2004
PHYSICS EDUCATION RESEARCH CONFERENCE

Previous Proceedings in the Series of Physics Education Research Conferences

Year	Title	Publisher	ISBN
2003	2003 Physics Education Research Conference	AIP Conf. Proceedings Vol. 720	0-7354-0200-0

To learn more about this title, or the AIP Conference Proceedings Series, please visit the webpage **http://proceedings.aip.org**

2004 PHYSICS EDUCATION RESEARCH CONFERENCE

Sacramento, California 4 – 5 August 2004

EDITORS
Jeffrey Marx
McDaniel College, Westminster, Maryland
Paula Heron
University of Washington, Seattle, Washington
Scott Franklin
Rochester Institute of Technology, Rochester, New York

SPONSORING ORGANIZATION
American Association of Physics Teachers (AAPT)

AMERICAN INSTITUTE OF PHYSICS

Melville, New York, 2005
AIP CONFERENCE PROCEEDINGS ■ VOLUME 790

Editors:

Jeffrey Marx
Department of Physics
McDaniel College
2 College Hill
Westminster, MD 21157
USA

E-mail: jmarx@mcdaniel.edu

Paula Heron
Department of Physics
Univeristy of Washington
Box 351560
Seattle, WA 98195-1560
USA

E-mail: pheron@phys.washington.edu

Scott Franklin
Department of Physics
Rochester Institute of Technology
Rochester, NY 14623-5603
USA

E-mail: svfsps@rit.edu

L.C. Catalog Card No. 2005932478
ISBN 0-7354-0281-7
ISSN 0094-243X

Printed in the United States of America

CONTENTS

INVITED PAPERS

PEER-REVIEWED PAPERS

Preface

With this year's Proceedings we begin a new tradtion: the Preface is written by the outgoing editor, who can reflect on changes that have occurred. The Proceedings are now four years old, and have shared many of the growing pains experienced by the Physics Education Research Community. We have grappled with the explosive growth in both quantity and breadth of research. Maintaining a high quality while allowing emerging areas the freedom to grow remains a critical issue both for the Proceedings and the PER community. One measure of this growth continues to be the increasing number of papers we receive; 51 this year, 30times as many as only two years ago.

The list of reviewers needed has also grown, and we thank the following seventy individuals who gave their time to help ensure that the high standards continue to be met: Wendy Adams, Rhett Allain, Alicia Allbaugh, Brad Ambrose, Gordon Aubrecht, Lei Bao, Bob Beichner, Joseph Beuckman, Scott Bonham, Eric Brewe, David Brookes, Kerry Brown, Alice Churukian, Andrew Crouse, Karen Cummings, Melisa Dancy, Dewey Dykstra, Andy Elby, Paula Engelhardt, Jerome Epstein, Eugenia Etkina, Noah Finkelstein, Tom Foster, Kara Gray, Richard Hake, Danielle Harlow, Kathy Harper, Kenneth Heller, Charles Henderson, Brant Hinrichs, Zdeslav Hrepic, Leon Hsu, Steve Kanim, Chris Kautz, Carolann Koleci, Rebbeca Lindell, Mike Loverude, Fran Mateycik, Tim McCaskey, Laura McCullough, Mark McKinnon, David Meltzer, Katherine Menchen, Donna Messina, Taha Mzoughi, Luanna Ortiz, Valerie Otero, K. K. Perkins, Steven Pollock, David Pritchard, David Rosengrant, Mel Sabella, Homeyra Sadaghiani, Cody Sandifer, Tim Seltzer, Peter Shaffer, Manjula Sharma, Chandralekha Singh, MacKenzie Stetzer, Robert Teese, Beth Ann Thacker, John Thompson, Alan Van Heuvelen, Stamatis Vokos, DJ Wagner, Rasil Warnakulasooriya, Aaron Warren, Carl Wieman, Edit Yerushalmi, and Xueli Zou.

There are still issues to resolve. While the electronic submission and distribution of papers should in principle make the reviewing process easier, in practice it has led to unacceptable delays. We apologize for the increasing length of publication time; it remains the single biggest issue I wish we could resolve. Instructions to reviewers is a smaller issue. Our goal is to have every paper reviewed to same standards; however, with seventy-three reviewers, ranging from graduate students to professor emeriti, this is a challenge.

When we first started the Proceedings, our hopes were to capture a snapshot of the PER community and address our belief that much good PER went unpublished. The Proceedings thus could fill a unique niche in PER literature, and be a valuable resource for the community. I hope we have succeeded, and wish the future editors the best of luck in the challenges ahead.

Scott Franklin
Outgoing Editor

PROGRAM

2004 PHYSICS EDUCATION RESEARCH CONFERENCE
SACRAMENTO, CALIFORNIA

Wednesday, August 4

4:00 - 6:00 pm: AAPT/PERC Bridging Session: Invited Talks & Panel Discussion

4:00 pm: Is Transfer Ubiquitous or Rare? New Paradigms for Studying Transfer
Jose Mestre, University of Massachusetts, Amherst

4:30 pm: Assessing Transfer of Conceptual Understanding
Karen Cummings, Southern Connecticut State University

5:00 pm: Measuring the Transfer of Mathematical Skills
Manjula Sharma, University of Sydney, Australia

5:30 pm: Panel Discussion
Discussant: To be announced

6:00 - 8:00 pm: Free Time

8:00 - 10:00 pm: Contributed Poster Session

9:00 - 10:00 pm: Round Table Discussion, *Parallel Sessions*

Thursday, August 5

8:00 - 8:15 am: Orientation

8:15 - 9:45 am: Workshops (W) & Targeted Poster (TP) Sessions - I

W-A: Getting Articles into Journals
Robert Beichner, North Carolina State University

W-B: The Physics Portal: Building a Self-Sustaining Internet-based Education Network
David Hestenes, Arizona State University; Bernard Haisch, ManyOne Network

W-C: Laboratory Math & Science for Cognitive Development - Dealing with the Real Level of our Physics Students
Jerome Epstein, Polytechnic University

TP-A: Challenges for the PER Community: Exploration of Common Assumptions, Open Questions & Current Controversies
Organizer: Paula Heron, University of Washington

TP-B: Beyond Student Transfer: Graduate, Postdoc, and Faculty Development & The Road to Sustainable, Scalable Inclusion of PER
Organizers: Noah Finkelstein, University of Colorado; Melissa Dancy, University of North Carolina

TP-C: Going Up? - Learning Transfer Among Students in Upper-Level Physics Courses
Organizer: Chandralekha Singh, University of Pittsburgh; Bradley Ambrose, Grand Valley State University

9:45 - 10:15 am: Break

10:15 am - 12:15 pm: Invited Talks & Panel Discussion

10:15 am: What Coordination has to say About Transfer
Andrea diSessa, University of California, Berkeley

10:45 am: Innovation and Efficiency in Transfer
Daniel Schwartz, Stanford University

11:15 am: When You don't See it -- Why not?
Zbigniew Dziembowski, Temple University

11:45 am: Panel Discussion
Discussant: Jose Mestre, University of Massachusetts, Amherst

12:15 - 1:45 pm: Luncheon Banquet & Talk
Duncan McBride, National Science Foundation

1:45 - 3:15 pm: Parallel Workshops (W) & Targeted Poster (TP) Sessions - II

W-A: Getting Articles into Journals
Robert Beichner, North Carolina State University

W-D: Transferring PER Results from the Domain of the Researcher to the Domain of the Practitioner
Kenneth Heller, University of Minnesota

W-B: The Physics Portal: Building a Self-Sustaining Internet-based Education Network
David Hestenes, Arizona State University; Bernard Haisch, ManyOne Network

TP-C: Going Up? - Learning Transfer Among Students in Upper-Level Physics Courses
Organizer: Chandralekha Singh, University of Pittsburgh; Bradley Ambrose, Grand Valley State University

TP-D: Issues in Studying Transfer of Problem Solving Skills
Organizers: Kathleen A. Harper, The Ohio State University; Thomas Foster, Southern Illinois University – Edwardsville; David P. Maloney, Indiana University, Purdue University, Fort Wayne

TP-E: Determining Transfer of Learning with Longitudinal Studies Using Grade & Demographic Data on Individual Students
Organizer: Wendell Potter, University of California, Davis

3:15 - 3:45 pm: Break

3:45 - 5:15 pm: Parallel Workshops (W) & Targeted Poster (TP) Sessions - III

W-D: Transferring PER Results from the Domain of the Researcher to the Domain of the Practitioner
Kenneth Heller, University of Minnesota

W-C: Laboratory Math & Science for Cognitive Development - Dealing with the Real Level of our Physics Students
Jerome Epstein, Polytechnic University

TP-A: Challenges for the PER Community: Exploration of Common Assumptions, Open Questions & Current Controversies
Organizer: Paula Heron, University of Washington

TP-B: Beyond Student Transfer: Graduate, Postdoc, and Faculty Development & The Road to Sustainable, Scalable Inclusion of PER
Organizers: Noah Finkelstein, University of Colorado; Melissa Dancy, University of North Carolina

TP-D: Issues in Studying Transfer of Problem Solving Skills
Organizers: Kathleen A. Harper, The Ohio State University; Thomas Foster, Southern Illinois University – Edwardsville; David P. Maloney, Indiana University, Purdue University, Fort Wayne

TP-E: Determining Transfer of Learning with Longitudinal Studies Using Grade & Demographic Data on Individual Students
Organizer: Wendell Potter, University of California, Davis

5:15 - 6:30 pm: Break

6:30 - 8:00 pm: Dinner Banquet

Poster Titles and Authors

The Design and Validation of the Colorado Learning Attitudes about Science Survey
Wendy Adams, Katherine Perkins, Noah Finkelstein, Carl Wieman, and Michael Dubson, University of Colorado, Boulder

Evaluating and Using BEMA (Brief Electricity & Magnetism Assessment)
Lin Ding, North, Robert Beichner, Ruth Chabay, and Bruce Sherwood, North Carolina State University

Understanding the MPEX 'Expert': A Comparison with Traditional Physics Faculty
Elizabeth Gire, Edward Price, and Barbara Jones, University of California, San Diego

A Conceptual Hierarchy of Lunar Phases?
Aaron Hines and Rebecca Lindell, Southern Illinois University - Edwardsville

Eliciting and Representing Hybrid Mental Models
Zdeslav Hrepic, Dean Zollman, and N. Sanjay Rebello, Kansas State University

Development of an Instrument for Evaluating Anxiety Caused by Cognitive Conflict
Yeounsoo Kim and Lei Bao, The Ohio State University

Using a Q-type Assessment Instrument to Study Correlation Between Teacher Attitudes and Student Perceptions of Physics Laboratories
Yuhfen Lin and Dedra Demaree, The Ohio State University; Xueli Zou, California State University – Chico; Gordon Aubrecht, The Ohio State University

Measuring Conceptual Change in College Students' Understanding of Lunar Phases
Rebecca Lindell, Southern Illinois University - Edwardsville

Student "Splits" Between Intuition and Scientist Answers*
Timothy McCaskey and Andrew Elby, University of Maryland

A Survey to Investigate Student Understanding of Quantum Tunneling
Jeffrey Morgan and Michael Wittmann, University of Maine

Correlating Student Attitudes with Student Learning using the Colorado Learning Attitudes about Science Survey
Katherine Perkins, Wendy Adams, Steven Pollock, Carl Wieman, and Noah Finkelstein, University of Colorado, Boulder

Student Understanding of Gauss' Law of Electricity
Chandralekha Singh and Paul Reilly, University of Pittsburgh

From Students' Perspectives: A Q-type Assessment Instrument*
Xueli Zou, California State University – Chico; Dedra Demaree, Yuhfen Lin, and Gordon Aubrecht, The Ohio State University

Rate of Change and Electric Potential
Rhett Allain and Robert Beichner, North Carolina State University

Diminishing Forces – Implications of a Misconception
Alicia Allbaugh, Rochester Institute of Technology

Assessing Student Understanding of Wave Amplitude and Intensity
Lei Bao and Dedra Demaree, The Ohio State University

Student Difficulties with Graphical Representation of Vector Products: Crossing and Dotting Beyond t's and i's
Warren Christensen, Ngoc-Loan Nguyen, and David Meltzer, Iowa State University

Identifying Student Concepts of "Gravity"
Roger E. Feeley R. Thompson, and Michael C. Wittmann, University of Maine

Student Understanding of Gravitational Potential Energy and Moving Objects
Michael Loverude, California State University - Fullerton

Student Understanding of Sound Propagation: Research and Curriculum Development
Katherine Menchen and John Thompson, University of Maine

Teacher and Curriculum Factors that Influence Middle School Students' Sense-Making Discussions of Force/Motion
Cody Sandifer, Towson State University

The Role of Evaluation Abilities in Student Learning & Performance
Aaron Warren and Alan Van Heuvelen, Rutgers University

Student Participation in Normative Behaviors in a Physics for Elementary Teachers (PET) Classroom*
Benjamin Williams, San Diego State University

Generated Analogies as Assertions of Categorization
Leslie Atkins, University of Maryland

Active versus Passive Learning
Florin Bocaneala and Lei Bao, The Ohio State University

Concerning Scientific Discourse about Heat
David Brookes; George Horton; Alan Van Heuvelen and Eugenia Etkina , Rutgers University

Introductory College Students' Explanations of Friction and Related Phenomena at the Microscopic Level
Edgar Corpuz and N. Sanjay Rebello, Kansas State University

Evidence of Transfer in Interview Data
Paula Engelhardt, Kansas State University

Model what you Preach: Explicitly Articulated Interactions for Transfer of Concepts by Physics Graduate Teaching Assistants
Cathy Ezrailson

Transfer between Paired Problems
Kara Gray and N. Sanjay Rebello, Kansas State University

Transfer: the Advantage of Simple Symbols
Andrew Heckler, Jennifer Kaminski, and Vladimir Sloutsky, The Ohio State University

Sample Exams and Transfer in Introductory Mechanics
Carol Koleci, Charles Chretien, and Warren Turner, Worcester Polytechnic Institute

Alternative Conceptions, Memory, & Mental Model in Physics Education
Gyoungho Lee, Seoul National University; Jiyeon Park; Yeonsoo Kim; Lei Bao, The Ohio State University

Investigating Students' Knowledge of Particle Structure of Matter in Different Cultures
Cui Lili, Dean Zollman, and N. Sanjay Rebello, Kansas State University

Student Descriptions of Refraction and Optical Fibers
Fran Mateycik, DJ Wagner, Rensselaer Polytechnic Institute; JJ Rivera; Sybillyn Jennings, Sage College

Transfer of Learning from Trigonometry to Physics
Darryl Ozimek, Paula V. Engelhardt, and N. Sanjay Rebello, Kansas State University

Analogical Scaffolding of Abstract Ideas in Physics
Noah Podolefsky, Wendy Adams, and Noah Finkelstein, University of Colorado, Boulder

Learning and Knowledge Transfer between Physics Problems
David Pritchard , Elsa-Sofia Morote, and Rasil Warnakulasooriya, Massachusetts Institute of Technology

Multiple Representations: A Quantitative Study on Students use of Free-Body Diagrams in Large Lecture Classes.
David Rosengrant, Alan Van Heuvelen, and Eugenia Etkina, Rutgers University

Student Learning of Quantum Mechanics
Homeyra Sadaghiani and Lei Bao, The Ohio State University

The Structure of Intermediate Mechanics Students' Physics and Mathematics Intuitions
Eleanor C. Sayre and Michael C. Wittmann, University of Maine

Students' Conceptions about Probability in a Double-slit Experiment for Electrons and Potential Well Problems
Pornrat Wattanakasiwich and Kenneth Krane, Oregon State University

SECTION I

Invited Papers

Is Transfer Ubiquitous or Rare? New Paradigms for Studying Transfer

Jose Mestre

Department of Physics
University of Massachusetts Amherst 01003

Abstract. Classic and emerging views of transfer are discussed; examples of research are provided to illustrate each.

INTRODUCTION

Transfer of learning has been an important research theme in educational/cognitive psychology for many years.[1] Considerable research, and even volumes reviewing transfer findings[2] have occupied researchers attempting to understand the conditions under which knowledge learned in one context transfers to new contexts. After decades of research and debate, no consensus exists on whether transfer is ubiquitous or rare. Clearly this question is important for an educational system such as ours built on the premise that what is learned in school is useful in other contexts in school, outside of school, and in the workplace. In this article I will review two views of transfer, the classic view and an emerging view, and discuss how the controversy over whether transfer is ubiquitous or rare depends on which of the two views one uses to interpret findings. I will then provide specific examples to illustrate the types of questions that can be answered using the two views.

TWO VIEWS OF TRANSFER

Classic View

The classic view of transfer is exemplified by the following types of definitions:

> *[Transfer is] defined as the ability to extend what has been learned in one context to new contexts* (ref 3, p. 74).

> *Central to traditional approaches to transfer is a dominant methodology that asks whether people can apply something they have learned to a new problem or situation* (ref 4, p. 67).

> *As a construct in educational psychology, [transfer] refers to ... a person carrying the product of learning from one task, problem, situation ... to another* (ref 5, p. 101).

> *[Transfer is] the degree to which a behavior will be repeated in a new situation* (ref 6, p. 4).

> *[Transfer is defined broadly] to mean the ability to apply knowledge or procedures learned in one context to new contexts* (ref 7, p. 3).

Definitions such as these suggest that transfer consists of measures of appropriate use of prior learning in new contexts, contexts which are of the researcher's choosing and based on the researcher's perception of similarities to the learning context. This view is reseacher-centered in that the researcher predefines the underlying concept that should transfer and designs an experiment to seek evidence for whether or not transfer occurs. Within this view, transfer is measured as an all-or-nothing event, and when measured in this way there is ample evidence that transfer is rare indeed.[4,6,8-9]

CP790, *2004 Physics Education Research Conference*, edited by J. Marx, P. Heron, and S. Franklin

To illustrate consider a well known example from an experiment[10] in which college students were given the following situation to study:

A general wishes to capture a fortress in the center of a country. There are many roads radiating outward from the fortress. All roads have been mined so that while small groups of men can pass over the roads safely, a large force will detonate the mines. A full-scale direct attack is therefore impossible. The general's solution is to divide his army into small groups, send each to the head of a different road, and have the groups converge simultaneously on the fortress.

Minutes later, after determining that students had understood this situation, they were given the following problem:

You are a doctor faced with a patient who has a malignant tumor in the stomach. It is impossible to operate on the patient, but unless the tumor is destroyed, the patient will die. There is a kind of ray that may be used to destroy the tumor. If the rays reach the tumor all at once and with sufficient high intensity, the tumor will be destroyed, but surrounding tissue may be damaged as well. At lower intensities, the rays are harmless to healthy tissue, but they will not affect the tumor either. What type of procedure might be used to destroy the tumor with the rays, and at the same time avoid destroying the healthy tissue?

It was expected that students would apply the underlying concept from the first situation to the second, but few students were able to solve the tumor problem. When explicitly told to use information from the first situation, over 90% of the students could solve the tumor problem. Findings like these led some researchers to argue that transfer is rare:

...most studies fail to find transfer ... [and] those studies claiming transfer can only be said to have found transfer by the most generous of criteria and would not meet the classical definition of transfer. ... in studies ... that claim transfer, transfer is produced by 'tricks' of one kind or another. ... Transfer is rare, and its likelihood of occurrence is directly related to the similarity between two situations. (ref 6, p. 15)

Emerging View

In the emerging view transfer is defined more broadly as a complex, dynamical process where knowledge is activated and applied in response to context. The focus is on studying what students do (both productively and unproductively) during a task. In contrast to the classic view, the emerging view is student-centered.[4,11] For example, Lobato's[12] "actor-oriented model of transfer" relies on "personal creations of relations of similarity" by the learner between learning and transfer contexts rather than similarities perceived by researcher.

Within this view, different types of questions can be studied, such as: What is the nature of the knowledge activated during a transfer task? How does the knowledge activated depend on context? How are representations of context created and of what do they consist? What can be said about the transfer dynamic (e.g. how reliable is it)? How can instruction be designed to promote flexible and positive transfer?

RESEARCH EXAMPLES FROM EMERGING VIEW OF TRANSFER

Images in Mirrors

In a simple yet compelling example of the context dependence of transfer, Hammer and his collaborators[13] describe a task in a graduate course for teachers where students worked in groups to decide what size mirror is needed to see an image of your entire body. One student, "Sherry," stated that you need a mirror the same size as your body because your whole body has to fit in the mirror. Other students in her group argued that a mirror only half the size of your body is needed, but Sherry still defended her position. Days later she told her group of a discovery: She had a bedroom mirror that was about half her size and she was able to see a reflection of her whole body daily at home. Despite this daily experience, this knowledge did not transfer to the context of a question about images posed in physics class.

Hammer et al., argue that what is happening in this example is complex. Had she simply recalled the fact that she saw her full image daily in a mirror half the size of her body, this might be considered simple recall rather than transfer. Sherry's reasoning revealed that she was not using her experiences with mirrors in answering the original question, but something else. Perhaps, Hammer et al. argue, it was her experience with doors, which require a size at least as big as your body to allow passing without obstruction; perhaps it was her experience with life-size images printed on surfaces that triggered her argument. She was clearly transferring something that was more abstract than her daily experience with mirrors. Examples like these has led Hammer and his collaborators to think of transfer in terms of activation of "resources" (fine-grained knowledge units) that an individual uses to frame the context. Different resources are activated to frame the question about mirrors and images in physics class, and viewing herself in the mirror in her bedroom.

Problem Posing

In my own problem posing work[14], I found that students transferred lots of knowledge learned in physics class, but did it piecemeal rather than holistically. This led to posing of problems that were locally coherent, but globally incoherent. For example, consider the following "concept scenario" around which students were asked to pose "textbook like" problems that matched the scenario:

Mechanical energy is conserved, followed by conservation of momentum, followed by conservation of mechanical energy with potential energy increasing and kinetic energy decreasing

One student posed the following problem:

A 3 Kg anvil is dropped from a height of 3 meters an (sic) falls on a lever that transfers its energy of the anvil to a tennis ball weighing 0.5 Newtons. At what height will the ball climb to. (See figure.)

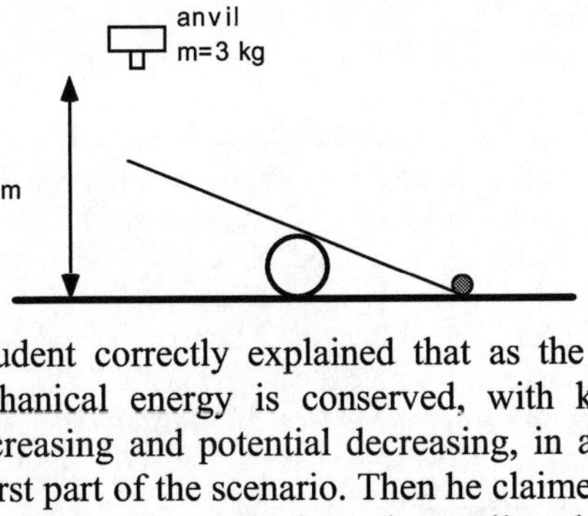

The student correctly explained that as the anvil falls, mechanical energy is conserved, with kinetic energy increasing and potential decreasing, in accord with the first part of the scenario. Then he claimed that energy is transferred exactly from the anvil to the ball in the collision, which is impossible given that the anvil loses considerable energy to the earth when it stops. Further, he wants to conserve momentum in the collision by setting the mass times the velocity of the anvil equal to the mass times the velocity of the ball. The last portion in which the ball rises matches the last portion of the scenario. Clearly this student matched different portions of the scenario, but did not create a consistent overall problem that matched the entire scenario.

From a transfer perspective, he was transferring lots of knowledge learned in physics class, yet his understanding was not refined enough to allow him to construct a coherent problem. The inconsistency in matching the momentum conservation portion of the problem and scenario has to do with the two portions that require conservation of mechanical energy at the beginning and the end of the scenario. This, as well as other students interviewed, showed a reasonable understanding of conservation of energy but only within the context of objects rising or falling in the Earth's gravitational field. In attempting to satisfy the two energy conservation portions of the scenario, the student had to reverse the direction of the object's motion from beginning to end, and hence the need for the see-saw arrangement that did not conserve momentum. Viewed from a classic perspective, transfer failed, yet viewed from the emerging perspective, transfer was ubiquitous.

Judging Realistic Motion

The last example is from our work[15] in which students judged which among several animations of balls moving along tracks is realistic. The apparatus used is shown below. Students were shown five animations of the ball on track B first, followed by the same five animations with both balls moving together in a race. A major finding was that the types of strategies that students used to make decisions about the realism of the ball's motion when a single ball moved along track B differed substantially from the strategies used to reason about the realism of both balls moving together in a race. In the one-ball motions, students relied on judging the realism of speed changes in the various portions of the tracks against their expectations of what the motion should be. In the two-ball animations, the expected race outcome played a major role in judging realism of motion. More specifically, one particular animation of the ball in track B moving by itself was rejected by all 50 students interviewed (26 enrolled in an educational psychology course and 24 enrolled in an honors physics course); in this animation, the ball's motion was unphysical in the upward section of the "V," namely the ball speed up while moving uphill.

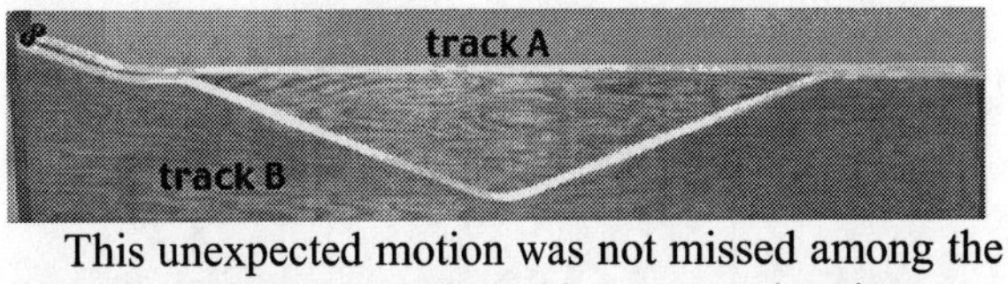

This unexpected motion was not missed among the student participants. When this same animation was run with both balls moving in a race, the result was a tie at the end of the race. Given that the expectation of a tie was very strong among the physics students (which they justified in terms of conservation of energy—the potential energy lost equals the kinetic gained, hence the balls have the same speed in the final flat shelf and hence they should tie), over 60% of the physics students (and none of the psychology students) selected this animation as most realistic. Clearly different types of knowledge were being activated and applied by the two groups of students, and in this instance the physics knowledge (conservation of energy) activated by the physics students in the two-ball race led to a tie selection despite one of the balls undergoing unphysical motion. This led us to posit that the type of knowledge that students activate and apply depends very strongly on both context and on the prior knowledge they possess. In this case, the more prior physics knowledge possessed by the physics students did not help them since that knowledge was not transferred appropriately.

DISCUSSION

Two views of transfer have been presented, the classic view from educational/cognitive psychology, and an emerging view. In the classic view the researchers designs experiments to test whether or not transfer of some body knowledge/learning occurs wholesale to another scenario. In the emerging view, transfer is explored in terms of the type of knowledge activated by students in response to context in order to make sense of the situations they are considering. In PER we are more interested in the types of questions that the emerging view of transfer allows us to answer, which revolve around the nature and functioning of the student's knowledge architecture, rather than questions about whether or not a particular piece or pieces of learning/knowledge was successfully applied at some instant in time in some new context.

ACKNOWLEDGMENTS

I would like to thank my collaborators Tom Thaden-Koch, Bob Dufresne, Bill Gerace, and Bill Leonard for their help and insights.

REFERENCES

1. Royer, J.M., Dufresne, R., & Mestre, J. "Foreword" in *Transfer of Learning from a Modern Multidisciplinary Perspective*, edited by J. Mestre, Greenwich, CT: Information Age Publishing, in press.
2. Detterman, D.K., & Sternberg, R.J. (Eds.). *Transfer on trial: Intelligence, cognition, and instruction.* Norwood, NJ: Ablex, 1993.
   Educational Researcher, **32** (No. 1), 2003.
   J. Mestre (Ed.), *Transfer of Learning from a Modern Multidisciplinary Perspective* Greenwich, CT: Information Age Publishing, in press.
3. Bransford, J.D., Brown, A.L. & Cocking, R.R. *How People Learn: Brain, Mind, Experience, and School.* Washington, DC: National Academy Press, 1999.
4. Bransford, J. & Schwartz, D. *Review of Research in Education.* **24**, 61-100, 1999.
5. Beach, K. *Review of Research in Education* **24**, 101-140, 1999.
6. Detterman, D.K. "The Case for the Prosecution" in *Transfer on trial: Intelligence, cognition, and instruction,* edited by Detterman, D.K., & Sternberg, R.J. .Norwood, NJ: Ablex. 1993, pp. 1-24.
7. Mestre, J. *Transfer of Learning: Issues and Research Agenda.* National Science Foundation Report #NSF03-212. 27pp.
8. Rebello, S., Zollman, D., Allbaugh, A.R., Engelhardt, P.V., Gray, K.E., Hrepic, Z., & Itza-Ortiz, S.F. "Dynamic Transfer: A Perspective from Physics Education Research" *Transfer of Learning from a Modern Multidisciplinary Perspective*, edited by J. Mestre. Greenwich, CT: Information Age Publishing, in press.
9. Schwartz, D., Bransford, J, & Sears, D. "Efficiency and Innovation in Transfer" *Transfer of Learning from a Modern Multidisciplinary Perspective*, edited by J. Mestre. Greenwich, CT: Information Age Publishing, in press.
10. Gick, M.L., & Holyoak, K.J., *Cognitive Psychology* **12**, 306-355, 1980.
11. Greeno, J.G., Smith, D.R., & Moore, J.L. "Transfer of Situated Learning." In *Transfer on trial: Intelligence, cognition, and instruction,* edited by D.K. Detterman & R.J. Sternberg, Norwood, NJ: Ablex, 1993, pp. 99-167.
12. Lobato, J., *Educational Researcher* **32**(#1), 17-20, 2003.
13. Hammer, D., Elby, A., Scherr, R., & Redish, E. "Resources, Framing and Transfer," in *Transfer of Learning from a Modern Multidisciplinary Perspective*, edited by J. Mestre. Greenwich, CT: Information Age Publishing, in press.
14. Mestre, J., *J. of App. Devel. Psych.* **23**, 9-50, 2002.
15. Mestre, J., Thaden-Koch, T., Dufresne, R., & Gerace, W. "The Dependence of Knowledge Deployment on Context Among Physics Novices" edited by. E. Redish, & M. Vicentini, *Proceedings of the International School of Physics "Enrico Fermi".* Amsterdam: ISO Press/Italian Physics Society, 2004, pp. 367-408.

How Do You Hit A Moving Target? Addressing The Dynamics Of Students' Thinking

David E. Meltzer

Department of Physics and Astronomy
Iowa State University

Abstract. From the standpoint both of research and instruction, the variable and dynamic nature of students' thought processes poses a significant challenge to PER. It is difficult merely to assess and characterize the diverse phases of students' thinking as they gain and express understanding of a concept. (We might call this the "kinematics" of students' thought processes.) Much harder still is uncovering the various factors (instructional method, student characteristics, etc.) that influence and determine the trajectory of students' thinking. (We could call this the "dynamics" of students' thinking.) The task of deciphering the mutual interaction of these factors adds to the challenge. I will outline some of the initial work that has been done along these lines by various researchers, and I will identify some directions for future research that I think might be fruitful for workers in PER.

INTRODUCTION

Our goal as educators is to better understand the *process* of student learning so as to be able to influence it more effectively. Students' learning of physics is characterized by a knowledge state that is a generally increasing function of time. Often, however, the inherent time-dependence of this process is given inadequate examination, in part due to the difficulty of investigating students' thinking at multiple time points during its evolution.

Characterization of a time-dependent process requires a bare minimum of two probes at different time points, while a varying rate requires three such probes. Alternatively, a probe may be carried out over a continuous (brief) time interval and variations during that interval observed. (This type of probe is characteristic of so-called "dynamic assessment" [1] and the "teaching experiment" [2].) In any case, such repeated probes of student thinking are logistically difficult to implement within actual classroom settings involving ongoing instruction.

In this paper I will outline some of the work that has been done by various researchers in exploring changes in student thinking over time, and I will identify some directions for future research that I think might be fruitful for workers in PER.

ASSESSING STUDENTS' MENTAL STATES AT A PARTICULAR TIME

It is useful to recall the complexity of a thorough probe of students' thinking at even a single point in time. Such a probe would require analysis not only of a students' ideas about a set of physics concepts and the relationships among them, but also of the ways in which the student perceives and implements the learning process itself.

Students' "Knowledge" State

At any given moment a student has a collection of ideas related to specific physics concepts, and a related set of ideas corresponding to the expressed or implied interconnections among those concepts. These ideas are in significant part dependent on context, that is, they often depend on the physical setting of a given problem, the form of representation employed in the problem, and so forth. One can try to assess this

CP790, *2004 Physics Education Research Conference*, edited by J. Marx, P. Heron, and S. Franklin

collection of student ideas by posing questions involving diverse contexts and a variety of representations [3-9]. In this fashion one can try to determine the "distribution function" of ideas (sometimes called the "mental model" [4,5]) characteristic of a particular student, or of a particular student population.

Students' "Learning State"

Another key component of students' thinking is the set of their ideas related to the practice of learning physics, along with the methods they actually employ to learn. This includes their study methods, their attitudes toward physics and physics learning, their motivation to learn, etc. One can attempt to assess these factors through a number of methods including observations of learning practices [10], attitudinal surveys [11,12], "dynamic assessment" [1], "teaching experiments" [2], etc.

CHARACTERIZING THE PROCESS OF STUDENT LEARNING

If we are to carry assessment beyond a single time point, we must determine the specific parameters needed for an assessment of the overall learning process. If we can obtain observable data corresponding to those parameters, we then need to determine how exactly to analyze those data.

Qualitative Parameters

The basic elements of a time-dependent analysis of student learning include *sequences* of the various parameters that characterize students' knowledge. These include the following: (1) The *sequence of ideas* and of *sets* of ideas (mental models) developed by a student during the process of learning a set of related concepts; (2) The *sequence of difficulties* encountered by a student during that learning process (difficulties are related to "*ideas*," but are not necessarily the same thing); (3) The *sequence of knowledge resources and study methods* employed by the student during that process; (4) The *sequence of attitudes* developed by a student during that process.

The fundamental assumption in this analysis is that all of the various elements may (and probably do) undergo change over time. There will always be a question of how rapidly this change occurs and, consequently, how frequently an assessment must be made in order not to overlook key stages of the process.

Quantitative Parameters

In addition to qualitative parameters, one can identify a number of potentially relevant measures to which numbers can be attached. These include the following: (1) The progression in *depth of knowledge* as measured by probability of correct response on a set of related questions (e.g., score S, range [0.00,1.00]); (2) The *average rate of learning* R of a set of related concepts (e.g., $R = g/\Delta t$ where g = normalized gain calculated using $S_{pretest}$ and $S_{posttest}$); (3) The *variations in the learning rate* V encountered by a student during that process (e.g., $V = \Delta R/\Delta t$); (4) The *time-dependent distribution function* characterizing the idea set of a student population. (This might be defined through a method analogous to that of Bao [4,5].)

Phase I: "Kinematics" Of Students' Thinking

The first level of investigation is to characterize the pattern of students' thinking as it *evolves* during the learning process. In principle the objective is to determine, at a number of different points in time, the set of students' ideas, difficulties, learning resources, etc. with respect to a well-defined concept or set of related concepts. (For instance, one might acquire data related to students' understanding of Newton's second law of motion.) Then, based on this time-series data, one can try to determine the *normal course of evolution* of those ideas and difficulties under a variety of standard learning situations.

Phase II: "Dynamics" Of Students' Thinking

The second phase of the investigation would be to determine the factors that *influence* the evolutionary pattern of students' thinking during the learning process. One might describe this objective as an attempt to answer the question, "What are the social and pedagogical forces that determine the path of a student's 'learning trajectory'?" More specifically, one could ask: What is the *relative* influence of (a) individual student characteristics (preparation, background, etc.) and (b) instructional method (including pedagogical techniques, classroom environment, etc.), on the observed sequences of ideas, difficulties, and attitudes? A crucial question would be to determine the extent to which the observed sequences might be altered due to efforts of the instructor and/or the students.

PREVIOUS WORK

A number of workers have investigated various aspects of the issues discussed in this paper. However, many related issues have been explored little or not at all. Here I will outline some of this previous work.

Sequence of ideas: A number of investigators have described shifts in mental models by analyzing the differences in typical student response patterns between pretests and posttests [4,5,7-9]. Savinainen et al. have also explored such patterns at mid-instruction points (between pre- and post-instruction) [8,9], while other workers have attempted to describe and characterize the sequence of ideas acquired by students during the learning process in a more detailed, step-by-step fashion [13-15]. Some workers (e.g., Thornton [3] and Dysktra [6]) have postulated the existence of specific "transitional states," which are well-defined sets of ideas occurring during the transition from novice to expert thinking.

Sequence of difficulties: The generalizability of *patterns* of learning difficulties is well established [16], but that of difficulty *sequences* has not been thoroughly investigated. In general, there has not been much detailed exploration into how the specific learning difficulties students encounter may change and evolve over the course of a semester or year.

Sequence of attitudes: There is evidence of regularities in attitude change during instruction [11], but also evidence that these regularities are dependent upon instructional context [12].

DYNAMIC ASSESSMENT

As an alternative to assessment of student thinking at a single instant (through a quiz, exam, etc.), a preplanned sequence of questions, hints, and answers may be provided and the students' responses observed throughout a time interval. This method has been formalized under the rubric "Dynamic Assessment" [1]. One first attempts to determine what types of problems the students can solve on their own, without additional assistance. One then continues by providing carefully measured and sequenced assistance through hints and answers, in order to assess the students' ability to respond to instructional cues with efficient learning. Among the assessment criteria are the amount of assistance required, the rapidity and depth of response, etc. A similar method is the "teaching experiment" [2], in which a mock instructional setting is used as a means to probe students' responses to various instructional interventions.

QUESTIONS FOR FUTURE WORK

Here I will list a number of questions that might serve as a basis for future investigations on these topics. For convenience, I will divide them according to whether they refer primarily to *characterizations* of the evolutionary process of students' thinking ("kinematics"), or to the factors that *influence* that process ("dynamics").

I. Kinematics

(1) Can one confirm the existence of *well-defined* "transitional mental states" related to learning of specific concepts, that is, sets of ideas concerning those concepts that are intermediate between those of a novice and those of an expert? If such transitional states do exist, do they vary among individuals according to differences in their background and preparation? Are different transitional states observed in traditional and reformed instruction?

(2) More broadly, one can ask: Does the individual "mental model" distribution function evolve according to some characteristic pattern? (This "distribution function" refers to the collection of student concepts related to a specific topic, as reflected for instance in the set of responses to a group of related diagnostic questions [4,5].) Is the evolution pattern correlated with individual characteristics (demographics, preparation, etc.) and/or with the nature of the instructional method?

(3) How does the population "mental model" distribution function evolve in general? (Here we refer to the average set of responses given by an entire class of students, or a number of similar classes.) Is the evolution pattern correlated with population demographics?

(4) Are there common patterns of variation in learning rates? For example, do learning rates typically increase or decrease monotonically throughout the course of a semester?

(5) Is the magnitude of the learning rate at an early phase of the process correlated with the long-term learning rate [17]?

(6) Is the picture of a student's learning trajectory provided by "dynamic assessment" (or teaching experiments) over a brief time interval more complete and accurate than that provided by a single standard quiz or exam?

II. Dynamics

(1) Can one trace back, in a causal fashion, the set of student ideas at a particular time, to the specific set of ideas and difficulties that had been acquired at an earlier time? More specifically: To what extent does the student's present set of ideas and difficulties determine the pattern of his or her thinking in the future?

(2) Are transitional states (if they exist) actually influenced by differences in students' preparation, and/or by the nature of the instructional method?

(3) Are the sequences of individual and population "idea distribution functions" (mental models) influenced by individual background and/or instructional mode?

(4) Are learning-rate variations influenced by individual background and/or instructional mode? More broadly, what are the factors that influence the trajectory of student learning, and what is the nature of the interaction among the various determining factors?

SUMMARY

The dynamic, time-dependent aspects of the student learning process are essential features of that process, and yet they are logistically difficult to observe and analyze. Future investigations in this area have the potential to yield valuable information that could help instructors increase the effectiveness of instruction in physics.

ACKNOWLEDGMENTS

The work of the Iowa State Physics Education Research Group has been supported by grants from the Iowa State University Center for Teaching Excellence, and from the National Science Foundation through the Division of Undergraduate Education, Division of Research, Evaluation and Communication, and Division of Physics.

REFERENCES

1. Lidz, C. S., *Practitioner's Guide to Dynamic Assessment*, New York: Guilford, 1991.
2. Engelhardt, P. V., Rebello, N. S., Corpuz, E., and Ozimek, D., "The Teaching Experiment: What It Is and What It Isn't," in *Proceedings of the 2003 Physics Education Research Conference*, edited by J. Marx et al., AIP Conference Proceedings, New York: American Institute of Physics, *in press*.
3. Thornton, R. K., "Conceptual Dynamics: Following Changing Student Views of Force and Motion," in *The Changing Role of Physics Departments in Modern Universities: Proceedings of International Conference on Undergraduate Physics Education*, edited by E. F. Redish and J. S. Rigden, AIP Conference Proceedings 399, New York: American Institute of Physics, 1997, pp. 241-266.
4. Bao, L. and Redish, E. F., *Am. J. Phys.* **69**, S45-S53 (2001).
5. Bao, L., Hogg, K., and Zollman, D., *Am. J. Phys.* **70**, 766-778 (2002).
6. Dykstra, D. I., "Why Teach Kinematics?" Parts I and II, preprint (2002).
7. Meltzer, D. E., *AAPT Announcer* **33 (4)**, 98 (2003).
8. Savinainen, A., *High School Students' Conceptual Coherence of Qualitative Knowledge in the Case of the Force Concept*, Ph.D. Dissertation, Department of Physics, Joensuu (Finland): University of Joensuu, 2004.
9. Savinainen, A., Scott, P., and Viiri, J., *Sci. Educ.* (*in press*).
10. Thornton, R. K., "Uncommon Knowledge: Student Behavior Correlated to Conceptual Learning," preprint (2004).
11. Redish, E. F., Saul, J. M., and Steinberg, R. N., *Am. J. Phys.* **66**, 212-224 (1998).
12. Elby, A., *Am. J. Phys.* **69**, S54-S64 (2001).
13. Hrepic, Z., Zollman, D. A., and Rebello, N. S., *AAPT Announcer* **33 (4)**, 134 (2003).
14. Itza-Ortiz, S. F., Rebello, N. S., and Zollman, D., *Euro. J. Phys.* **25**, 81-89 (2004).
15. There are also a number of "case studies" in the literature in which very detailed, diary-like expositions follow small numbers of students on almost a week-to-week basis. It is hard to know whether and to what extent such descriptions might be generalizable to broader student populations.
16. For example, see the many studies referenced in McDermott, L. C., and Redish, E. F., *Am. J. Phys.* **67**, 755-767 (1999).
17. Meltzer, D. E., *AAPT Announcer* **27 (4)**, 89 (1997).

Twenty Questions for PER: How Does It All Fit Together?

Edward F. Redish*, Michael C. Wittmann†

*Department of Physics, University of Maryland, College Park, MD 20742-4111, USA
† Department of Physics and Astronomy, University of Maine, Orono, ME 04469-5709, USA

Abstract: In physics education research (PER), we have focused our attention for many years on finding ways to improve our instruction and have achieved some notable successes. In this paper, we suggest that the time has come to embed this activity in a more complete and scientific view of PER, one that builds a coherent understanding of the system of teaching and learning in addition to improving the practice of our instruction. We outline five broad topics of interest for PER and discuss questions that need to be addressed in each topic over the next few years. The topics are: the model of the participants, the model of the contexts, the model of the content, the engineering of instruction, and the overall epistemology of PER — How do we decide when we think we know something?

INTRODUCTION

Little of what we do in physics education research (PER) resembles the complex and productive interplay of theory and experiment common to other sciences. We suggest some broad general questions to help put research in PER in perspective. Our goal is to suggest paths of action that synthesize seemingly disparate research areas.

We often say that as physics education researchers, we are applying the methods of science to help us understand how our students learn and do not learn physics. Often, we restrict our "application of the scientific method" to observing what our students do and trying to correlate their learning with instructional changes that seem, intuitively, to make sense to us. It is more like "seat-of-the-pants" engineering than like physics; more like Edison's search for the proper filament for a light bulb than like the current attempt to understand spintronics in order to (some time in the future) build a better microchip.

In scientific research, experiment and theory perform an intricate dance, with theory taking the lead at one time and suggesting experiments, with experiment taking the lead at another time and producing results that demand new theoretical explanations. A good example of this is the development of the theory of electricity. Although Benjamin Franklin was much concerned with saving houses from lightning strikes, his research into electricity involved much more than engineering better lightning rods. He did extensive experiments and proposed theories of the structure of electricity, including demonstrating the two charge model and proposing the theory of charge conservation. Coulomb's later detailed and accurate experiments relied heavily on the theoretical two-charge model built up as a result of earlier careful experimental and theoretical developments.[1]

In PER, we rarely use a serious theoretical frame. We are rarely able to generate reliable predictions in situations we haven't observed, and we often do not understand what our experimental results are telling us about theory. To move forward into a more productive and scientific mode of research, we must understand and model complicated educational systems and situations. We need to model the participants, both students and teachers. We need to model the way they interact, both with their overall environment and with the specific environments that happen in their classrooms. We need to model the content they are learning, both physically and pedagogically. Clearly, these models of participants, contexts, and content are not independent; they interact strongly. Contexts play a crucial role in determining the structure and response of individuals. The properties and responses of individuals play a crucial role in determining how group interactions take place. The way individuals think determines how content is seen, organized, and learned (and even discovered).

CP790, *2004 Physics Education Research Conference*, edited by J. Marx, P. Heron, and S. Franklin

These three strongly interacting research areas suggest that our theories need to overlap physics, cognitive, and behavioral science. In each of these areas, much is known both inside and outside the PER community. We need to consider what we can learn from successful engineering of instructional environments and how our models of participants, contexts, and content can inform the engineering of our classrooms. In this brief paper, we will not summarize what is known in any of these areas. Rather, we will raise some broad general questions that appear to us to be a useful guide to synthesis in PER.

PER: WHAT DO WE KNOW AND HOW DO WE KNOW IT?

Our first question concerns the epistemology of PER: How do we decide when we know something? In physics, we are accustomed to having mathematical predictions and quantitative experiments. People, like individual electrons, appear to have free will; people's behavior is predictable only statistically. But people, unlike electrons, are not identical or interchangeable. This leads to numerous questions of "How do we know? Why do we believe?"

We might call the broad question for this section: *The Epistemology Question — What are appropriate tools, qualitative and quantitative, for deciding we have learned something in PER?*

There are many questions one could ask to determine how we decide to know something in PER. The first comes from the fact that we listen to our students, interview them, and monitor their behavior on homework, quizzes, and tests. What do we learn from this?

1. What information do we need in order to infer what's happening in an individual's mind?

If we are to follow the route suggested by scientific progress, we should make sense of student thinking in terms of mechanisms. Those mechanisms are likely to be inferred rather than observed. Inferring what students and teachers are thinking from their behavior only expresses a part of what's happening. It's rather like inferring quarks inside a nucleon. How do we decide if a mental model or a social response is "really there"? Does it suffice to invoke Gell-Mann's early hedge and say "It doesn't matter if it's really there or not, this hypothesis is a useful way of thinking about the phenomena we are trying to describe"? What experimental data and triangulations do we need to decide a mental description is reliable and useful?

Other questions probe the issues of comparisons between individual and large-population studies and hidden assumptions in statistical studies.

2. How many students do we need to study before we can be satisfied that we have seen "most of" the behaviors we can expect from a large population?
3. When do large-scale statistical studies give us appropriate information about the learning of individuals and when do they mislead us?

As physicists, we tend to prefer large-scale studies where we can use familiar mathematical tools to generate numbers representing our uncertainty. But these tools and our interpretation of the results rely heavily on the underlying assumption that we have a homogeneous population that is normally distributed. This is rarely the case. As in medical research, we must question the law of large numbers that assumes we only have many random fluctuations. Saying, "this drug is successful in treating this disease in 99% of the cases" is insufficient if in the other 1% of the cases patients go into anaphylactic shock. We first must identify subpopulations at risk before applying statistics.

We also have new technologies, both experimental and theoretical, which are beginning to probe how people think. Do these methods have value for us?

4. What useful information about mental structures relevant for PER do brain-scanning techniques (fMRI, MEG, PET, EEG) give us?
5. Does modeling with networks, either in the small (neural networks attempting to model specific tasks) or in the large (statistical network analysis showing emergent structures) tell us anything useful about how to model thinking in physics?

MODELING THE PARTICIPANTS

Students and teachers are the "electrons" and "nuclei" of our many-body classroom system. If we want to understand how our classroom works, we must understand how our students and teachers "work" – if we want to understand the properties of an atom, we have to understand the properties of electrons and nuclei. We learn about electrons and nuclei by studying them through interactions. The same is true of students and teachers. We can build models of how people think, but how those thoughts interact with and are affected by outside events, both physical and socio-cultural, is an essential component of this study. Synthetic models of thinking appropriate for PER have begun to be formulated. See, for example [2].

We might call the broad question for this section: *The Ontology Question—What are appropriate mental structures for describing how people think and learn physics?* This might be divided into two: statics (description) and dynamics (change). Note, however, that all brain processes are dynamic, so even calling our descriptive questions "statics" is a bit misleading.

We pose three broad questions that refine the static and descriptive question: What are appropriate mental structures for describing how students think about and interact with their educational environment?

6. Some researchers have described students' naïve knowledge as robust, others as volatile. Does one description dominate pre-expert students in physics? If not, what kind of description is appropriate and when?
7. Some researchers have described students' knowledge as stored and recalled, others as generated on the spot dynamically depending on context. Does this make a difference for us? If so, how do we determine which is right and when?
8. Students often appear not to use in our classrooms knowledge that they know and use in other contexts. How can we describe, understand, and control this phenomenon?

Of course we want to not just describe how students think: we want to change how they think, i.e., we want them to learn. We pose three broad questions related to the question: How do mental structures change in response to experience, i.e., how do people learn?

9. What mechanisms describe sense-making in physics?
10. What level of mental structure is appropriate to use to describe learning in physics?
11. What does it mean to make a transition from a novice to an expert, and how do we describe this transition and the transition process?

MODELING THE CONTEXT

Individuals build their cognitive structures through their response to and interaction with both their physical and socio-cultural environment. Our understanding of an individual's cognition and learning has to dovetail with our understanding of both how their physical and the socio-cultural worlds function. We need to develop a descriptive understanding of how social environments like a classroom function and an understanding of how students respond to and are shaped by these environments. For a discussion and review of the literature on these issues see [3].

We might call the broad question for this topic: *The Socio-Cultural Question – How do students' experiences and interactions with their environments affect what they bring in to the class and how they respond to learning environments?* As before, our questions concern both static and dynamic issues.

We pose three questions related to statics:

12. What are appropriate ways of thinking about and describing how classrooms function or small groups interact?
13. What aspects of group interactions give us information about physical reasoning in the group?
14. What are appropriate mental structures for describing how individuals are affected by and affect an educational environment?

We pose three questions related to learning:

15. What are the feedback loops that connect individual actions, group dynamics, and larger scale environmental factors?
16. How do mental structures change in response to experience, i.e., how does context affect how people learn?
17. How does the learning of an individual and a group differ, and when do these differences matter?

MODELING THE CONTENT

As physicists, we often have a tendency to focus our classroom attentions solely on the content of the class: What should we cover and how? As physics educators, we often have a tendency to focus on what our students bring to the classroom that either does or does not match the way professionals think about a topic. But it is important to remember that physical laws are not given from the outside; they are the human brain's attempt to find ways to think about the physical universe that make sense to the human mind.

As we think about how to teach particular topics in physics, we need to consider what we might call: *The Cognitive Physics Question – Does thinking about how people think change the way we choose to talk (and think) about physical laws and principles?*

When we consider how we teach physics, we can't just take everything we have learned and create a class that works for ourselves. We know so much more than our students – some of it being so deeply learned that we don't even realize we know it or are using it. When you read a story to a 4 or 5 year old with the goal of helping her learn to read, you might not appreciate that she might not have any idea that the symbols on the page represent the words you are saying. Do we have a similar problem with the physics we are trying to teach? Perhaps understanding thinking and learning can help us better design learning goals for our courses.

18. Does understanding how people think and learn help us "deconstruct" our knowledge of physics so as to better understand what kinds of problems our students might have?
19. How does understanding how our students think and learn change the physics content we want to teach and how to orient it for them?

We already understand enough about thinking and learning to appreciate that learning physics is learning a way to think, not just learning facts about the physics. There is "content beyond the content," as well. Physics is not just about the laws of how the world works, it's about ways to think about how the world works and ways to find out new knowledge.

20. How can we teach more general way of thinking at the same time we are teaching specific content topics? What is particularly valuable for our students to know?

DESIGNING ENVIRONMENTS

As mentioned above, an essential goal of PER is figuring out how to teach physics more effectively. We can call this: *The Engineering Question – How can we create learning environments that respond most effectively to help particular populations of students learn physics in a variety of situations?*

This question is the one that has been most successfully addressed by the PER community and therefore we don't elaborate on it by adding specific questions.

CONCLUSION

The outline of our future work can be seen in our past success as primarily a research-based design community (*i.e.*, not a pure-research community). Researchers have studied specific issues in student learning. Curricula have been designed, implemented, evaluated, and refined. Issues of classroom design have been as important (though often less officially discussed) as individual learning. Thinking through the core concepts of the physics has led to excellent research and evaluation questions and has often led to researchers developing a deeper understanding of the material. Results have been presented at conferences and published in journals.[4] As a community, we understand these methods, agree on their relevance, and build on past history. We may each emphasize different elements, but we must keep in mind the interconnectedness of the research results when we ask ourselves, "What do I do in class on Monday?"

We recognize that our questions have been deeply researched in other areas that study human behavior. (See for example, [5].) We suggest that to begin to make more sense of what we do in PER, we will have to relate what we are learning to what is known in other fields such as general education research, cognitive science, neuroscience, and sociology. We intend that our questions all be interpreted in the context of physics learning. This leads to our final, broad overall question: *The Discipline Question — To what extent is thinking about and learning physics (and other scientific disciplines) different from the more general learning and thinking that tends to be studied in cognitive science and education research?*

In the next decades we need to build on what is known both in our own field and in other relevant research areas. By doing so, we can make more sense of our own research so that we can truly begin to create a coherent and useful synthesis.

ACKNOWLEDGMENTS

This work is supported in part by a grant from the National Science Foundation (#REC-0087519).

REFERENCES

[1] E. T. Whittaker, *A History of Theories of Aether and Electricity* (Dover Press, 1990).
[2] E. F. Redish, in *Proceedings of the Enrico Fermi Summer School, Course CLVI*, E. Redish and M. Vicentini, eds. (Italian Physical Society, 2004) pp. 1-64
[3] V. Otero, *ibid.* pp. 447-472.
[4] L. C. McDermott and E. F. Redish, *Am. J. Phys.*, **67**, pp. 755-767 (1998).
[5] J. Bransford, A. Brown, and R. Cocking, *How People Learn: Brain, Mind, Experience, and School* (National Academies Press, 2000).

Seeding Change: The Challenges of Transfer and Transformation of Educational Practice and Research in Physics (Part I)

Noah Finkelstein

Department of Physics
University of Colorado, Boulder

Edward Price

Center for Teaching Development
University of California, San Diego

Abstract. Academia appears to do a remarkable job at producing the next generation of research faculty. The long-anticipated shortage of well-qualified researchers has not appeared [1]. At the same time, while there are calls to reform educational practices in college and university classrooms, we are not widely preparing our future faculty to develop or implement these research-based educational practices. What mechanisms exist to foster the development of such practices and promote the field of PER more generally? These coupled papers examine the interrelated problems of supporting the development of the field, and the 'transfer' of what is known from PER to the more general populace of physics instructors. The first of these coupled papers outlines a framework of professionalism and faculty attitudes that is applied in the second paper in order to examine two programs designed to address the challenges of including education into the broader physics culture.

INTRODUCTION

In physics there is an asymmetry between research and education. This asymmetry is evident in the practice of current physicists and the preparation of future ones, in the attitudes and beliefs of physicists, and in the hiring and promotion practices at the most prestigious physics departments. In these papers we characterize this asymmetry and explore ways to reduce it. We focus on the role of graduate school, which we identify as a critical experience in the preparation of future physicists and a key point of leverage in attempts to integrate teaching and education research into the broader physics culture.

In Part I, we discuss physics graduate school in terms of preparing physics professionals. Using the notion of *professional*, we examine the relative roles of teaching and research in physics graduate school. We also consider faculty attitudes and beliefs about education and about research on education, and how these shape graduate experiences and the practices of the community. In Part II, we discuss two programs designed to integrate education into the core practice of physicists and examine the impact of these programs on the participants.

PHYSICS PROFESSIONALS

Graduate and postdoctoral programs in physics are designed to prepare professional physicists. Providing we agree upon what *professional* means, few would argue with such a statement. Professionalism as a concept embodies far more than a mastery of given content domain, such as theories of early star formation, or the trapping and manipulation of atoms. While there are many aspects to the professional life of physicists, here we focus on two critical elements: research and teaching. Though teaching may not be seen as within the traditional corpus of physicists' professional practice, we argue that, indeed, it is. Physicists may or may not be in professional environments that require explicit teaching (such as the university, college or pre-college environments); however, most physicists engage in various forms of teaching when broadly considered (whether it includes training within the work place, or sharing work with others in the form of presentations). More subtly, by including teaching and education within our conception of the core practice of physicists, we explicitly state the value and importance of education. Education is the process by which we produce more physicists and the means by which we share what is

CP790, *2004 Physics Education Research Conference*, edited by J. Marx, P. Heron, and S. Franklin

known and useful in our discipline with the rest of society (including those that vote, fund, and set policies for our work and society). While universities adequately prepare graduate students for their professional roles as researchers, our efforts at preparing these same students in teaching and educational practice in physics are lagging. By more thoroughly defining the characteristics of professionals, we highlight the achievements and shortcomings of our efforts in the preparation of graduate students and postdocs as physics professionals. In Part II, we examine two programs that successfully augment the traditional training that graduate students and postdocs receive; we also consider the issues surrounding making these programs integrated and sustainable.

Six Characteristics of Professionals

Shulman considers a profession to 'describe a special and unique set of circumstances for deep understanding, complex practice, ethical conduct and high-order learning, circumstances that define the complexity of the enterprise and explain the difficulties of prescribing both policies and curriculum in this area' [2]. Professionalism has both technical components and ethical or moral components. It is an evolving rather than prescriptive domain of practices. Shulman characterizes six critical (necessary but not sufficient) conditions of professional learning: service, understanding, practice, judgment, learning, and community [2]. We examine these six characteristics of professional learning to explore how we might include education more thoroughly within the preparation of future physicists.

Service: Service is the goal of a profession – to augment and contribute to society using specialized skills, knowledge, and judgment bound to a particular community and domain of inquiry. Professional physicists serve society (by characterizing and shaping our world). Our graduate students, with thorough preparation in research are well trained to serve in this manner, by advancing the field of physics and expanding what is known and practiced. Service in *teaching* physics is a muted theme in our graduate programs – it is represented by paltry teaching requirements, limited or absent training programs, and an area of minimal supervision or feedback. Graduate school is complete when a student contributes research to advance the field (in service to the sub-domain and society at-large). Teaching is either considered an obstacle, or an addendum.

Understanding: The body of knowledge, the scope of what is known, tested, debated, and developed is one of defining characteristics of the profession. Understanding Beowulf, while critical to some scholarly and professional pursuits, is generally *not* considered within the body of understanding necessary for the professional physicist. Through courses in particular domains of inquiry, and particularly through laboratory and research apprenticeships, graduate students develop an understanding of the research domain of physics. When it comes to teaching, in general, graduate programs provide remarkably few opportunities to promote student understanding of teaching, what is known in the growing field of physics education research, or the rich literature of educational practice.

Practice: The practice of physics may be considered the purpose, or ends, of knowledge. To apply, to test, and to further what is known in the field requires the *practice* of physics. Ultimately, the practice of research is the hallmark of a graduate program – through a fairly explicit model of apprenticeship, graduate students become professional researchers through practice. Teaching, in contrast, remains highly variable in terms of opportunity to practice (requirements vary between no and 3 terms of teaching, generally). It remains largely de-coupled from other aspects of professional learning and the lack of feedback and support of teaching turns the practice of teaching from a form of professional preparation to a stunted form of practice, a mechanism to fund graduate pursuits or fulfill requirements.

Judgment: Judgment bridges understanding and practice. Judgment is based on the ideological commitments of a profession, and distinguishes between prescriptive labor (following defined tasks) and professional activity. One of the defining characteristics of professional physicists is their ability to define problems and simplify questions in order to identify appropriate physical mechanisms that shape the investigation of our world. Through coursework, participation in research labs, colloquia, conferences, and professional journals, graduate students develop a sense of judgment about what research questions are appropriate, and how and when to revisit and revise a particular research approach. Meanwhile, little or no coaching on judgment in teaching is present, and too often teaching is approached *ad hoc* (for graduate students have little understanding upon which to base their practice).

Learning from experience: Learning from practice serves to inform academic understanding, practice, and judgment. This cyclic process in physics allows physicists to revise and revisit theories, challenge paradigms, and develop new forms of inquiry.

Research experiences and feedback, in the form of group meetings, conferences, and papers, shape graduate student preparation in research. In teaching, relatively few mechanisms for reflecting on educational practice exist. Students tend to teach as they were taught without much reflection on the process. Only in the exceptional cases is teaching recognized in graduate preparation (with teaching prizes, or punishment). In the normal course of events, there are no opportunities to reflect upon teaching practice other than to recognize that these duties have been discharged.

Community: A community of professionals has distributed expertise across and responsibility for a domain. This community holds the reigns to validate new forms of understanding, practice, and judgment. In physics, when these community values are breached (e.g. when data are falsified), it is a scandal and the community rejects the work and the individual conducting the work. The research community is well defined for graduate students. They participate as a cohort of students taking classes, then in a research group, as conference participants and members of a discipline, and ultimately as contributors to the field (through presentation and publication). Teaching has no such established or built-in community in graduate preparation, despite it being a widespread practice of the professionals they seek to become.

FACULTY ATTITUDES

The asymmetry between teaching and research is not limited to graduate preparation; it is also reflected in faculty attitudes and practices. These two elements of university education are interrelated: faculty attitudes influence the focus of graduate school, while the attitudes of future faculty are shaped by the graduate experience. We first consider physics faculty attitudes and beliefs about teaching and learning, and then turn to attitudes and beliefs about *research on* teaching and learning. In Part II, we focus on the relationship between attitudes and graduate preparation.

Faculty attitudes about teaching and learning

It is generally acknowledged that instructors' attitudes and beliefs have a major impact on classroom practices and consequently on student learning [3-6]. For instance, White and Gunstone note that university teachers' views of the structure of physics were overwhelmingly the dominant criteria for deciding curriculum and pedagogy [7]. Instructors' beliefs about teaching and learning are formed by personal experience, experience with schooling and instruction, and experience with formal knowledge, including pedagogical knowledge [6]. Many beliefs about teaching and learning are established early and are strong and enduring [4,6].

Instructors' attitudes and beliefs about teaching and learning are particularly relevant in physics, where those attitudes and beliefs may constrain widespread use of the findings of PER. While PER provides guidance on effective instructional practices, it rests on a particular set of assumptions about teaching and learning that may be at odds with instructors' beliefs. Despite evidence that interactive engagement methods are more effective than traditional instruction [8], enthusiasm for using interactive techniques varies widely among instructors. We attribute this, in part, to differing attitudes and beliefs about teaching and learning. (Other constraints might include lack of resources, institutional support, or awareness [9].) For example, interactive techniques generally aim to improve the performance of all students, and particularly middle and low achieving students. If the instructor believes the purpose of the course is to 'find out who's good at physics' (as opposed to bringing all or most of the students to a particular level of competence), then he or she may feel that interactive techniques undermine the instructional goals. Additionally, if an instructor believes that physics is either something you 'get' or you do not, then he or she may not feel interactive techniques are worthwhile. We consider instructors' beliefs and attitudes about teaching and learning of central importance to widespread implementation of physics education research.

Faculty attitudes about research on teaching and learning

Physics faculty are experts at using research methodologies to study physical phenomena. This expertise is developed through a lengthy graduate school apprenticeship; as discussed above, no analogous experience exists for education. At the same time, physicists readily seem to transfer their physics skills to new domains such as biophysics, finance, and consulting [10]. In fact, such flexibility is a point of pride among physicists. One might expect that this research expertise could be applied to education as well. While some physicists have done this and others make calls to do so [11], outside of PER, research on teaching and learning, and scholarship of teaching and learning remain rare.

In this section we consider some possible origins of resistive attitudes towards education research. We begin by considering the role of a framework for understanding education, the context of education and research in physics, and the culture of physics.

A framework for understanding education

Physics education research provides a framework for investigating physics education and shaping effective educational practice. It is unclear how much physics faculty know about PER [9], but a lack of pedagogical content knowledge and its practice seems widespread. As noted above, White and Gunstone found that faculty approaches to teaching physics were dominated by their views on the structure of the material [7]. The failure to consider presenting the material in a way that facilitates student learning suggests a view of teaching that focuses on the topic and the teacher, as opposed to learning and the student.

We find, anecdotally, that even when made aware of PER, many traditional physics faculty question whether such an approach to education is valid, productive, or within the purview of physicists. We attribute this rejection, in part, to attitudes and beliefs about teaching and learning. Furthermore, we believe these attitudes and beliefs are an expression of physics culture more broadly.

Context of education and research in physics

By definition, graduate preparation occurs in an institutional context of Ph.D. granting universities. This context informs the goals and practices of graduate students and faculty, but may be a mismatch for graduates bound for institutional settings with other goals and practices. We do not dispute the need for institutions that emphasize research, but wish to emphasize that the *minority* of graduate students become faculty at institutions similar to those in which they were trained [1]. Such a focused emphasis on research to the exclusion of other professional characteristics has consequences for the entire physics community.

In addition to institutional context, we can consider the domain of problems investigated by graduate students. Generally graduate students in physics develop their skills as researchers in the context of a single domain (studying physical phenomena). Is it unreasonable to think physicists could see how these skills can be applied to educational issues? Physicists, as a group, have had little difficulty applying research methodologies to other scientific fields (e.g., biophysics) [10]. This suggests that physics graduate preparation is not context bound *per se*, but that the support of using these research skills depends on the target discipline. Our claim is that many (perhaps most) of the skills of analysis, reasoning, and experiment that are developed in graduate research may be productively levied in educational practice and research, *if* such pursuits are valued.

Culture of physics

The physics community at large considers teaching and research to be separate endeavors; ones that do not inform each other. Ultimately, this attitude can be viewed as a cultural feature of physics. This attitude is manifest in norms, practices and institutional structures such as the asymmetric treatment of education and research during graduate preparation. How and why this view becomes part of physics culture (and that of higher education generally) is a complex question that we begin to address by considering the process of graduate preparation. In Part II, we focus on the question of how to create a culture where teaching is included in the core practice of physicists and considered a researchable endeavor.

REFERENCES

1. Committee on Science Engineering and Public Policy (COSEPUP). Enhancing the Postdoctoral Experience for Scientists and Engineers. Washington DC: National Academy Press (2000).
2. L.S. Shulman, Professing the Liberal Arts, in *Education and Democracy: Re-imagining Liberal Learning in America*, edited by Robert Orrill. New York: College Board Publications, 1997
3. S. De Souza Barros, and M.F. Elia, "Physics Teachers' Attitudes: How do they affect the reality of the classroom and models for change?", in *Connecting Research in Physics Education with Teacher Education*, ed. Tiberghien, A., et al., ICPE (1998)
4. M. F. Pajares, Rev. Ed. Res, 62, p307 (1992)
5. P.L. Peterson, et al., Cog. Inst., 6, 1-40, (1989)
6. V. Richardson, "The role of attitudes and beliefs in learning to teach," in *Handbook of research on teacher education, 2nd ed.*, edited by J. Sikula. New York: Macmillan, 1996, pp. 102-119.
7. R.F. Gunstone, and R.T. White, "Teachers' Attitudes About Physics Classroom Practice", in *Connecting Research in Physics Education with Teacher Education*, edited by A. Tiberghien, et al., Pub ICPE, 1998.
8. R. Hake, *Am. J. Phys.*, 66, p64 (1998) and *Conservation Ecology* 5(2), 28, 2002
9. M. Dancy and C. Henderson, these proceedings.
10. See articles in *The Industrial Physicist*, e.g.: March/April, 1997, p32; September, 1997, p46; April/May, 2003, p20, and December, 1999, p9.
11. For example, P.S. Shaffer and L.C. McDermott, Am. J. Phys. 60, 994-1003 and 1003-1013 (1994); E.F. Redish, Am. J. Phys. 62, 796-803 (1994).

Seeding Change: The Challenges of Transfer and Transformation of Educational Practice and Research in Physics (Part II)

Edward Price

Center for Teaching Development
University of California, San Diego

Noah Finkelstein

Department of Physics
University of Colorado, Boulder

Abstract. These coupled papers examine the interrelated problems of supporting the development of PER, and the 'transfer' of what is known from PER to the more general populace of physics instructors. Two programs are examined to highlight these interrelated issues: the Postdoctoral Fellowships in Mathematics Science Engineering and Technology Education and the Preparing Future Physics Faculty Program. Data on successes and failures of these programs are presented and analyzed from a perspective of professionalism and attitudes and beliefs outlined in the first paper. We claim such programs set the seeds for inclusion of education and education research in the broader culture of physics.

INTRODUCTION

In Part I of this paper we examined physics graduate school in terms of preparing physics professionals. We also considered physics faculty attitudes and beliefs about education and research on education. In both areas we noted an asymmetry between the emphasis placed on research and education. In Part II, we discuss two programs designed to address these coupled issues.

TARGETING GRADUATE AND POST-DOCTORAL EDUCATION

During graduate school, physicists engage in authentic research experiences [1], but most often there is no corresponding apprenticeship regarding teaching and learning. In Part I we analyzed this discrepancy in terms of the professionalization of physicists. By extending the focus of physics graduate school to include structured attention to education, we may begin to give education greater prominence and validate education research in physics, by physicists. In this way, we can begin to shift the culture of physics to include education in the core practice of physicists.

We consider two preparation programs, one for physics graduate students and one for postdocs. Both programs focus on increasing the prominence of educational issues. After a description of the programs, we consider their impacts, potential for being sustained, and interaction.

Preparing Future Physics Faculty

In 1998, the American Association of Physics Teachers (AAPT) funded Preparing Future Physics Faculty (PFPF), a graduate program designed to augment traditional training in research. PFPF was a discipline-specific version of Preparing Future Faculty, a program initiated by the Council of Graduate Schools and the Association of American Colleges and Universities. PFPF and PFF were responses to calls for increased emphasis on preparation in the areas of teaching and professional development by the Association of American Universities [2] and the National Academy of Sciences [3]. The University of California, San Diego was one of the sites chosen for a PFPF program. One of the authors [NF] was involved with establishing the program; the other [EP] is a former participant and current director. The program ran with external support until 2000, and has since continued with the

CP790, *2004 Physics Education Research Conference*, edited by J. Marx, P. Heron, and S. Franklin

support of the physics department and UCSD's campus-wide Center for Teaching Development. Over the six years of its existence, the program has undergone substantial changes and evolved to address four goals:

- Preparing graduate students for their future responsibilities as educators by promoting awareness and understanding of PER;
- Raising awareness of differences in the needs and opportunities at different academic institutions (i.e., community colleges, bachelor's granting institutions, and regional and research universities);
- Providing physics graduate students with professional and career development in areas such as conducting a job search and writing grant proposals;
- Creating an environment where physics graduate students can discuss issues in the physics community.

Graduate students participating in PFPF attend weekly seminars on topics relating to the goals discussed above. The sessions cover a range of topics, but can be grouped into four areas: physics education, academic settings, professional/career development, and exploration of the physics community. In addition to weekly seminars, graduate students are encouraged to participate in a range of practice-based activities: researching, developing curricula, and teaching. Research projects include graduate students engaging in PER-based studies of local practice (such as examining instructor beliefs about teaching). Curricular development often takes the form of graduate students appropriating PER-based activities and adopting them for local practice – for example, graduate students have augmented the complement of Interactive Lecture Demonstrations (ILDs) running in the introductory sequence by building an RC circuit ILD (and testing its effectiveness in the algebra-based course). Finally, teaching practice is heavily emphasized. All students are encouraged to conduct a 5-10 minute micro-teach (presenting a single topic to the rest of the PFPF seminar). Subsequently, students engage in observations and guest lectures in local introductory courses and at partner institutions (community and teaching colleges). Ultimately, several students have become instructors- of- record, taking responsibility for designing and implementing a full term class at these partner institutions. All of these activities are supervised both locally by the PFPF supervisors and at the host institutions by practicing faculty. These activities ground the seminar discussions in practical experience, making both more meaningful. The scope of engagement (ranging from guest lecturing to teaching a course as instructor-of-record) depends on the participant's interests and constraints. Guest lecturing is valuable experience with a small time commitment. On the other hand, teaching a course provides a more comprehensive experience but is a demanding undertaking. Our most successful participant activities combine the best of both approaches by including a group planning component and a modular workload. By involving multiple participants, these programs achieve a significant impact, yet require only modest effort from individual graduate students.

Postdoctoral Fellowships in Science, Mathematics, Engineering and Technology Education

Several years earlier, the NSF instituted the Postdoctoral Fellowships in Science, Mathematics, Engineering and Technology Education (PFSMETE) in order to support postdocs as they built the expertise necessary to create and to succeed within hybrid science – education environments. The primary objective of PFSMETE was the preparation of individuals with recently awarded doctorates in science for leadership roles in science, mathematics, engineering, and technology (SMET) education through the development of expertise in science education research [4]. Further analysis of the program guide suggests a number of subordinate objectives primarily related to PFSMETE research. The NSF expected that PFSMETE research would also promote: 1) the communication of scientific knowledge to a variety of learners, 2) the assessment of teaching and learning programs, 3) the application of inter-disciplinary knowledge to curriculum development, and 4) the use of technology in SMET education.

During the three years of its existence, NSF granted 63 Fellowships to allow those who had recently been awarded doctorates in a scientific discipline to pursue education research of their own design. It is notable that the program adopted a student- directed (Fellowship) model rather than an institution- directed (Research Associate) model. That is, awardees brought funding with them to an institution to engage in research of their design, rather than engage in predefined research directed by others. In 1999, citing budget constraints, the program was cancelled, meaning that NSF would honor current awards, but offer no more Fellowships. The Fellows were distributed around the country and represented a host of disciplines, including mathematics, physics, geology, biology, chemistry, and engineering.

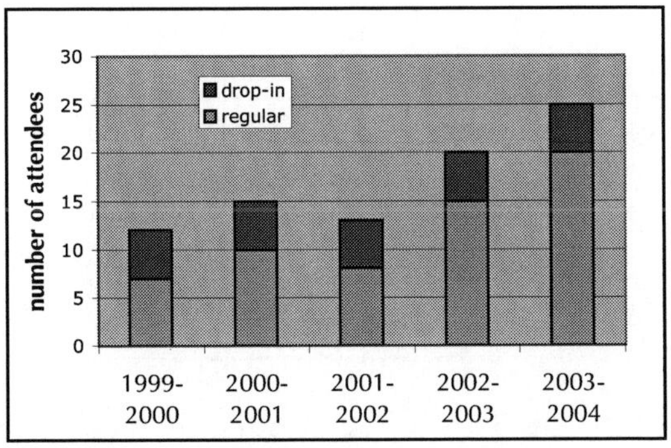

FIGURE 1 . Physics graduate program weekly attendance

RESULTS

While it should be clear that affecting students' choices and preparation is a long-term endeavor, we may assess the preliminary impact of these programs. First, it is worth considering whether students choose to participate in these voluntary programs. In the graduate program, participation has increased since its inception, as shown in Figure 1. Over this period, the UCSD physics department has averaged about 115 graduate students at any one time. Given an average of 5-6 years of graduate study, it can be seen that a significant fraction of the department's graduate students have participated in the PFPF program.

In the postdoctoral program, of the 63 participants, 12 Fellowships were awarded in physics. They were given to: Alicia C. Alonzo, Bruce B. Birkett II, Scott W. Bonham, Andrew R. Elby, Noah Finkelstein, Scott V. Franklin, Elaine Su-Eng Fu, Apriel K. Hodari, Leonardo Hsu, Beth Hufnagel, Catherine Ishikawa, and Gregory R. Schultz. The majority of these Fellows continues in an academic or educational setting, making up roughly 1/4 of the current cohort of self-identified junior faculty in PER [5].

As measured by surveys of the participants, each of these programs has been successful at building bridges between physics and education, and infusing physics education research into traditional practices in physics. We have surveyed PFPF participants on their attitudes about the importance of teaching and what they have learned from the program. The results are summarized in Table 1 on the following page. While it is unsurprising that the graduate students consider education a substantive part of their future career (they have self-selected to participate in the program), it is striking how few felt education was valued by the physics research community. Undoubtedly this reflects the almost exclusive research emphasis of a traditional physics Ph.D. program. Encouragingly, a very high fraction of participants report being more aware of the results of physics education research and plan to use those results in their own teaching, indicating that we have been preliminarily successful in achieving our goals of transferring traditional research skills to PER and to preparing students for teaching. Former participants who are now teaching report following through on these intentions.

Similar results were found in a survey of the PFSMETE participants issued in 2001. Postdocs report being satisfied with the Fellowship (9.5 out of 10 point scale) and well trained by the program (8.5 out of 10 points). Furthermore, nearly all participants report more interaction with educational programs as a result of the Fellowship, as shown in Figure 2.

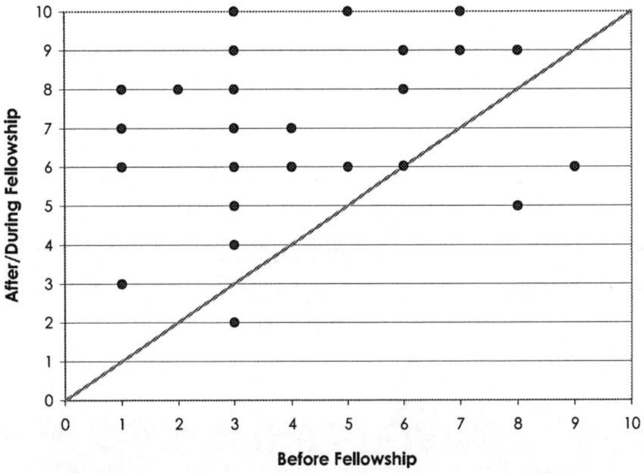

FIGURE 2. Contact with education programs as a result of PFSMETE. 0 represents no contact, 10 maximum of contact. (Notably two participants had joint degrees in education and science prior to PFSMETE)

PROFESSIONALIZATION REVISITED

One may revisit the characteristics of profession offered in Part I and observe that both programs, PFPF and PFSMETE, begin to develop a profession of education within the sciences. The role of these programs within the physics and education communities is illustrated in Figure 3. The PFPF and PFSMETE programs seek to extend the region of overlap between the PER and traditional physics communities. This may be seen as attempt to blur disciplinary boundaries and help foster cultural change – to produce a hybrid culture and introduce tools and practices from one intellectual domain into another. It

TABLE 1. Survey of PFPF Participants.

33/38 students surveyed considered education a substantive part of their future career. Of those 33:

Statement	Percent agreeing
Feel education is valued by the physics research community	18%
After participation in PFPF	
More aware of the results of PER	82%
Planning to incorporate PER results in teaching	94%
View PER as a legitimate research activity with in physics community	88%

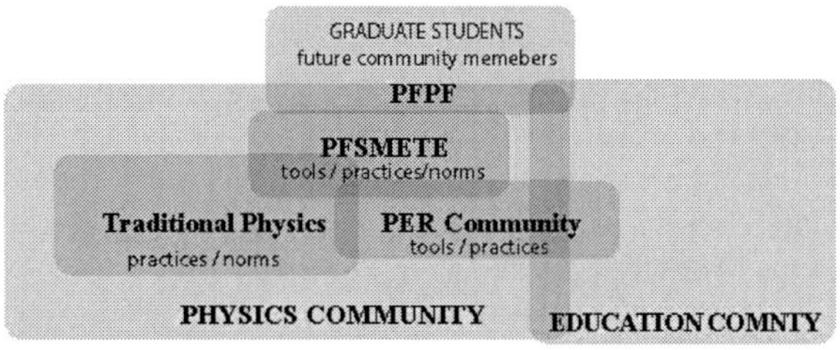

FIGURE 3. Interactions between members of the physics and education communities, including the PFPF and PFSMETE programs.

is worthy of note that the PFSMETE Fellowship directly supported the development of the PFPF program, as one of the authors and architects of the PFPF program [NF] was supported by PFSMETE at the time the PFPF was established.

SUSTAINABILITY

The establishment of these programs may be viewed as the creation of *settings*, new and sustained relationships between individuals [6]. PFPF and PFSMETE's structures, explicit (and implicit) goals, history, and surrounding context reflect their initial intent to sustain the particular relationships developed by participation in the programs and to infuse education into physics. Sarason contends that in developing settings, a program's success or failure depends upon two critical factors: the initial structure of the program and the adaptation of that structure to local conditions [6]. These two programs included these critical features – they coupled the program structures to the surrounding institutional structures while simultaneously building flexibility and autonomy of practice that allowed individuals and programs to adapt. This perspective provides insight into the continued success of UCSD's PFPF program; despite the loss of external funding, the program has been supported by the local institution and adapted to local interests and needs. Similarly, while the PFSMETE program was cancelled due to external factors (funding constraints), the program's impact continues via the contributions of former Fellows.

Using this idea of the creation of settings and Shulman's characterization of professional education, we posit that two key characteristics shape the success of both the programs and the experiences of the individuals within them: 1) a balance between academic autonomy (independent approach and voluntary participation) and institutionalization (directed work and inclusion); and 2) building potential for crossing established academic cultural boundaries (particularly between traditionally conceived disciplines and education). Again, the balance of program autonomy and institutionalization allow for inclusion by the surrounding institution, and allows the program to adapt and evolve to local concerns. The second critical feature, that of bridging, embodies the specific goal of these programs– to effect cultural change.

ACKNOWLEDGMENTS

These programs were supported by the National Science Foundation and American Association of Physics Teachers. The authors are grateful for members of the Physics Education Research Groups at UCSD and UC, Boulder as well as the students and postdocs in these programs.

REFERENCES

1. J.S. Brown, et al. *Ed. Res.*, **18**, p32. (1989)
2. Association of American Universities, Committee on Graduate Education Report and Recommendations, Oct. 1998. http://www.aau.edu/reports/GradEdRpt.html
3. National Academy Of Sciences Committee On Science, Engineering, And Public Policy Reshaping the Graduate Education of Scientists and Engineers, 1995, http://www.nap.edu/readingroom/books/grad/brief.html
4. National Science Foundation, Postdoctoral Fellowships in Mathematics, Science and Technology Education, document number: NSF 97-42. (1997)
5. As listed by the Y-PER Physics Education Research– listserve of junior faculty.
6. Sarason, S.B.. *The Creation of Settings and the Future Societies*. SF, CA: Jossey-Bass Publishers. (1989)

Transfer of Learning in Quantum Mechanics

Chandralekha Singh

Department of Physics, University of Pittsburgh, Pittsburgh, PA, 15260

Abstract. We investigate the difficulties that undergraduate students in quantum mechanics courses have in transferring learning from previous courses or within the same course from one context to another by administering written tests and conducting individual interviews. Quantum mechanics is abstract and its paradigm is very different from the classical one. A good grasp of the principles of quantum mechanics requires creating and organizing a knowledge structure consistent with the quantum postulates. Previously learned concepts such as the principle of superposition and probability can be useful in quantum mechanics if students are given opportunity to build associations between new and prior knowledge. We also discuss the need for better alignment between quantum mechanics and modern physics courses taken previously because semi-classical models can impede internalization of the quantum paradigm in more advanced courses.

INTRODUCTION

Transfer of learning is defined as the application of knowledge and skill acquired in one context to another [1]. Cognitive theory suggests that transfer can be difficult especially if the source (from which knowledge is to be transferred) and target (to which knowledge is to be transferred) do not share surface features. The failure to transfer is attributed to the fact that knowledge is encoded in memory with the context in which it was acquired and the features of the target may not trigger the recall of relevant resources in memory about the source even if the two share deep features. Transfer in physics is often challenging. There are very few concepts and principles in physics and learning requires unpacking and understanding their applicability in diverse physical situations. For example, waves can be mechanical, electromagnetic or quantum mechanical, all of which share deep features but are superficially very different. However, transfer is essential for a good grasp of physics because physics is hierarchical and new knowledge builds on prior knowledge. In order to improve the transfer, it is important to help students de-contextualize learning and store it in memory at an abstract level. This can be facilitated by learning the same concept in different contexts.

It is clear that some appropriate transfer does occur from other courses to quantum mechanics (QM). Here, we will focus on identifying learning that is difficult to transfer appropriately for students in QM and discuss possible reasons. This type of study is important for designing effective instructional strategies to improve transfer [2]. Our investigation involves administering and analyzing written tests and conducting individual interviews. In this paper, we will outline the general difficulties found and their possible sources. Detailed statistics will be presented elsewhere.

Sources of major difficulties with QM include its novel paradigm, abstractness and mathematical sophistication. The quantum paradigm does not conform to our everyday experience. In order to transfer previous learning, e.g., waves or probability concepts learned in classical contexts, students must first learn the basic structure of quantum mechanics and then contemplate how the previously learned knowledge applies to this novel framework. Also, most students take a one or two semester course in modern physics before QM in which quantum mechanics is covered. Below, we discuss why there should be better co-ordination between the teaching of modern physics and quantum mechanics, and more effort channeled on helping students understand the limits of the semi-classical models and perhaps the time alloted to such models should be reduced.

INAPPROPRIATE TRANSFER FROM CLASSICAL MECHANICS

The written free-response questions and individual interviews show that some students have not acquired or retained relevant knowledge from previous courses (e.g., superposition principle). Here, we will focus on evidence pointing to the difficulty in appropriate transfer due to lack of robust knowledge hierarchy (schema) for quantum mechanics. For example, one common difficulty is assuming that an object with a label "x" is orthogonal to or cannot influence an object with a label "y". This is evident from responses such as "The magnetic field

CP790, *2004 Physics Education Research Conference*, edited by J. Marx, P. Heron, and S. Franklin

is in the z direction so electron is not influenced if it is initially in an eigenstate of S_x" or "Eigenstates of S_x are orthogonal to eigenstates of S_y". In introductory physics, x, y and z are indeed conventional labels for orthogonal components of a vector. Unless we give students an opportunity to understand the structure of quantum mechanics, such difficulties will persist. Students must be given opportunity to build schemas that allows them to understand that although an electron in an external magnetic field pointing in the z direction is in a real physical three dimensional space of the lab, to make predictions about the measurement performed in the lab using quantum mechanics, one needs to map the problem to an abstract Hilbert space in which the state of the system lies and where all the observables of the real physical space get mapped on to operators acting on states.

The conflict between the classical and quantum paradigm also forced students to incorrectly conclude that "Successive measurements of continuous variables of a particle, e.g., position, produce "somewhat" deterministic outcomes whereas successive measurements of discrete variables, e.g., spin, can produce very different outcomes." For example, during the interview, one students said the following regarding successive measurements on a single electron: "If an observable has a continuous spectrum...the next measurement won't be very different from the first one. But if the spectrum is discrete then you will get very different outcomes." When asked to elaborate the student added "For example, imagine measuring the position of an electron. It is a continuous function so that time dependence is gentle and after a few seconds you can only go from A to its neighboring point (pointing to an x vs. t graph that he sketches on the paper) you cannot go from this without going through this intermediate space." When asked to elaborate his views about observables with discrete spectrum, the student said "Think of discrete variables like spin...they can give you very different values in a short time because the system must flip from up to down. I find it strange that such (large) changes can happen almost instantaneously. But that's what quantum mechanics predicts..." This example again illustrates the need for helping students build schemas consistent with quantum mechanics.

We performed a study in which we asked students: "If the period $T = 2\pi/\omega$ of a Quantum Harmonic Oscillator (QHO) is known, how would you verify it experimentally?" A common response was that one must figure out some way to perform two successive measurements of the position of a particle separated by a time T to make sure they are the same. The problem is that according to the postulate of quantum mechanics one cannot measure the position twice for the same particle and hope to get the desired information because the first measurement will collapse the state. Further probing during interviews showed that students were transfering the classical notion of period using the logic that the position of the QHO is deterministic. Interviews clarified that students were not referring to the "probability" of position measurement repeating itself after a time T or to the coherent states of a QHO. Few students provided correct response along the line that one can perform measurements of position on ensembles of identically prepared systems, e.g., measure position on half of the systems at time $t = 0$ and half after time T. The expectation values of position obtained in the two cases should be the same. Another correct response involving an experiment to obtain the energy separation between the adjacent levels to find ω and using it to determine T was rare.

Written responses and individual interviews also show that students have difficulty distinguishing between classical and quantum bound states. Classically a particle is bound if the energy of the particle is such that there are two turning points. Quantum bound state is different due to the quantum mechanical tunneling and the energy of the particle in a bound state must be less than the potential at both $-\infty$ and $+\infty$. A majority of the students transfered the classical notion and looked for classical turning points to determine quantum bound states.

INAPPROPRIATE TRANSFER FROM MODERN PHYSICS

Experts understand the limits of the models they employ and use more refined models when simpler ones fail. Modern physics courses often emphasize semi-classical models without helping students understand their limitations. For example, WKB approximation is used as a back-of-the-envelope calculation for 1-D Time-Independent Schroedinger Equation (TISE). Within this approximation, instead of an extended wave function with varying probability density in regions with different potentials, one often uses language such as "The particle spends more "time" in this region since its total energy is fixed but if the potential energy is higher then the kinetic energy is lower". Students who do not understand the limits of these models take the word "time" literally and believe that there is a non-trivial time-dependence even in a stationary state. These notions from semi-classical model of what "time-development" means makes learning time-development in QM even more challenging. Further misconceptions that develop from this include ideas such as "There must be work done on a particle when it tunnels through a barrier because the particle was initially on one side of the barrier and later on the other side so it must dig a hole to get through" [3]. Modern physics texts often use a similar "imprecise" language when describing an electron in the stationary state of a hydrogen atom. They write that the electron spends more "time" in one place and less "time" in another instead of explanation involving the probability of finding it at different distances from the nucleus. The notion that the particle is either here or there but spends different amount of time in different places is inconsistent with the indeterministic nature of position in quantum mechanics. It breeds many misconceptions that become very difficult to rectify in QM.

Our research also shows that students have trouble distinguishing between individual measurements and expectation values of physical observables. Modern physics courses often avoid discussion of the probability of measuring individual outcomes and only discuss expectation values. Many students conclude that for a physical observable, its expectation value and individual measurement are the same or that each individual measurement of the observable will yield the expectation value. Individual interviews also show that many students believe that one cannot perform individual measurements in quantum mechanics despite the fact that they have learned about double slit experiments with single photon or single electron. In such experiments, interference pattern builds up gradually as particles hit the screen one by one. It is clear from such experiments, that individual measurement of position of the particle on the screen cannot be predicted ahead of time and individual measurement is very different from the expectation value of position. Our research also shows that students often do not understand that expectation value refers to the average value of the observable obtained after performing a large number of measurements on identically prepared systems and have difficulty understanding the active nature of measurement in quantum mechanics.

Students in QM learn about incompatible observables (position-momentum, position-energy etc.) and the uncertainty relation between them. After relevant instruction, we asked students about the measurement of the distance from the nucleus of an electron in the ground state of a hydrogen atom. Many students claimed it will always yield the same value and some explicitly cited the Bohr model.

DIFFICULTY IN LATERAL TRANSFER

Quantum postulates are very different from everyday experience and there is a "quantum leap" in the mathematical sophistication required. Therefore, it is difficult for students to de-contextualize concepts and apply them in contexts other than the ones in which they were acquired. For example, the superposition principle which is learned previously in many classical contexts is also crucial for understanding the wave character of particles, in particular for grasping how the wave-packets are formed. In the interviews, we found that even those QM students who can explain the superposition principle in classical contexts have difficulty in discerning its relevance to "particle" waves. Many believe that for a given potential, the stationary states are the only wave functions possible. They have trouble understanding that any smooth wave function meeting the boundary condition is acceptable and wave packets are formed by the linear superposition of stationary states. The Time-Dependent Schroedinger Equation (TDSE) is mentioned in passing in most modern physics texts and the idea of linear superposition of wave function is usually ignored. In modern physics, the concept of wave packet is introduced in an ad hoc way when discussing the semi-classical models and wave-particle duality but almost never after the quantum mechanical model is introduced (except in the context of tunneling through barriers where little justification or reasoning is provided for why wave-packets are valid states of the system). Unfortunately, TDSE is severely deemphasized even in QM courses.

We posed to QM students a question about whether a particle in a 1-D infinite square well potential of width "a" can be in a state $sin^3(n\pi x/a)$ at a given time. Students had learned in the lectures and homeworks that the particle can be in the states $Asin(\pi x/a) + Bsin(2\pi x/a)$ or $Ax(a-x)$ at a given time since they represent linear superposition of stationary states and satisfy the relevant boundary conditions. Many students claimed that $Asin(n\pi x/a)$ were the only possible solutions. When we explicitly posed a question asking whether they agree with a student who claims that all allowed wave functions satisfy $H\Psi = E\Psi$, roughly one third said "yes". Some explicitly claimed that this is true "by definition". Even when the question was posed explicitly in the context of hydrogen atom about whether the stationary states were the only possible wave functions, the response was similar. Transferring superposition ideas from one type of potential to another type (e.g., from infinite square well to QHO) was also challenging for students.

Time-development of wave function is an important concept for understanding among other things, magnetic resonance imaging, interaction of light with matter and quantum computing. In a linear superposition of stationary states, each term evolves in time via its own phase factor, e.g., $Aexp(-iE_1t/\hbar)\Psi_1 + Bexp(-iE_2t/\hbar)\Psi_2$. Students have difficulty in transferring these concepts learned in one context to another because the instruction does not help them recognize the deep features involved in quantum dynamics and the role of the system Hamiltonian in this process [2]. Our research shows that the most common mistakes in the questions asking for the wave function after time t, given the wave function at $t = 0$, are responses such as $(A\Psi_1 + B\Psi_2)exp(-iEt/\hbar)$ or $Ax(a-x)exp(-iEt/\hbar)$.

These expressions are not solutions of TDSE and cannot represent valid wave functions. During the interviews, when students who provided such responses were asked about the meaning of "E", the common response was that it is the energy of that "state". This statement is incorrect because these are not states of definite energy.

In another study, we asked students to explain whether the probability density should depend on time if the wave function of a particle in an infinite square well at time $t = 0$ was $A\Psi_1 + B\Psi_2$. Interestingly, even those who recognized that each stationary state evolves according to different time-dependent phase factors, often said that the probability density should be time-independent. They claimed that the time dependence will cancel out when one takes the absolute square of the wave function. Their explanations show that they lost the interference effects because they were adding the probabilities of being in each stationary state ($|A\Psi_1|^2 + |B\Psi_2|^2$) instead of adding the probability amplitudes at time t and then squaring. It is important to give students an opportunity to think through the consequences of this analysis in the context of concrete experiments, e.g., a double slit experiment.

Students often over-generalize results related to "position" to other observables. For example, quantum measurement hypothesis says that the measurement of an observable will collapse the state into an eigenstate of the corresponding operator. Our research shows that many students believe that the measurements of all physical observable, e.g., energy, will collapse the wave function into a position eigenstate (a delta function in position). Students also struggle with qualitative ideas behind the Fourier transform. We asked students the following question: "Given that at time $t = 0$, the wave function $\Psi(x)$ of a particle incident on a potential barrier is a localized wave packet approximated by a delta function, what if anything can you say about the momentum space wave function $\tilde{\Psi}(k)$? Draw a rough sketch with your explanation." The common responses claimed that $\tilde{\Psi}(k)$ should also be localized about the same location, but perhaps be slightly broader or that one should just find the "k_0" corresponding to the "x_0" where the particle is localized (some sketched square barriers in k-space right below the barrier in x space).

Our investigations also show that students do not automatically transfer the formalism of quantum mechanics learned in the 2-D Hilbert space of spin 1/2 system to the infinite dimensional Hilbert spaces. Even if students have internalized the formalism for spin 1/2, connection with the infinite dimensional space is challenging for at least two reasons. First of all, the mathematical complexity increases manifold. Secondly, the infinite dimensional Hilbert space dealing with the orbital degrees of freedom causes greater interference with students' cherished classical ideas. Quantum mechanical spin of particles is abstract and students do not have much "feel" for it. Therefore, concepts such as "the spin of the particle does not in general have a definite value" or "different components of spin cannot be measured simultaneously" do not produce as much conflict with the existing schemas. On the other hand, it is much more difficult to internalize that position, momentum or energy of the particle do not in general have definite values or that they cannot be measured simultaneously. The existing classical schemas related to these observables interferes with the counter-intuitive quantum mechanical tenets.

SUMMARY

In order to help students transfer concepts from the previous courses to QM or within QM from one context to another appropriately, it is important to give them opportunity to build schemas consistent with the postulates of quantum mechanics and help establish lateral connections between useful concepts learned previously and the new knowledge. Our research also shows the need for better alignment between instruction in modern physics and QM. In particular, the manner and extent to which semi-classical models are covered in modern physics should be re-evaluated.

One encouraging finding from the perspective of developing effective instructional strategies is that despite the abstractness of quantum mechanics, the types of difficulties students have are universal. Students over-generalize classical and semi-classical ideas in ways that can be classified in a few categories. We are developing and assessing Quantum Interactive Learning Tutorials (QuILTs) that take into account student difficulties and incorporate paper-pencil tasks and computer simulations. The research-based tutorials that provide scaffolding. They are designed to help students build robust schemas consistent with quantum mechanics principles while helping them discern how their prior knowledge resources can be utilized in appropriate situations. They help students organize their knowledge hierarchically and represent it at a more abstract level in memory to facilitate de-contextualization, categorization and recognition based upon deep quantum mechanics laws.

ACKNOWLEDGMENTS

We are grateful to the NSF for award PHY-0244708.

REFERENCES

1. M. Gick and K. Holyoak, NY, Academic Press (1987).
2. C. Singh, Am. J. Phys., **69(8)**, 885-896, (2001).
3. J. Morgan et. al., AAPT announcer, 2003.

A Repeat Performance? Challenges In Developing Robust Conceptual Understanding in Quantum Mechanics

Bradley S. Ambrose

Department of Physics, Grand Valley State University
Allendale, MI 49401, ambroseb@gvsu.edu

Abstract. Ongoing physics education research (PER) at Grand Valley State University (GVSU) is guiding the development of instructional materials for teaching modern physics. Results from this project can help address the question: To what extent should we expect upper-level physics students to be able to apply concepts previously covered in class—even those addressed through PER-based instruction at the advanced level—to different situations? Extensive research at the introductory level has already revealed that such transfer is extremely difficult for beginning students to do on their own. Preliminary results from this project suggest that, even among upper level students, specific conceptual and reasoning difficulties must be addressed explicitly and at multiple instances during instruction.

INTRODUCTION

Physics education research (PER) at the introductory level has revealed that most students lack a working knowledge of basic physics concepts after standard lecture instruction, whether at or beyond the introductory level. [1] Results from this ongoing project yield insight on the following question: After PER-based instruction of a given concept (or concepts) in an initial context, to what extent do upper level students correctly apply that same concept (or concepts) in a different context? This question will be explored in the context of stationary bound state wave functions, partly due to the growing research base in student learning of quantum mechanics, [2-6] and also because the qualitative analysis of such wave functions is possible with careful reasoning about probability density and deBroglie wavelength. [7]

CONTEXT OF INVESTIGATION

The primary context of investigation is the sophomore level modern physics (MP) course at Grand Valley State University (GVSU). The course format consists of three 50-min lectures and a 3-hr lab each week. Most students enrolled (8–14 each year) pursue majors or minors in physics. The course covers special relativity, photon model of light, wave-particle duality, and basic topics in quantum mechanics.

The author served as either the lab instructor or the instructor of the entire course for the GVSU classes to be discussed. The lab portion of each of those classes consisted of a mixture of traditional labs and small-group inquiry-based activities based on new or existing materials from *Tutorials in Introductory Physics.* [8] Written qualitative questions were administered in class as tutorial pretests and post-tests.

Supplemental data will also be presented from a junior-level course in quantum mechanics (QM) at the University of Washington (UW). Unless otherwise specified, the relevant instruction in that course was delivered in traditional lecture format.

STUDENT DIFFICULTIES WITH BOUND STATE WAVE FUNCTIONS

A research task that has been fruitful in probing student ideas about the qualitative behavior of wave functions is referred to here as the "stepped well" question. The students are asked to consider a particle in the potential shown in Fig. 1. They are told that the

CP790, *2004 Physics Education Research Conference*, edited by J. Marx, P. Heron, and S. Franklin

wave function $\psi_n(x)$ corresponds to a highly excited energy state (*i.e.,* has many nodes between $x = 0$ and $x = L$). Students are asked to compare regions I and II according to (a) the amplitude of the wave function and (b) the spacing between consecutive nodes of the wave function. The students were required to explain their reasoning and to sketch a qualitatively correct graph of the wave function $\psi_n(x)$.

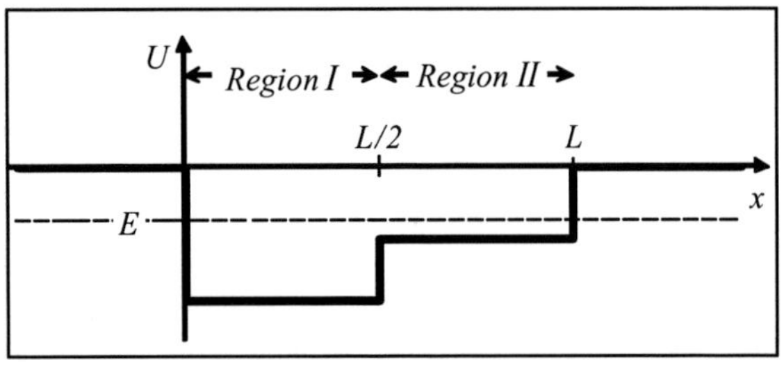

FIGURE 1. Potential energy graph from the "stepped well" question.

Although students did not have the time or requisite math background to solve for the wave function analytically, a qualitative analysis [7] is sufficient. The kinetic energy of the particle is greater in region I than in region II. Semi-classically, the amplitude of the wave function is larger in region II because that is where the particle "spends more time" per unit distance. Also, the (local) deBroglie wavelength is inversely proportional to momentum, so the nodes of the wave function are more tightly spaced in region I, where kinetic energy is greater.

Measuring Conceptual Transfer Of DeBroglie Wavelength and Probability

The "stepped well" question was given to QM students at UW ($N = 61$, Fall 2001) and to MP students at GVSU ($N = 10$, Winter 2002, designated here as "MP class #1"). The UW class had already completed a tutorial on electron interference and deBroglie wavelength [8]. GVSU class #1 had a similar tutorial as well as another tutorial on classical probability.[1] These tutorials have proven effective in enhancing student understanding in the contexts that they explicitly cover. [9, 10]

Even after completing the previous tutorials, however, very few students correctly answered all parts with correct reasoning (see Fig. 2). The low success rates, particularly on the amplitude question, suggest that few students could spontaneously extend their understanding of the requisite ideas to the new context of wave functions. Careful analysis of responses revealed several persistent conceptual difficulties, two of which are discussed below.

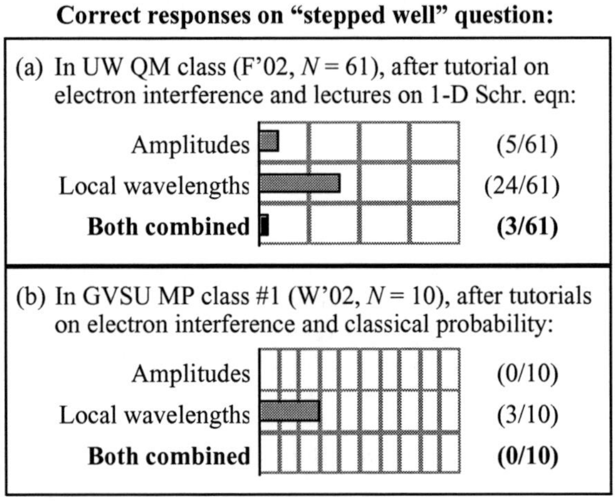

FIGURE 2. Results from "stepped well" question (a) by QM students at UW and (b) by MP class #1 at GVSU.

Belief That Particle "Spends More Time" In Regions Of Lower Potential Energy

Most students incorrectly believed the amplitude of the wave function would be larger in region I, where the potential energy is lower. (See example of an incorrect student graph in Fig. 3.) Many articulated the idea that the particle, as one student stated, "will want to 'fall' into the lower potential." Students have been shown to hold similar incorrect intuitions about classical situations, such as a block undergoing simple harmonic motion. [10]

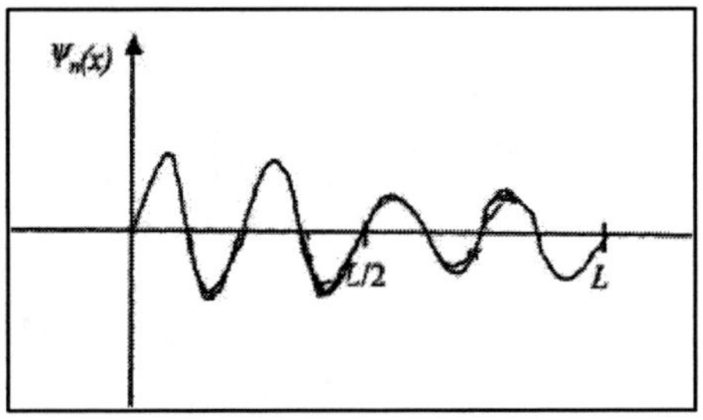

FIGURE 3. Example of incorrect student graph of the wave function on the "stepped well" question. Note the incorrect behavior of both the amplitude and local wavelength.

[1] These tutorials, and others, were pilot-tested at GVSU as part of the FIPSE project P116B000300, "Redefining the Teaching of Applied Quantum Mechanics through the Dissemination of a Proven Reform," R. N. Steinberg and M. C. Wittmann (2000-2004).

Failure To Relate DeBroglie Wavelength To Kinetic Energy

Many students also failed to recognize that local deBroglie wavelength varies with kinetic energy. Some associated wavelength with total energy. As one student explained, "the spacing between nodes in region I and II is equal because energy is constant." The graph in Fig. 3 illustrates this particular error.

ADDRESSING STUDENT DIFFICULTIES: THE NEED FOR MULTIPLE INTERVENTIONS

The results presented above revealed the need to address persistent student difficulties specifically in the context of wave functions. Preliminary versions of new tutorials were tested in the MP classes that took place in Winters 2003 and 2004 at GVSU, referred to here as classes #2 and #3.

Three additional tutorials, described very briefly here, have been tested to date. The first, *Square well potentials (SqW)*,[2] is designed to help students compare and contrast bound state wave functions for infinite and finite square "well" potentials. It also guides students to recognize that the absolute square of $\psi_n(x)$ yields the classical probability density in the limit of large *n*, thus motivating the correspondence principle. The next tutorial, *Reflection and transmission of matter (R&T)*,[3] focuses on scattering from potential "steps" and tunneling, with particular emphasis on relating local deBroglie wavelength to potential energy. The third tutorial, *Relating classical mechanics to quantum mechanics (RCQ)*,[3] helps students apply both local deBroglie wavelength and the correspondence principle in the context of more complicated potentials, including the "stepped well" and harmonic oscillator potentials.

Assessment Of Tutorials Square Well Potentials *And* Reflection and Transmission of Matter

The "stepped well" question was used as a tutorial post-test after *SqW* in class #2 and after both *R&T* and *SqW* in class #3. The results from these classes are summarized in parts (a) and (b) of Fig. 4. Despite the small class sizes, interesting trends can be noted.

Correct responses on "stepped well" question:

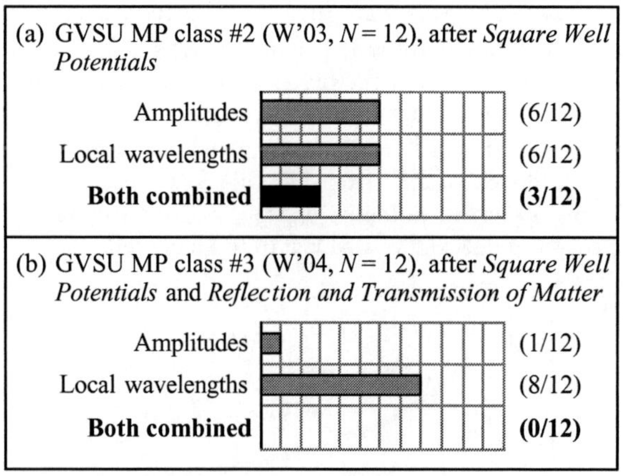

FIGURE 4. Results from GVSU modern physics classes #2 and #3 on the "stepped well" question.

In class #2 half of the students correctly answered on the amplitudes question. The modest level of success could be attributed to the *SqW* tutorial and to its emphasis on the correspondence principle.

Most of the students in class #3 were successful on the wavelengths question, which can be attributed to the *R&T* tutorial. However, very few answered the amplitude question correctly, even though *SqW* was the tutorial completed most recently before administering the "stepped well" question. Many students seemed to make a naïve association between a potential "step" (as discussed in *R&T*) and the "step" inside the stepped "well." One student explicitly used reasoning regarding reflection and transmission at the "step" to explain why the amplitude would be larger in region I. Others, possibly remembering their experience matching boundary conditions in the scattering problem for the "step," incorrectly answered instead that the amplitudes were equal in both regions. More in-depth investigation is needed to pinpoint the reasoning underlying these errors. Nevertheless, despite the fact that *SqW* and *R&T* address amplitude and wavelength issues in the context of wave functions, students did not develop a sufficiently robust understanding of those concepts in the context of a more complicated potential "well."

Assessment Of Tutorial Sequences That Included Relating Classical Mechanics To Quantum Mechanics

The results summarized above motivated the need for the third tutorial *(RCQ)* as well as a new research task, the "unknown potential well" question, to serve as a post-test for this tutorial. On the post-test students were shown a graph of the probability density (see

[2] Initially developed at GVSU.
[3] Initially developed at UW and subsequently modified at GVSU.

Fig. 5) for a classical particle in an unknown 1-D potential. Students were first asked to rank locations x_1, x_2, and x_3 according to potential energy. They were also asked to sketch the behavior of a highly excited quantum mechanical energy state of the same potential. To answer correctly, students needed to reason "backwards" from the classical probability density graph to potential energy, and then correctly relate local deBroglie wavelength to kinetic energy.

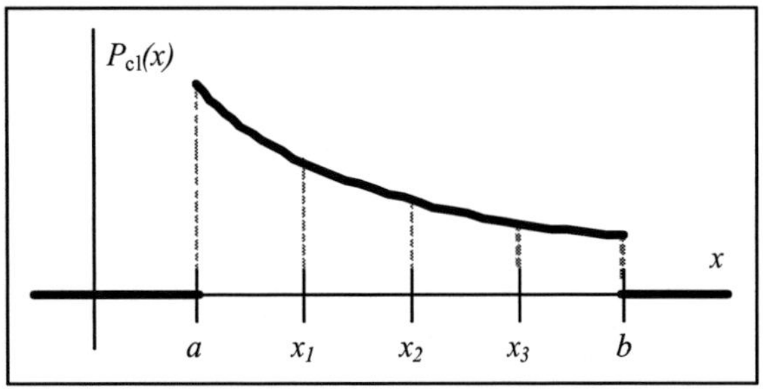

FIGURE 5. Classical probability density graph from the "unknown potential well" post-test, designed for the tutorial *Relating classical mechanics to quantum mechanics.*

The *RCQ* tutorial and the "unknown potential well" post-test (as a final exam question) were used in classes #2 and #3 at GVSU (N = 24). The results from both classes were similar even though class #2 did not complete the *R&T* tutorial. The combined results are shown in Fig. 6. About two-thirds of the students answered all parts correctly with correct reasoning.

Correct responses on "unknown well" post-test

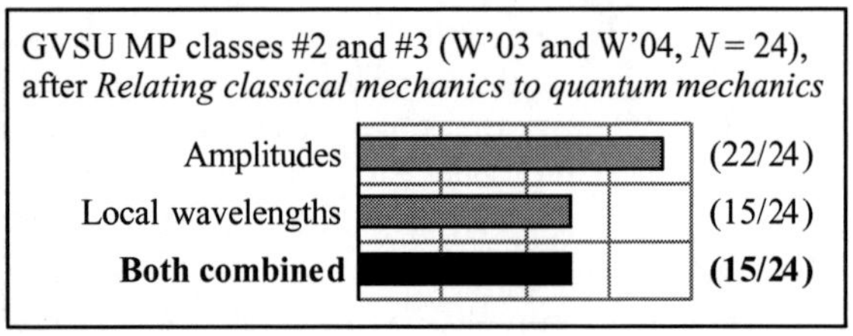

FIGURE 6. Results from GVSU modern physics classes #2 and #3 on "unknown well" post-test. Both classes completed the tutorial *RCQ* as well as one or both of the tutorials *SqW* and *R&T.*

IMPLICATIONS FOR INSTRUCTION AND FUTURE RESEARCH

The promising results from the "unknown well" post-test suggest that the preliminary tutorials on wave functions can enhance student understanding because they provide students with the opportunities necessary to confront and resolve their incorrect ideas and intuitions. Addressing conceptual difficulties arising in earlier contexts (with deBroglie wavelength in the context of electron interference, and with probability density in the context of classical motion) did not remedy difficulties with those same concepts in the later context of wave functions.

Future research is needed to probe the ability of students to distinguish between the behavior of highly excited and low-lying energy state wave functions, as well as their understanding of probability, probability amplitude, and probability current in both classical and quantum mechanical situations. Nevertheless, the results presented here indicate that ongoing research and curriculum development in upper level courses, despite their typically low enrollments, can yield valuable insight into student learning.

ACKNOWLEDGMENTS

Special thanks are due to Lillian McDermott, Peter Shaffer, and Andrew Crouse for ongoing collaboration and cooperation, particularly for the collection of some of the data presented here. Thanks are also due to Chandralekha Singh for co-organizing the 2004 PERC targeted poster session TP-C, where this report was first presented.

REFERENCES

1. McDermott, L. C., and Redish, E. F., *Am. J. Phys.* **67**, 755-767 (1999).
2. Thacker, B. A., *Am. J. Phys.,* **71**, 599-606 (2003).
3. Cataloglu, E. and Robinett, R. W., *Am. J. Phys.* **70**, 238-251 (2002).
4. Wittmann, M. C., Steinberg, R. N., and Redish, E. F., *Am. J. Phys.* **70**, 218-226 (2002).
5. Bao, L. and Redish, E. F., *Am. J. Phys.* **70**, 210-217 (2002).
6. Singh, C., *Am. J. Phys.* **69**, 885-895 (2002).
7. French, A. P. and Taylor, E. F., *Am. J. Phys.* **39**, 961-962 (1971).
8. McDermott, L. C., Shaffer, P. S., and the Physics Education Group at the University of Washington, *Tutorials in Introductory Physics*, Upper Saddle River, NJ: Prentice-Hall, 2002.
9. Vokos, S., Shaffer, P. S., Ambrose, B. S., and McDermott, L. C., Phys. Ed. Res., Am. J. Phys. Suppl. **69**, S24-S35 (2001).
10. Ambrose, B. S., "Student understanding of probability in the classical and semi-classical regimes," at the National Meeting of the American Physical Society, Philadelphia, PA, 2003 (unpublished).

Student Learning In Upper-Level Thermal Physics: Comparisons And Contrasts With Students In Introductory Courses

David E. Meltzer

Department of Physics and Astronomy
Iowa State University

Abstract. We found that students in an upper-level thermal physics course were in general quicker than introductory students at grasping and applying fundamental concepts. Upper-level students seemed, in general, more receptive to employing qualitative reasoning using multiple representations, and capable of using it more effectively than introductory students. In addition, upper-level students were better able to utilize guided-inquiry curricular materials in the sense of reasoning with greater depth and grasping more subtle issues. However, although the overall level of preparation and ability was higher in the upper-level course, the broad range of preparation represented among the students presented various practical challenges to implementing active-learning instructional strategies. Moreover, even quite capable upper-level students would falter unexpectedly and unpredictably on various conceptual difficulties that are common among introductory students. The unpredictable and inconsistent nature of this effect demonstrated that instructors must always be prepared to detect and address such difficulties in upper-level courses.

INTRODUCTION

We have been engaged in an ongoing investigation of student learning of thermal physics in introductory courses [1-3]. In the course of this project, we have probed students' reasoning regarding heat, work, the first law of thermodynamics, calorimetry, and related topics. Based on this work, we have developed and tested preliminary versions of guided-inquiry curricular materials.

During Fall semester 2003, I taught a junior-level thermal physics course targeted at physics majors and other advanced students. In this course, many instructional methods were used that are often characterized as "interactive engagement" [4] or "active learning" [5]. Fourteen students were enrolled, mostly junior and senior physics majors along with several students majoring in chemistry or engineering. This course provided an opportunity to compare introductory and advanced students regarding learning of similar topics, using some identical curricular materials and methods. In this paper, I will discuss some of the main features of this experience.

METHODOLOGICAL ISSUES IN UPPER-LEVEL CLASSES

Students taking upper-level physics courses are certainly not representative of the overall student population enrolled in introductory general physics courses. Only a small percentage of students in the introductory course have a specific interest in physics as a major field, and most would be far more likely to take upper-level engineering courses than to enroll in advanced-level physics courses. For this reason, we must assume that observations regarding learning and transfer in advanced-level physics courses are characteristic only of a highly selected subsample of students enrolled in a typical introductory course. It is also important to remember that small class sizes (common in upper-level courses) are associated with a relatively high probability that any *one* particular class will not be fully representative of other, similar classes [6].

When evaluating students' performance in upper-level courses, two distinct factors come into play: (1) students' knowledge of material previously covered in introductory courses, and (2) students' *learning* of new

CP790, *2004 Physics Education Research Conference*, edited by J. Marx, P. Heron, and S. Franklin

material during the advanced course. In order to make some determination of students' knowledge of material learned in previous courses, it is essential to administer pretests on the relevant material to the students before those topics are covered in the advanced course. This poses a logistical challenge, because extensive pretesting ordinarily requires substantial amounts of class time. Because pretests are administered before instruction on the relevant topics, they can not be graded for course credit. Moreover, even when pretest data are acquired, an important dimension of student learning is missed, that is, the degree to which students may have increased their *readiness* to learn new topics as a result of their experiences in the introductory courses.

Although students may give incorrect answers or inadequate explanations on pretest questions related to specific topics, it may well be that their exposure to those topics in their previous courses had provided them with a basis for rapidly and effectively assimilating a second round of instruction on those same topics, as is often provided in advanced courses. Determining the students' state of "readiness" for new learning is an extremely challenging problem for researchers who develop and administer diagnostic pretests. Ordinary tests assess students' knowledge at a particular point in time and do not provide a measure of the rate at which such knowledge might be *changing*, nor of its *susceptibility* to change.

When considering "transfer" of learning from introductory courses, we are interested in two distinct (but hard to separate) factors: (1) *application* of knowledge previously learned, and (2) *synthesis* of new concepts based on knowledge elements learned or learning skills developed in previous courses. In the case of "application" of knowledge, we might try to determine the degree to which students can apply knowledge of concepts and techniques acquired in previous courses to the solution of more complex problems that make use of those same concepts or techniques. However, we have an equal or greater interest in the degree to which students can learn *new* concepts or techniques in the advanced course, and effectively apply those new concepts in problem solving. One assumes that an important product of instruction in introductory courses is the preparation of students for learning of new concepts in more advanced courses. However, it is probably much more difficult to determine students' ability to *learn* new concepts, than it is to determine their ability to *apply* previously learned concepts in new contexts.

STUDENTS' INITIAL KNOWLEDGE

On the first day of class, a small set of diagnostic questions related to calorimetry and the first law of thermodynamics was administered to provide information regarding students' initial knowledge. Overall performance on these questions was superior to the average post-instruction performance of students in the introductory physics courses reported in Refs. 1-3, although a broad range of knowledge levels was found.

During the present (Fall 2004) semester, I am again teaching the thermal physics course. On the first day of class, a larger set of diagnostic questions was administered to the students. Two (out of a total of 15) of these questions are shown in Fig. 1.

This *P-V* diagram represents a system consisting of a fixed amount of ideal gas that undergoes two ***different*** processes in going from state A to state B:

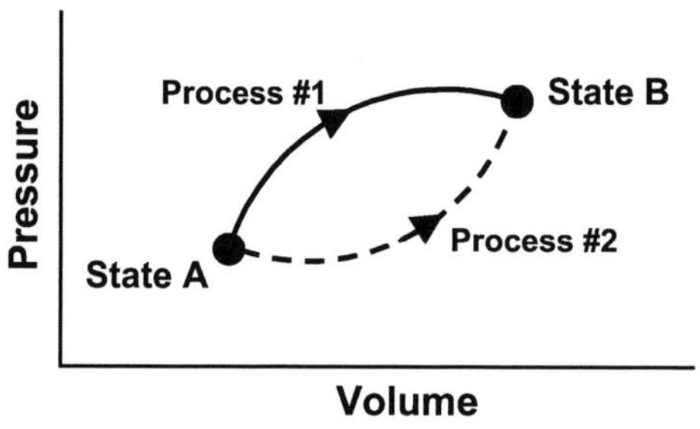

[In these questions, ***W*** represents the work done ***by*** the system during a process; ***Q*** represents the heat ***absorbed*** by the system during a process.]

1. Is *W* for Process #1 ***greater than***, ***less than***, or ***equal to*** that for Process #2? Explain.
2. Is *Q* for Process #1 ***greater than***, ***less than***, or ***equal to*** that for Process #2? Please explain your answer.

FIGURE 1. Two of the questions posed to students in both introductory and upper-level physics courses. Answers: (1) *greater than*; (2) *greater than.*

Most of the questions in this second diagnostic set had been administered (after instruction was completed) to students in the introductory calculus-based general physics course during the investigations reported in Refs. 1-3.

Among the 21 upper-level students responding to these questions, a wide range of initial knowledge levels was evident. Some students showed good ability to apply first-law concepts, while others showed little or none. On some questions, average performance was clearly superior to the post-instruction performance of students in the introductory course, while on other questions performance was virtually indistinguishable from that of students in the lower-level course.

On Question #1 shown in Figure 1 (the "work" question), about one quarter of students in both introductory and upper-level courses answered incorrectly that the work done by the system in Process #1 would be the same as that done in Process #2. By contrast, on Question #2 (the "heat" question), 38% of students in the upper-level course gave a correct or nearly correct answer with an acceptable explanation, compared to only 15-20% of students *after instruction* in the introductory courses. On questions related to cyclic processes, thermal reservoirs, and isothermal processes, performance of students in the upper-level course was comparable to the post-instruction performance of a self-selected sample of interview volunteers from the introductory course whose course grades were well above the class average [2].

COMPARISONS AND CONTRASTS WITH INTRODUCTORY STUDENTS

Students in the upper-level course demonstrated a number of important learning skills that were significantly better developed than among students in the introductory course. At the same time, even very able students in the advanced course periodically demonstrated a vulnerability to learning difficulties similar or identical to those found among students in the introductory course.

Upper-Level Students Demonstrated Superior Learning Skills

Learning skills displayed by upper-level students were superior in a number of respects to those of students in the introductory course. For example, they demonstrated an ability to make use of qualitative reasoning, multiple representations, and guided-inquiry curricular materials that was generally beyond that of the introductory students.

In covering similar material, upper-level students were quicker to generalize over specific contexts with a unifying concept. By contrast, introductory students tended to focus on pattern matching, recognizing commonalities among different problems without necessarily extracting a unifying physical theme. Despite having superior mathematical skills, upper-level students relied *less* on purely mathematical calculations and arguments than did introductory students in working identical problems. They were less likely to simply point to an equation as an explanation, and more likely to use arguments based on proportional reasoning.

Upper-level students found it easier than did introductory students to interpret the meaning of diagrams, bar charts, and other graphical material, even in novel contexts. They were more comfortable in making use of multiple representations (verbal, diagrammatic, etc.) to express their own thinking, and they showed less reliance on purely mathematical forms of reasoning. Even upper-level students with relatively less preparation demonstrated facility with multiple representations.

Upper-level students made effective use of guided-inquiry worksheets originally developed for use with introductory students. Typically, upper-level students worked through problems faster and more thoroughly, and required less guidance from instructors, than did students in the introductory course. Moreover, they were less likely to become bogged down in problem minutiae such as instructions or descriptions of apparatus, and they showed less confusion in interpreting instructions. These students worked well in groups, usually had productive discussions, and helped each other effectively. They showed a willingness to devote extra time to the resolution of confusing points.

Common Reasoning Difficulties Were Shared by Upper-Level Students

Even students receiving the highest overall grades would sometimes encounter conceptual difficulties that were the same as or similar to those observed among introductory students. The appearance of these learning difficulties among the upper-level students was intermittent and unpredictable, but recurrent (although the *same* difficulties did not generally recur). *Providing* they were addressed directly, these difficulties appeared to be resolved efficiently and thoroughly with few observable remnants.

Notable examples of conceptual difficulties encountered included the following: (1) Several students had substantial difficulty in applying the state-function property of entropy to conclude that ΔS would be equal for a free-expansion process and an isothermal process sharing identical initial and final states. In general, invoking state-function properties in contexts involving *entropy* seemed to be more difficult for most students than in the context of internal *energy* [7]. (A similar finding was recently reported by Kautz [8].) (2) Many students were slow in learning to compare engine and refrigerator efficiencies to the Carnot efficiency in order to check compliance with the second law. In addition, there were difficulties in making the correct identification of heat and work inflows *to* and outflows *from* the system in these problems. (3) When working through a guided-inquiry worksheet using diagrams that depicted a cyclic process, some students initially concluded that net work done by the system during the process had to be zero. Similar difficulties had been prevalent among students in the introductory course [2] and were evident among the upper-level students on the first-day pretest. (4) Many students displayed considerable difficulty in distinguishing between systems that had identical temperatures but different internal energies, and vice versa. (This is related to the classical confusion between heat and temperature, long recognized as a recurring learning difficulty in teaching thermodynamics to diverse student populations.)

Challenges and Difficulties

Consistent with observations made among students in introductory courses, both highly favorable and highly unfavorable reactions toward interactive-engagement techniques were displayed by upper-level students. The 10-15% unfavorable rating on evaluations matched that found in the introductory algebra-based course. Use of guided-inquiry worksheets during class (instead of in a separate recitation section) created logistical difficulties due to the broad range of speeds with which students worked. Insufficient pretesting and lack of previous relevant research made optimal course planning difficult.

SUMMARY

Students' performance on qualitative and quantitative problems throughout the course (on homework, quizzes, and exams) provided substantial evidence of effective learning in the context of this "active-learning" environment. There was some evidence of transfer of learning from previous courses, in that students seemed able to make use of (sometimes fragmentary) ideas acquired during previous instruction in the process of synthesizing an improved overall grasp of the subject. However, there was also substantial evidence suggesting that instructors must be attentive to sudden and unpredictable appearances of standard learning difficulties even among upper-level students.

When presented with unfamiliar concepts, upper-level students appeared to learn and apply them more efficiently than did introductory students. Experience with other advanced courses and a willingness to do substantial amounts of homework apparently contributed to significant learning gains. Upper-level students demonstrated, on the average, greater motivation; however, it is difficult to separate motivational factors from skill factors with respect to their relative significance in the production of observed learning gains.

The fundamental problem regarding analysis of transfer in the context of upper-level courses is the difficulty in answering this question: When learning is observed in upper-level courses, does it represent (a) transfer of *knowledge* acquired in introductory courses, (b) application of *learning skills* acquired in introductory courses, or (c) knowledge and/or skills possessed by the student all along, perhaps even before beginning introductory courses? (Or, perhaps, all three?) It is likely that extensive longitudinal investigation with diverse courses and student populations will be required to apportion the proper weights among the various relevant factors.

ACKNOWLEDGMENTS

Ngoc-Loan Nguyen has made significant contributions to this report, based on his experiences as a teaching assistant in both the introductory course and the upper-level thermal physics course. This work was supported in part by NSF DUE-#9981140 (T. J. Greenbowe, Co-P.I.) and NSF PHYS-#0406724.

REFERENCES

1. Meltzer, D. E., "Student Reasoning Regarding Work, Heat, and the First Law of Thermodynamics in an Introductory Physics Course," in *Proceedings of the 2001 Physics Education Research Conference*, edited by S. Franklin, J. Marx, and K. Cummings, Rochester, NY: PERC Publishing, 2001, pp. 107-110.
2. Meltzer, D. E., *Am. J. Phys.* **72**, 1432-1446 (2004).
3. Christensen, W. M., Nguyen, N.-L. P., and Meltzer, D. E., *AAPT Announcer* **34 (2)**, 138 (2004).
4. Hake, R. R., *Am. J. Phys.* **66**, 64-74 (1998).
5. Knight, R. D., *Five Easy Lessons*, San Francisco: Addison Wesley, 2002, pp. 46-62.
6. This is because the variance (about the population mean) for mean values of small subsamples drawn from a large population is relatively large, compared to the variance for larger subsamples.
7. Ref. 2, Sec. V B.
8. Kautz, C., *AAPT Announcer* **34 (2)**, 120 (2004).

Tracing Difficulties With Relativistically Invariant Mass To Difficulties With Vector Addition Of Momentum In Newtonian Contexts

Andrew Boudreaux

Western Washington University
Department of Physics and Astronomy
Bellingham, WA 98225-9064
boudrea@physics.wwu.edu

Abstract. For effective transfer of knowledge, it is necessary to break the transfer of conceptual difficulties. In physics courses that include special relativity, students are expected to relate the invariant mass of a system to the energy and momentum of the individual particles that make it up. Many have difficulty doing so. Student responses indicate that some difficulties stem from a failure to treat energy-momentum as a four-vector. Introductory students experience related difficulties in a purely non-relativistic context: many fail to take the vector nature of three-momentum into account when relating the momentum of a system to the momenta of its constituents. Results suggest that these difficulties are widespread, and not necessarily resolved through the study of advanced topics.

INTRODUCTION

Many instructors – in physics and other disciplines – expect their students to be able to transfer ideas they have learned to new situations. Research on student learning in a variety of domains indicates, however, that this ability is often not a result of instruction.[1,2] An experienced physics instructor will recognize that introductory students encounter difficulty when asked to apply a concept or line of reasoning to a situation different from that in which it was learned.[3] Explicit guidance is often required for them to do so. In contrast, facility in applying ideas flexibly in diverse contexts is characteristic of physics experts working within their own discipline. The following questions thus arise: Does traditional instruction in advanced topics promote facility in transfer? If not, what specific difficulties with transfer occur?

In the study of relativity beyond the introductory course, students are expected to apply the four-vector in the analysis of problems in kinematics and dynamics.[4] It is reasonable to expect that students will assimilate and gain competence with this new tool by extending their understanding of the already familiar three-vector. The energy-momentum four-vector, an essential tool in the analysis of relativistic collisions, is built from quantities familiar from Newtonian physics: energy (a scalar), and momentum (a three-vector). To secure a meaningful understanding of the conservation of energy-momentum in relativistic collisions, the student must first transfer his or her understanding of energy and momentum to the new context, and then extend this understanding to encompass more complicated situations. Research results described below indicate that despite instruction this transfer often does not occur. In some cases, a transfer of debilitating difficulties prevents meaningful learning.

The first section of the paper identifies specific difficulties that advanced students encounter as they study special relativity. The following section traces these difficulties to difficulties with basic, underlying ideas.

IDENTIFYING DIFFICULTIES WITH ENERGY-MOMENTUM

Two written questions have been used to investigate student understanding of relativistic

CP790, *2004 Physics Education Research Conference*, edited by J. Marx, P. Heron, and S. Franklin

energy-momentum. These questions – together with overviews of student performance – are described first. Individual student responses are then presented as evidence for specific difficulties.

Non-Interacting Particles Question

In this question, students are shown three systems, each consisting of two non-interacting identical particles. Each particle has mass m and momentum either zero or of magnitude p_o (see Figure 1). Students are told that the moving particles have relativistic speed in the frame shown. They are then asked to rank the systems, first according to the magnitude of total momentum, and then according to mass.

System A

System B

System C

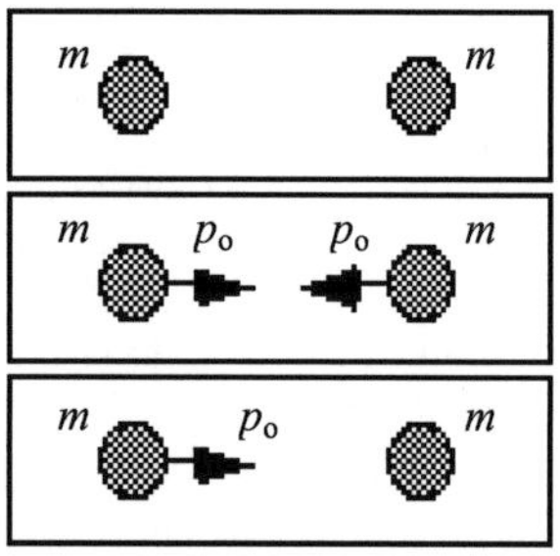

FIGURE 1. Two-particle systems used in the *Non-interacting particles* question.

To rank the systems correctly by momentum, a student can, for each system, sum the corresponding three-momentum components of the individual particles. The resulting system momentum components can be used to determine magnitudes: $P_A = P_B = 0$, and $P_C = p_o$. To rank the systems by mass, the student must first apply $m^2 = E^2 - p^2$ to each particle in order to find its energy in terms of m and p_o.[5] The energies and momenta of the particles can then be summed component-wise to find the energy-momentum four-vector of each system. Finally, the magnitudes of the system four-vectors can be compared to rank the systems by mass: since system B has greater energy but smaller momentum than C, $M_B > M_C$. To compare A and C, a brief calculation must be performed:

$$\begin{aligned} M_C^2 &= [m + (m^2+p_o^2)^{(1/2)}]^2 - p_o^2 \\ &= 2m^2 + 2m(m^2+p_o^2)^{(1/2)} > 4m^2 = M_A^2. \end{aligned} \quad (1)$$

The rankings are thus $P_C>P_A=P_B$ and $M_B>M_C>M_A$.

The *Non-interacting particles* question has been administered as a written quiz to 13 students in a junior-level relativity course at the University of Washington. The students had received traditional instruction on energy-momentum. On the momentum task, 8 of 13 answered correctly. The remaining five students all gave the same incorrect ranking: $P_B > P_C > P_A$. Only two students ranked the masses correctly. Six students gave a variety of incorrect rankings while five left the mass ranking blank.

Photon Question

In this variation students were asked to consider systems of two massless photons, rather than two massive particles. Students were told that each photon has energy E. The momentum directions differ from those given in the original problem (see Figure 2).

System A

System B

System C

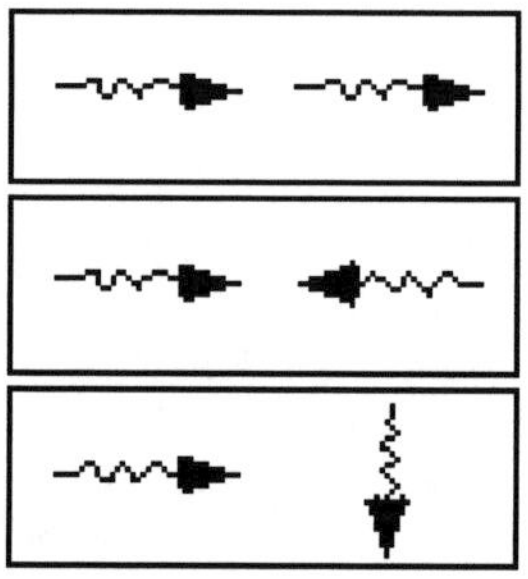

FIGURE 2. Systems used in the *Photon* question.

Students are asked to determine the following in terms of E: (i) the energy-momentum four-vector of each photon, (ii) the energy-momentum four-vector of each system, (iii) the magnitude of the total momentum of each system, and (iv) the mass of each system. To answer correctly, a student can first recognize that for massless particles the magnitude of the momentum is identical to the energy. This allows the energy-momentum four-vector of each photon to be determined. Then reasoning similar to that outlined above can be applied to yield the following results:

TABLE 1. Correct answers to the *Photon* question.

System	Energy-momentum four-vector of system	Magnitude of total momentum of system	Total mass of system
A	(2E, 2E, 0)	2E	0
B	(2E, 0, 0)	0	2E
C	(2E, E, –E)	$E(2)^{1/2}$	$E(2)^{1/2}$

The *Photon* question was administered to 14 students in the same relativity course described above. It was given as a written homework assignment approximately one week after the quiz on which the original question was administered. When determining the system four-vectors, the majority of students responded correctly (12/14 correct for system A, 11/14 for system B. and 10/14 for system C). Correct responses for the momentum and mass, however, were less frequent:

TABLE 2. Number of correct responses on the *Photon* question.

System	$\|P_{system}\|$	M_{system}
A	10/14	13/14
B	11/14	7/14
C	7/14	7/14

Discussion Of Specific Difficulties

On both the *Non-interacting particles* question and the *Photon* question, students found determining the invariant mass of the system challenging. Below are described certain types of reasoning that seemed common among the incorrect responses.

Difficulties summing energies and momenta as four-vector components in determining the four-vector of the system. For some students, the failure to determine the system mass correctly seemed related to a failure to treat individual energies and momenta as components of a four-vector. On the *Photon* question one student recognized that $|\boldsymbol{p}| = \mathrm{E}$ for an individual photon, but answered (2E, 2E) for the energy-momentum four-vector of system A, (0, 0) for system B, and (E, E) for system C. These answers for the system energy, correct for system A but incorrect for B and C, suggest a tendency to combine energies inappropriately as signed quantities. Furthermore, the incorrect momentum components given for system C indicate a failure to recognize that the total three-momentum vector must include at least two non-zero components. It is as if the student has applied a simplistic rule of "complete reinforcement" for photons propagating in the same direction, "complete cancellation" for opposite directions, and "non-combination" for orthogonal directions.

The incorrect energy-momentum four-vectors led the student to the incorrect answer of zero total mass for each system. In all, 4 out of the 14 student responses indicated that the masses of all three systems are the same.

Tendency to confuse mass and momentum. One student gave correct energy-momentum four-vectors for each photon system A-C. He went on, however, to answer incorrectly for the total momentum: "$P_A^2 = (2\mathrm{E})^2-(2\mathrm{E})^2 = 0$; $P_B^2 = (2\mathrm{E})^2$; $P_C^2 = [(2\mathrm{E})^2-\mathrm{E}^2-\mathrm{E}^2]$." The student confuses the invariant system mass for the total three-momentum. Another student, who like the first obtained correct energy-momentum four-vectors for each system, gave a slightly different incorrect response for momentum: "$P_A^2 = P_x^2-0^2-0^2$; $P_B^2 = 0^2-0^2-0^2$; $P_C^2 = P_x^2-P_y^2-0^2$." This response does not incorporate the energy part of the four-vector, as the previous response did, but inappropriately applies the algorithm for computing the magnitude of a four-vector to determine the magnitude of three-momentum.

TRACING DIFFICULTIES WITH ENERGY-MOMENTUM TO DIFFICULTIES WITH THREE-MOMENTUM

The inability of some students to determine the invariant system mass seemed related to difficulties with concepts and reasoning not specific to the relativistic context. On the *Non-interacting particles* question, for example, 5 of 13 students gave the incorrect system momentum ranking $P_B > P_C > P_A$. One of these students reasoned that "the magnitude of $P_B = 2p_o$, the magnitude of $P_C = p_o$, and the magnitude of $P_A = 0$." Another explained "B > C > A; the total momentum is $p_1 + p_2$ and has nothing to do with direction."

This tendency to sum vector magnitudes when finding the momentum of a system – a serious difficulty with a foundational idea – prompts consideration of the following possibilities: 1) students capable of correctly applying the momentum concepts in non-relativistic contexts are failing to transfer their understandings to the new, more complex situation, and/or 2) difficulties with momentum can be traced back to non-relativistic contexts and are not fully resolved through advanced study. We thus must ask: To what extent do students experience related difficulties with momentum in non-relativistic contexts?

To address this, an additional written question was administered. Students were told that two identical objects – with momenta of the same magnitude –

collide and stick together to form a new, composite object. Students were then asked to compare p_o, the magnitude of the momentum of the original objects, with P_C, the magnitude of the momentum of the composite object.[5] To answer correctly, a student can add the momentum vectors of the two original objects to determine that of the composite. This leads to the comparison $p_o < P_C < 2p_o$.

The question has been administered to 304 students in two sections of the introductory calculus-based mechanics course at the University of Washington. Students had received traditional instruction on conservation of momentum, including quantitative homework problems on collisions in two dimensions. Performance in the two sections was comparable: about 30% gave the correct response, while the most common incorrect response, $P_C = 2p_o$, was given by about 35%. The following explanations are typical of students who gave the most common incorrect response:

> *"Magnitude doesn't consider direction. Since both particles have equal mass and magnitude of momentum, $P_C = 2p_o$."*

The student explicitly considers direction, but indicates that it does not have bearing on the magnitude of the total momentum. This type of incorrect reasoning is similar to that used by more advanced students in the relativistic context of the *Photon* question. The difficulties with momentum that prevented advanced students from answering correctly about invariant mass can be traced to similar difficulties in a Newtonian context.

A variation of the question has been administered to 53 physics graduate students and advanced undergraduates in the teaching seminar at the University of Washington. Although about 75% answered correctly, 9 out of the 53 students gave incorrect responses similar to the $P_C = 2p_o$ response given by introductory students on the original version of the question. This suggests that the tendency to add magnitudes when summing momentum vectors is not fully resolved through advanced study. Students may be transferring this unresolved difficulty to situations involving relativistic energy-momentum as they progress to more advanced material.

Another domain in which the ability of advanced students to transfer knowledge has been examined is relativistic electrodynamics. In a study involving over 100 students in two sections of the honors introductory course at the University of Washington, student ability to relate current as well as electric and magnetic forces in multiple frames was probed. More than half of these students failed to recognize that the current in a wire maintains the same value under a Galilean transformation. Many students seemed unable to apply the operational definition of current in a frame moving relative to the laboratory.

CONCLUSION

As physics students engage with the study of advanced material, they must bring basic ideas to bear in new, more complex situations. This requires first a solid foundation of understanding of the underlying ideas, and then a vertical transfer to the more complicated situation. The examples presented here suggest that in some cases students instead transfer a deficient understanding of the underlying concepts. In order to promote meaningful learning of advanced topics, difficulties with basic ideas must be addressed.

ACKNOWLEDGMENTS

The author is grateful to past and present members of the Physics Education Group at the University of Washington for their support of and assistance with the research described here. Special thanks are due to Stamatis Vokos for productive discussions and critical feedback on this manuscript.

NOTES

1. Barnett, S.M., and Ceci, S.J., "When and where do we apply what we learn? A taxonomy for transfer" in *Psychological-Bulletin* **128 (4)**, 612-637 (2002).
2. Bransford, J.D., and Schwartz, D., "Rethinking transfer: A simple proposal with multiple implications," in *Review of Research in Education,* ed. A. Iran-Nejad and P.D. Pearson, Washington, D.C.: AERA, 1999, pp. 61-100.
3. See, for example, Arons, A.B., *A Guide to Introductory Physics Teaching*, New York: Wiley and Sons, 1990.
4. For a lucid presentation of the ideas at the sophomore level, see Taylor, E.F. and Wheeler, J.A., *Spacetime Physics*, New York: W.H. Freeman, 1992.
5. Mass is identified with the magnitude of the energy-momentum four-vector; units in which c=1 are used.
6. For more details, see Boudreaux, A., "An Investigation of Student Understanding of Galilean Relativity," Ph.D. dissertation, Department of Physics, University of Washington, 2002 (unpublished).

Transfer of Learning through Gender and Ethnicity

Austin M. Calder, Emily L. Ashbaugh

Department of Physics, University of California, Davis, California 95616

Abstract. This brief report compared the performance by gender and ethnicity of 6720 students in an introductory course for the life science majors: Physics 7A and 7B. We compared performance between ethnicities and genders using Z scores taken by quarter. We also performed a binary analysis with achievement of a high grade in 7B as the dependent variable. The results indicate that on average males score higher than females in every ethnic group, and that the only statistically significant ethnic differences in our binary analysis were White and African American. The model indicated that being female reduced odds of achieving a high grade in 7B by one half. Odds were reduced by more than half for African Americans and increased by three halves for White. We also compared gender equity over 18 quizzes. Equity favored quiz questions that are more open ended; this is consistent with some earlier findings in studies of gender equity in introductory physics courses.

INTRODUCTION

At UC Davis, Physics 7 is an algebra-based course for life-science majors designed using an active-learning curriculum [1]. Our student population is about 65% female and 60% minority. 7A, the first course in the series, covers conservation of energy, thermodynamics, and fluid mechanics. 7B covers forces, kinematics, and Newton's laws. We have detailed data on over 6,500 students who have taken both 7A and 7B. This study is part of a larger research endeavor involving longitudinal studies [2] of transfer of learning. We are investigating what factors affect performance in introductory physics and which skills transfer to different courses such as mathematics, chemistry, and biology.

METHODS AND RESULTS

We collected data from 25 quarters of Physics 7 at UC Davis. From each quarter, we have students' scores on each of the four to eight quizzes, and each of the four to nine individual final exam questions. The data set includes records from over 8500 students from 7A and over 8000 students from 7B, with 6720 students having taken both quarters. We collected gender and ethnicity information for each of these 6720 students, resulting in a large longitudinal database.

To more easily compare performance across multiple quarters, we converted grades from quizzes, finals, and overall grades to Z scores, by quarter. We converted to Z scores before paring the dataset down to the 6720 students with data from both quarters, so although the mean by quarter is zero for the original data, the mean by quarter in our set of 6720 students may deviate from zero.

With a dataset including performance metrics, ethnic data, and gender, we began our analyses by testing the equity of grades by ethnicity and gender. To test the equity with regards to ethnicity, we plotted the mean course grades from 7A and 7B by ethnic group (Fig. 1). UC Davis allows self-identification to 15 ethnicities, but after viewing trends amongst the ethnic groups, we combined groups to make a more manageable six ethnicities. We split these six by gender to view trends by gender and ethnicity (Fig. 2).

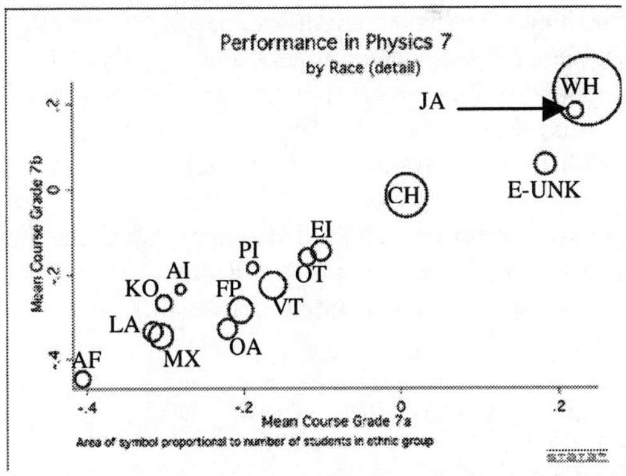

FIGURE 1. Here ethnic groups are plotted (weighted by population) in detail. Key: AF-African American CH-Chinese JA-Japanese AI-American Indian LA-Latino MX-Mexican OA-Other Asian FP-Filipino VT-Vietnamese KO-Korean PI-Pacific Islander OT-Other EI-East Indian / Pakistani E-UNK-European Unknown WH-White

CP790, *2004 Physics Education Research Conference*, edited by J. Marx, P. Heron, and S. Franklin

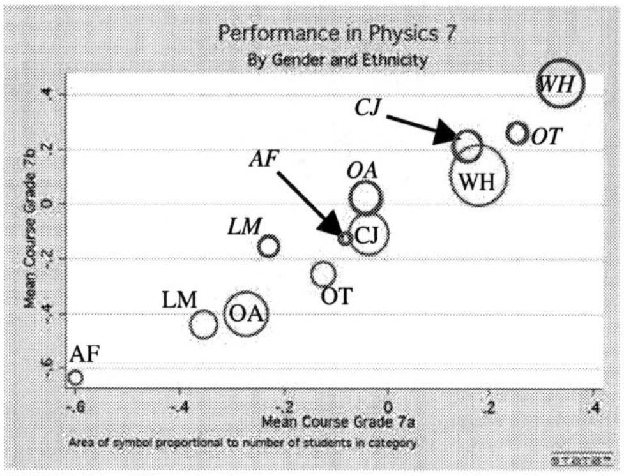

FIGURE 2. Six ethnic groups are split with male students labeled in italics. The ethnicity ordering is preserved with the exception on African American males. Key: AF-African American CJ-Chinese or Japanese LM-Latino or Mexican OA-Other Asian OT-Other WH-White

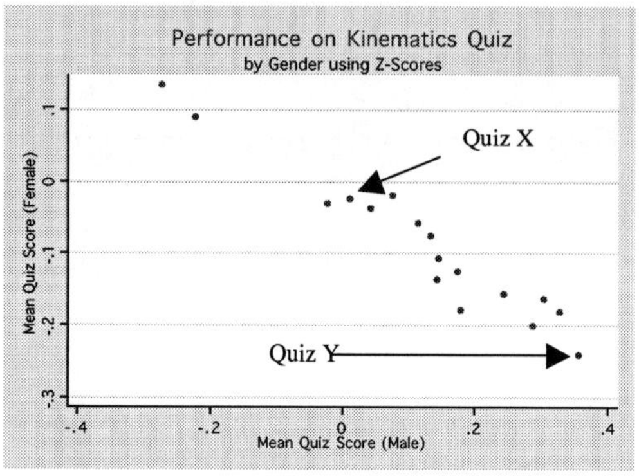

FIGURE 3. Performance on quizzes from 17 academic quarters is compared by gender. The extreme two points on the left are during summer sessions and have significantly smaller student samples. Two example quizzes, one equitable (Quiz X) and one inequitable (Quiz Y), are shown below.

To further analyze the gender equity, we scrutinized kinematics quiz data from Physics 7B. Though different quizzes are given each quarter, Quiz 3 (of four) always evaluates students' ability to analyze position, velocity, and acceleration graphs. Of the 25 quarters of available data, seven quarters had four quizzes (biweekly as opposed to weekly), and we only examined these quarters (5827 students). Quiz 3 is traditionally considered the least equitable quiz from 7B, and a t-test confirmed our prediction, with 10 of the 18 quarters having significantly different means by gender at the 95% confidence level. In each statistically significant case the males outperformed the females. Figure 3 shows the mean scores for females and males on Quiz 3. Because we are plotting deviations from the mean, correlations necessarily must be negative. Our graph is more non-linear than might be predicted because we were obliged to drop several students (mostly summer session students attending other schools) who took only 7B. One particularly skewed quarter was dropped from the plot for this reason.

For a non-graphical method of analyzing the importance of gender and ethnicity in our data, we performed a binary logistic odds regression (fit to Equation 1).

$$\text{High7} = C_{L7A}(\text{Low7A}) + C_{H7A}(\text{High7A}) + C_F(\text{Female}) + C_{WH}(\text{WH}) + C_{AF}(\text{AF}) + C_{OA}(\text{OA}) + C_{CJ}(\text{CJ}) + C_{LM}(\text{LM}) \quad (1)$$

Here "High7" is the logarithm of the odds of achieving in the upper third of students for 7B, "Low7A" =1 when the student's grade is in the lowest third of 7A (zero otherwise); "High 7A" is 1 for students in the highest third of 7A. "Female" is 1 if the student is female; "White" is 1 if the student is white and 0 otherwise, etc.

The coefficients of the fit are given in Table 1. Performance in 7A, gender, and being white or African American are the only statistically significant factors in determining success. Additionally, the magnitudes of the coefficients confirm that grades in 7A (-1.73 for low 7A and 1.85 for high 7A) are more important to success in 7B than gender or ethnicity, since the coefficient of fit for being female is only –0.688 and the largest magnitude for ethnicity is –0.810 for African American students.

DISCUSSION

The strong linearity in Figures 1 and 2 demonstrates that performance in 7A correlates strongly with performance in 7B across all ethnic groups. White students outperform all minority groups except the Japanese, but note that the scale of the graph exaggerates the differences, since the overall difference between the extremes is just over half a standard deviation. Figure 2 splits the ethnicities by gender, and in all 6 ethnic groups the males outperform the females. Males in the African American and Other categories exceed the females by over half a standard deviation, compared to 0.2 of a standard deviation in the remaining ethnicities, but note that these two ethnicities are the smallest (146 AF and 492 OT). By drawing a line through Mean7A=Mean7B, we can see that males generally perform better in 7B than 7A, and females perform better in 7A.

The two quizzes from Fig. 3 are given below. We noticed that the quiz with more gender equity appears to be the more difficult of the two. This coincides with some recent data in gender equity [3]. The males out performed the females on Quiz Y, which was mathematical in scope and deplete of physical intuition. A student could do well on Quiz Y without true functional understanding of the concept. Again this is in agreement with our previous observations on gender in introductory physics.

TABLE 1. Binary analysis coefficients

	Coef.	Std. Error	P>|z|	$e^{Coef.}$
Low 7A	-1.730	0.116	0.000	0.177
High 7A	1.853	0.067	0.000	6.377
Female	-0.688	0.065	0.000	0.503
WH	0.423	0.128	0.001	1.527
AF	-0.810	0.314	0.010	0.445
LM	-0.244	0.172	0.157	0.784
OA	-0.182	0.140	0.194	0.834
CJ	0.120	0.139	0.391	1.127

Quiz X

Imagine you are riding in a bumper car moving at a constant velocity. You collide with a stationary bumper car, which slows you down, but you continue moving in the same direction. The other bumper car moves off in the same direction as you are moving, with a speed intermediate between your initial and final speeds. Assume all speeds are constant except during the actual collision. [Not shown: axes labeled x(t), v(t), and a(t)]

a) Sketch position, velocity, and acceleration graphs of the motion of the car **you are riding in**, covering the period before the collision, the collision, and after the collision. Make sure the time axes are the same for all three graphs. You will have to make choices regarding which physical direction you represent as being positive and where the physical origin is located.

b) On the same set of axes, **but using a dashed line**, sketch position, velocity, and acceleration graphs of the motion of the car **you are NOT riding in**, covering the period before the collision, the collision, and after the collision.

c) What is the direction of the net force on the car **you are riding in** (i) before, (ii) during, and (iii) after your collision with the other car? Give the direction with respect to your motion. Explain how this force is consistent with your graphs.

Quiz Y

The horizontal motion of an object is described by the v(t) velocity versus time plot shown (Fig. 4) below.

a) Calculate the values of acceleration for each time interval of the v(t) plot shown above.

$a_I(t)$=____ m/s^2; $a_{II}(t)$=___m/s^2; $a_{III}(t)$=___m/s^2; $a_{IV}(t)$=___m/s^2.

b) Complete the properly scaled a(t) acceleration versus time plot that is consistent with the v(t) plot shown above. Use the dotted lines to align the time intervals on these plots.

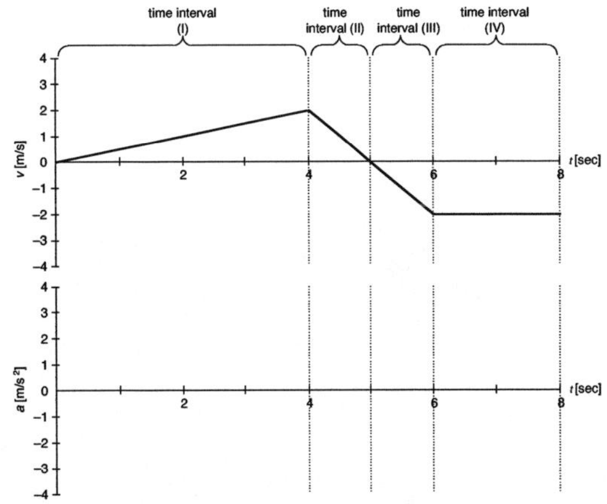

FIGURE 4. Figure for kinematics question on Quiz B

c) If the initial position x = 0.0 m, what is the position x of the object at t = 8.0 s

d) State the time interval(s) in which this object is speeding up in the negative direction, and explain how you know this from the v(t) graph above.

Future Research Directions

We are currently obtaining more information on all of our students, including grades from other university courses. With detailed information on a large sample of students we will be able to see more longitudinal trends. Some possible questions we seek to answer are:

1) How do course grades and timing of mathematics courses affect performance on mathematical tasks in physics?

2) Which other courses affect particular tasks in physics (e.g. how strongly does a course grade in Chemistry affect performance in thermodynamics?)

3) Is gender equity achievable at the introductory college physics level?

It is hoped that further research will detail some of these issues.

REFERENCES

1. W. H. Potter, C. J. De Leone, and L. B. Coleman, "What is the Physics 7 Course?" Paper presented at the 2000 NARST Annual Meeting
2. W. H. Potter, "Evidence that a Reformed Physics Course has Lasting Impact on Student Learning and Performance" Talk given at the Joint Meeting of APS California Section and AAPT N. California/Nevada Section, November 14-15, 2003, Lawrence Berkeley National Lab and University of California, Berkeley
3. Conversation with Wendell Potter, July, 2004

SECTION II

Peer-Reviewed Papers

The Design and Validation of the Colorado Learning Attitudes about Science Survey

W. K. Adams, K. K. Perkins, M. Dubson, N. D. Finkelstein and C. E. Wieman

Department of Physics, University of Colorado, Boulder, Colorado 80309

Abstract. The Colorado Learning Attitudes about Science Survey (CLASS) is a new instrument designed to measure various facets of student attitudes and beliefs about learning physics. This instrument extends previous work by probing additional facets of student attitudes and beliefs. It has been written to be suitably worded for students in a variety of different courses. This paper introduces the CLASS and its design and validation studies, which include analyzing results from over 2400 students, interviews and factor analyses. Methodology used to determine categories and how to analyze the robustness of categories for probing various facets of student learning are also described. This paper serves as the foundation for the results and conclusions from the analysis of our survey data [4][5].

INTRODUCTION

Over the last decade, researchers in science education have identified a variety of student attitudes and beliefs (ABs) that shape and are shaped by student classroom experience[1]. Over the last year at Colorado, we have developed and validated an instrument, the Colorado Learning Attitudes about Science Survey, CLASS[2], which builds on existing surveys (MPEX, VASS, EBAPS)[3]. This survey probes student's ABs and distinguishes the ABs of experts from novices. The CLASS was written to make the statements as clear and concise as possible and is readily adapted to use in a wide variety of science courses. Students are asked to respond on a Likert-like (5-point agree to disagree) scale to statements such as: "I study physics to learn knowledge that will be useful in life.", or "After I study a topic in physics and feel that I understand it, I have difficulty solving problems on the same topic.", or "To learn physics, I *only* need to memorize important equations and definitions." In this paper we will discuss the methods used to validate the survey. We will also discuss the subtleties of choosing categories of statements and list the seven categories we have chosen. The survey has generated some very interesting results which are discussed briefly here and in depth in the companion papers by Perkins et al.[4] and Pollock[5].

DESIGN AND ADMINISTRATION

The CLASS was designed for use with a broad population, takes only ten minutes to complete and covers many areas of student's ABs about physics. To make it suitable for a variety of courses serving non-science majors, physics majors or graduate students words such as "domain" or "concepts", which are not prevalent in a typical introductory student's vocabulary, were avoided. Every effort was made to avoid statements that include two different statements. Finally, one of the most difficult tasks was creating statements that were interpreted in only one way by both faculty and students.

Students (though perhaps not physicists) use the word physics in at least three ways: a particular course, the scientific discipline, or the physics that describes nature. We designed the survey to embrace a single meaning of the word physics to avoid confusion. We focused the statements on physics that describes nature; noting this sense sometimes overlaps with physics as a discipline. By taking this approach, it made the statements meaningful even if a student had never taken a physics course.

This survey has been administered before (pre) and after (post) instruction to 2400 students in 10 courses over the past year either online or paper and pencil. Scoring is done by determining the percentage of a group of students who agree with the experts' view. We calculate the overall score and then in the five categories listed in Table 2. Each category consists of

CP790, *2004 Physics Education Research Conference*, edited by J. Marx, P. Heron, and S. Franklin

three to eight statements that correlate with one another and target a specific attitude or belief about science.

VALIDITY AND RELIABILITY

Validation was done in three steps: First experts were interviewed and then took the survey; second students were interviewed to confirm the clarity and meaning of statements; and finally a detailed factor analysis was performed to create and verify existing categories of statements.

Three experts underwent a series of interviews on the initial draft of CLASS V.1 (Version 1 – Fall 2003). Their comments were used to hone the statements and remove any that could be interpreted more than one way. When this process was complete, seven experts took the survey. Their answers confirmed the expert point of view used in scoring. These experts were physicists who have extensive experience with teaching introductory courses and worked with thousands of students. Some of these experts are involved with physics education research; others are simply practicing physicists interested in teaching. The experts provided consistent responses to all statements in V.2 (Version 2 – Spring 2004) except to three statements, none of which are included in the final five categories.

Student interviews were carried out by obtaining 34 volunteers from six different courses at a mid-size multipurpose state university (MMSU) and a large state research university (LSRU). Care was taken to acquire a diverse group of interviewees. Interviews consisted of first having the student take the survey with pencil and paper. Then, during the first ten minutes, students were asked about their major, course load, best/worst classes, how they study, class attendance and future aspirations to characterize the student and their interests. After this, the interviewer read the statements to the students while the student looked at a written version. The students were asked to answer each statement using the 5-point scale and then talk about whatever thoughts each statement elicited. If the student did not say anything, he/she was prompted to explain his/her choice. After the first five or six statements, the students no longer required prompting. If the students asked questions of the interviewer, they were not answered until the very end of the interview.

Interview results showed students and experts had consistent interpretations on nearly all of the statements. A few statements were unclear or misinterpreted by some of the students. Some of these were reworded or removed for V.2. and the remainder were addressed with V.3. (Version 3 – Fall 2004). Finally there were statements that elicited unexpected student ideas, which will be used for further refinement of the survey.

Statistical analyses were used to test the validity of the sub-groupings of statements into categories. We performed a factor analysis, a data reduction technique that groups similar statements using correlations

TABLE 1. Reduced Basis Factor Analysis of Categories - CLASS Version 2

Predetermined Categories	FA Results	Optimum categories	Emergent Categories	FA Results	Optimum categories
Independence	MF	Conceptual Understanding	Category 1	SS*	Real World Conn. and Personal Interest
Coherence	PC	Conceptual Understanding	Category 2	SS*	Real World Conn. and Personal Interest
Concepts	MF	Conceptual Understanding	Category 3	BQ	Conceptual Understanding
Reality World View	SS	Real World Connection	Category 4	WF	Dropped
Reality Personal View	SS	Personal Interest	Category 5	NS	Dropped
Math	SS	Math Physics Connection	Category 6	SS	Sense Making/Effort
Effort	PC	Sense Making/Effort	Category 7	WF	Dropped
Skepticism	PC	Dropped			

SS = Strong Single Factor; BQ = Better w/ 1 or 2 different statements; WF = Weak Factor; NS = Statements didn't make sense together; MF = Multiple factors; PC = Poorly Correlated

* This category is a single factor; however, even stronger when split into two

between statement responses. We used the principle components extraction method along with a direct oblimin rotation and performed both an exploratory and a confirmatory factor analysis. For more detail on factor analysis see reference [6].

First, we did an exploratory factor analysis, which analyzes the results from all statements and then groups statements that were answered similarly by students into independent factors. The exploratory factor analysis was performed with V.2 of the survey on three sets of data from a calculus-based physics I course (N=416): pre-test results, post-test results and the shift from pre to post. The results from this exploratory factor analysis provided a set of independent, emergent categories. These provide an oblique basis set that best spans the space of student responses.

Another useful perspective is to look at specific groups of statements that probe facets of learning that the physics professor can directly address. Following this idea we chose our original categories based on the categories used by the MPEX and the VASS and expanded upon them slightly during the first two phases of validation of the CLASS (predetermined categories). Such categories emphasize what a physicist believes is useful for learning physics (expert perspective) rather than emphasizing the way students think (student perspective). This means that some of these categories may not be independent of one another; however, if statements are properly designed, these categories are still self-consistent and provide useful information.

Next, we performed a reduced basis factor analysis using the predetermined categories and the emergent categories. With this technique a factor analysis is performed using a basis set that is limited to those statements we believe should be in the category plus a small number of additional statements that are candidates for the category based on their correlations. After carrying out a reduced basis factor analysis, we evaluate the scree plots, correlation coefficients, and factor loadings to determine the categories "robustness". Multiple iterations of this analysis and adding/subtracting statements are used to optimize the categories. After determining robust categories in this fashion, we evaluate the statements not included in any category and search for new categories by looking specifically for correlations with those statements. Table 1 above illustrates the categorization process. Results of this analysis on CLASS V.2 provided the five very robust, albeit not completely independent, categories shown in Table 2.

TABLE 2. CLASS Version 2 Categories

Category	Robustness
Personal Interest	7.75
Reality World Connection	7.38
Math Physics Connection	6.51
Conceptual Understanding	6.11
Sense Making/Effort	5.89

Robustness Ratings done for calculus-based Physics I students at LSRU.

Reliability studies were conducted in Calculus-based physics I at LSRU which is offered every semester with an enrollment over 500 students. During the 2003-2004 school year the course was taught by the same professor, who allowed us to administer the survey to his course pre and post, both Fall and Spring semester. The pre and post results for the two semesters were not statistically different for statements that were the same on both surveys. See Table 3 for overall scores, Real World Connection and Math Physics Connection for both Fall and Spring semesters.

TABLE 3. Reliability Data

Category	Pre	Post	Std. Err.
Fall			
Overall	66%	67%	1%
Real World Connections	74%	77%	2%
Sense Making/Effort	77%	68%	2%
Spring			
Overall	68%	70%	1%
Real World Connections	74%	79%	1%
Sense Making/Effort	78%	71%	1%

Note: There are six questions, which changed between CLASS V.1 and CLASS V.2 that are included in the overall scores; however, the questions in the categories listed are identical.

APPLICATIONS

There are several useful ways to use the scores from the CLASS. One can look at the pre-test results and their influence on student learning or retention. One can also look at the change in attitudes over a semester to determine what effect instruction had on students' ABs. In Table 4 we show results for six courses covering a range of introductory physics courses. We see that students' *incoming* 'Personal Interest' increases with level of physics course. Thus, students who make larger commitments to studying physics tend to be those who identify physics as being relevant to their own lives. As seen with other surveys, the CLASS shows student ABs deteriorate after instruction; unless, ABs are explicitly addressed by the instructor. We see in Table 4 that in the courses at LSRU, which explicitly attended to ABs, the overall scores did not deteriorate; however, in the courses at MMSU there was a substantial decline in ABs. A companion paper by Perkins et al.[4] goes into more detail on these courses and also carefully looks at

TABLE 4. Evident correlation between favorable 'Personal Interest' and physics course selection

Course Type	School Type/Term	Dominant student population	N	Overall %favorable		Personal Interest %favorable on Pre-test (Std. Error of Mean)
				Pre	Post	
Non-Sci-I	LSRU/Fa03	non-sci	76	57%	58%	54% (4%)
Non-Sci-II	LSRU/Sp04	non-sci	36	71%	72%	69% (5%)
Alg-I	MMSU/Fa03	pre-meds	35	63%	53%	55% (3%)*
Calc-I	LSRU/Fa03	engineers	168	65%	67%	70% (2%)
Calc-I	LSRU/Sp04	engineers	398	68%	70%	72% (1%)
Calc-I	MMSU/Fa03	physics maj	38	65%	57%	74% (4%)

I=1st semester, II=2nd semester; typical standard deviation for 'Overall' is ~16%;*Average of 76 students enrolled at start.

correlations of students' ABs with their learning gains. They show that students with large learning gains have a greater positive shift in ABs while students with lower learning gains show a deterioration in ABs.

CONCLUSIONS AND FURTHER WORK

This paper describes the philosophy and methods behind the development of the CLASS. In addition we detail the validity and reliability studies for this survey. We also define a process of selecting categories of statements and determining their robustness. This paper serves as the foundation for the results and conclusions from the analysis of our survey data and future applications of the survey.

Analysis and refinement of the CLASS is still in process. Over the next year we plan to perform a factor analysis of the results for other courses, a final revision of the current statements and creation of statements to target other categories that were not adequately addressed by the current version of the survey. The survey will also be altered slightly to be appropriate for use in Biology, Math, Astronomy and Chemistry and administered to these courses this Fall. Finally we would like to step beyond simply characterizing groups of students to identifying individual student characteristics.

Note added in proof: CLASS Version 3 and a listing of questions by category can be found at http://cosmos.colorado.edu/phet/survey/CLASS/.

ACKNOWLEDGMENTS

Thank you to Steven Pollock and Courtney Willis for many helpful discussions and the members of the PER@C group. This work is supported, in part, by the NSF.

REFERENCES

[1] Bransford, J.D., Brown, A.L., and Cocking, R.R. (2002). *How People Learn* Washington D.C.: National AcademyPress.
Hammer, D. (2000) Student resources for learning introductory physics, *American Journal of Physics, 68,* S52-S59.
Redish, E.F.,(2003). *Teaching Physics with Physics Suite*, John Wiley & Sons. Seymour, E. and Hewitt, N.,(1997). *Talking about Leaving*, Westview Press.

[2] A copy of the CLASS V.3 can be found at http://cosmos.colorado.edu/phet/survey/CLASS/

[3] Redish, E., Saul, J.M. and Steinberg, R.N. (1998). Student Expectations in Introductory Physics *American Journal of Physics, 66,* 212-224. www.physics.umd.edu/perg/expects/index.html
Halloun, I. A. "Views About Science and Physics Achievement: The VASS Story." In The Changing Role of Physics Departments in Modern Universities: Proceedings of the ICUPE, E.F. http://modeling.asu.edu/R&E/Research.html
Elby, A., Epistemological Beliefs Assessment for Physical Science. http://www2.physics.umd.edu/~elby/EBAPS/home.htm

[4] Perkins, K., Adams, W., Finkelstein, N. and Wieman, C. (2004). Correlating student attitudes with student learning using the Colorado Learning Attitudes about Science Survey, *submitted to PERC Proceedings 2004.*

[5] Pollock, S.(2004). No Single Cause: Learning Gains, Student Attitudes, and the Impacts of Multiple Effective Reforms, *submitted to PERC Proceedings 2004.*

[6] Crocker, L. and Algina, J., (1986). *Introduction to Classical and Modern Test Theory,* Fort Worth: Hold, Rinehart and Winston, Inc.
Kachigan, S. K., (1986). *Statistical Analysis,* New York: Radius Press.
Kim, J. and Mueller, C. W., (1978). *Factor Analysis – Statistical Methods and Practical Issues,* Beverly Hills: Sage Publications.

Development of an Instrument for Evaluating Anxiety Caused by Cognitive Conflict

Yeounsoo Kim and Lei Bao

Department of Physics, The Ohio State University, 174W. 18th Avenue, Columbus, OH 43201, USA

Abstract. Physics learning situations often involve many cognitive conflicts between a student's present understandings and new information being learned. Cognitive conflict is known as an important factor in conceptual change. Therefore, it is important to help physics teachers and students develop skills and knowledge for more effective conflict management. However there is no readily available method to monitor the existence and features of cognitive conflicts that students may encounter during their learning. We focus the study on student anxiety caused by cognitive conflict so that we can improve student motivation. This study is targeted to develop an easy-to-use instrument that can be implemented in the classrooms to monitor student anxiety in cognitive conflict situations and the effects on student motivation. In this paper, we will discuss the structure of this instrument and show results from using this tool in our Physics by Inquiry class.

INTRODUCTION

Students' conceptual changes have been a dominant area of research in science education for more than two decades. Many researchers of constructivism in science education have argued that cognitive conflict is an important factor in conceptual change even though there are still questions about its positive and negative effects on science learning [1-6].

During learning, cognitive conflicts among ideas or opinions are inevitable. Courses such as Physic by Inquiry (PBI) [7] are especially designed to help students construct knowledge from seeing and resolving conflicts among peer students, among students and instructors, and between a student's present understandings and new information being learned. When conflicts occur, methods for effective management of such conflicts are crucial to learning. Depending on how conflicts are managed, they can lead to constructive or destructive outcomes.

However, there is no readily available method by which to identify the types of constructive or destructive cognitive conflict that students may reveal in their learning. Therefore we need to develop an easy-to-use instrument that can be implemented in classrooms to monitor the status of students' cognitive conflict and the effects on students' motivations in learning.

THE STRUCTURE OF THE INSTRUMENT

Based on a review of research on cognitive conflict [2-6] and our previous studies [8,9], we have designed and tested an instrument to monitor students' status of their anxiety caused by cognitive conflict. It is called the in-class Conflict and Anxiety Recognition Evaluation (iCARE). Table 1 shows the structure.

The iCARE is comprised of four parts: situation (types of cognitive conflict) [10], experience (factors of cognitive conflict) [5], measurement of anxiety, and types of anxiety [8, 9]. The full version of iCARE can be obtained from www.modelanalysis.net/ICARE/.

The first part was designed to identify the specific situations that cause a cognitive conflict. During a class, a student may encounter situations that caused differences (a) between student's preconceptions and a result of demonstration or experiment, (b) between a student's opinions and those of others in the group, (c) between a student's opinions and those of the instructor.

CP790, *2004 Physics Education Research Conference*, edited by J. Marx, P. Heron, and S. Franklin

Table 1. The Structure of the iCARE.

Category	Factor
1. Situations (Types of conflict)	Individual conflict - Idea & Result - Idea & Idea Social conflict - Student & Student - Student & Teacher
2. Experiences (Factors of conflict)	Surprised Interested Trying to pay attention
3. Measurement of anxiety	Confused Uncomfortable Upset
4. Types of anxiety	Low anxiety High anxiety

The measurement of the different situations that cause the conflicts is important since they may have different effects on cognitive development [10].

The second part identifies the experiences that students may have in cognitive conflict situations. Lee et al. [6] developed the cognitive conflict process model to explain the cognitive conflict. In this model the cognitive conflict occurs when a learner (a) recognizes an anomalous situation (e.g., "The differences surprised me."), (b) expresses interest (e.g., "The differences increased my interest in the topic.") or anxiety about resolving the cognitive conflict, and (c) engages in cognitive reappraisal of the situation (e.g., "The differences made me want to pay more attention to the topics and spend more time to work on it."). In second part, we give students the above example sentences and ask them to identify their experiences in conflict situation.

The third part measures the anxiety caused by cognitive conflicts. We used three items based on the Cognitive Conflict Level Test [5]. The items are as follows. (a) "The result of this experiment confused me." (b) "Since I can't resolve the differences, I am uncomfortable." (c) "I am upset because I cannot understand the reason for the result." All items were on a 5-point Likert scale (1= "not at all true", 5= "very true"). If the total scores are less than 9, the student is considered to have a low level of anxiety, and higher total scores were labeled as a high level of anxiety [9].

The forth part identifies the types of anxiety related to student behavior in cognitive conflict situations. Based on our previous research, we suggest eight types of anxiety according to the level of anxiety, which are shown in Table 2. According to Kim [8] and Cho, Kim, and Kwon [9], among these types, 'confidence in preconception' of high anxiety and 'revision of current theory' of low anxiety promote the learning of conceptual change. In general, students who have 'confidence in preconception' can distinguish between their preconceptions and the scientifically consistent models. Students in the category of 'revision of current theory' can often suggest explanatory hypotheses as a trial to correct preconceptions in order to recognize and resolve conflicting situations. The students in the remaining six categories need more external interaction with instructors.

Table 2. Types of anxiety caused by cognitive conflict

Level	Type	Example
Low	Multiple prediction	"Before the experiment, I predicted multiple possible outcomes. From the experiment, I have seen one of my predictions proved.
	Revision of current theory	"I was confident that by reevaluating my previous beliefs, I would be able to find an explanation without others' help."
	Dependence on others	"Without my effort, I only accepted what instructors or my classmates had said."
	Using past experience	"I made my predictions for this experiment by thinking about my past experience."
High	Confidence in preconception	"Before the experiment, I was highly confident in my previous understandings of the subject."
	Reviewing additional variables	"After I saw the experiment's outcome, I tried to explain it by considering things that I might have ignored as I was making the predictions."
	Lack of confidence	"But right now I don't think I have learned enough physics to build a good explanation yet."
	Conflict with past experience	"On this experiment, the results are inconsistent with what I expected based on my experience"

RESEARCH CONTEXT

This data collection was conducted during one quarter of the PBI course at The Ohio State University. This 10-week course covers the topics on electric circuits. Thirty students were enrolled in this class.

The PBI course is a group-learning environment that implements an elicit-confront-resolve model of learning. In the learning process, there are many situations in which students may encounter conflicts. Accompanying the inquiry method is a system of formative assessment and feedback through

Table 3. Topic of section and iCARE used

Section	Topic	iCARE
1	Single-bulb circuits	1st
2	A model for electric current	1st
3	Extending the model for electric current	1st
4	Series and parallel networks	2nd
5	Kirchhoff's first rule	2nd
6	Equivalent resistance	2nd
7	Multiple batteries	2nd
8	Kirchhoff's second law	2nd
10	Ohm's law	2nd

checkpoints, questions of the day, pretests, homework, exams, and journal entries.

The iCARE was given to students as the post-evaluation to each section (except for section 9 which was not evaluated due to scheduling problems). Table 3 lists the sections to which we used iCARE as the post-evaluation. From section 1 to section 3, we used first version of iCARE focused on anxiety. From section 4, an improved version was used that includes measurement for identifying the type of conflict and experiences in conflict situations.

The structure shown in Table 1 reflects the second version. On average, students took about 5 minutes to complete the evaluation in class right after they finished the work of one section. This can help students report with 'fresher' experience about their learning during the section.

RESULTS

From the data, we have identified a number of interesting results.

First, most of the students (average 93.3%) experienced cognitive conflict in every section. However, the situations that cause conflicts are different. Figure 1 shows the percentage of the situations against total reported situations by each

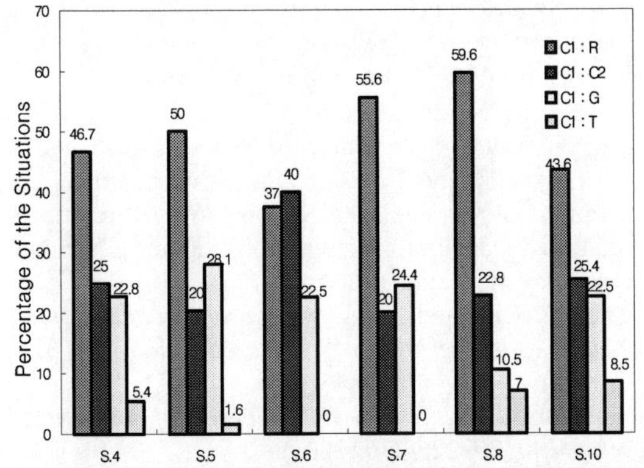

Figure 1. Percentage of the Situations by Each Section.

section. Among the possible situations, the difference between the student's preconception (or prediction) (C1) and the result of an experiment (R) remains the highest in all sections (from section 4 to section 10). In Figure 1, C2 means the student's preconception of a different experiment, G means another group member's opinion, and T represents the teacher's opinion. The pattern C1:R represents a conflict between a student's preconception and the result of an experiment.

In addition, students also demonstrate a variety of reactions to the conflicts; students can be 'surprised', 'interested', and 'trying to pay more attention' to the topic. Figure 2 indicates the percentage of different types of experiences against total reported experiences by each section. It shows that from section 5 to section 10 the reaction of 'interested' decreases, and the reaction of 'trying to pay more attention' to the topic increases. The reason could be related to the increasing difficulty of the content.

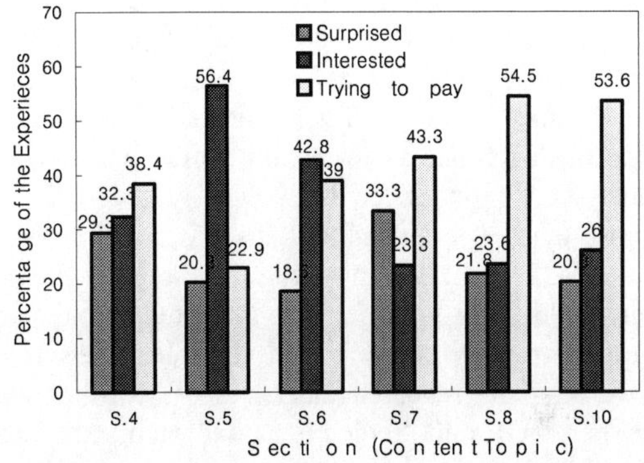

Figure 2. Percentage of Students by Experience.

Second, we found that students can have different levels of anxiety over different sections (content topics).

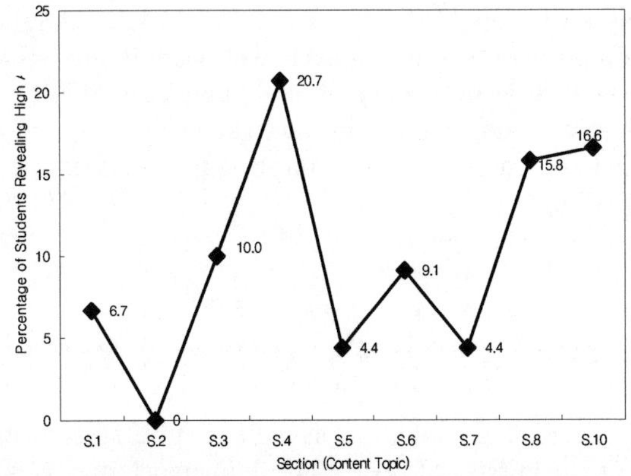

Figure 3. Percentage of Students Revealing High Anxiety by Each Section (excluding Section 9).

Table 4. The t- test results of variance for students' first midterm exam scores by levels of anxiety

	Anxiety level	N	M*	SD	t	p
Exam scores	Low	23	89.04	6.03	2.432	.022**
	High	6	81.16	10.45		

* Maximum score: 100, **: $p < .05$

Figure 3 shows the percentage of students revealing a high level of anxiety by section.

In particular, in section 4 (series and parallel networks), 41% of students experienced cognitive conflicts when working with exercise 4.4 (the experiment that asks students to rank the current through the battery for a list of series and parallel combinations). Students reported that this experiment made the most impression on them.

Further analysis suggests that the students who experienced high anxiety in section 4 also had lower first midterm exam scores than those who experienced low anxiety. The students took the exam a week after doing section 4. The exam consisted of the problems related to series and parallel networks. As shown in Table 4, the t-test results indicate there was a significant difference in the exam scores between the students who experienced high anxiety ($M = 81.16$) and those who experienced low anxiety ($M = 89.04$), $t(27) = 2.432$, $p< .05$. In particular, among the six students exhibiting high anxiety, two students showed the type of 'reviewing additional variables' (see Table 2). Two students revealed the types of 'conflict with past experience' and 'reviewing additional variables'. One student had 'lack of confidence' and 'confidence in preconception'. And the last one had 'reviewing additional variables' and 'confidence in preconception'. These details can provide useful information when implementing strategies to help students resolve conflicting situations.

In this study, instructors follow the typical PBI instruction methods. We didn't try to use additional strategies to help the students who were found to have a high level of anxiety. In our future studies, we will develop and implement new strategies to address students' anxiety based on the measurement from iCARE.

CONCLUSIONS AND IMPLICATIONS

From this study, we have seen interesting results from using the iCARE. The implementation is also practical so that it can serve as a standard tool to evaluate students' status of anxiety in cognitive conflict situations. Such a tool will be a crucial element for further research on strategies to help students resolve cognitive conflicts and control anxiety. In addition, we can use the iCARE to inspect how specific curriculum components affect student learning in terms of triggering cognitive conflicts and causing anxiety. From students' learning behaviors and reactions to conflict situations, we can also obtain additional assessment of students' preparations and learning styles. Further, we can construct a more systematic study in which all these behavior-based results and students' performance data can be correlatively analyzed to yield a better picture of student learning.

ACKNOWLEDGMENTS

We thank all members of the PER group at OSU, especially Florin Bocaneala, and of the PER group at KNUE for their help in conducting this research. This work is supported by Korea Research Foundation Grant (KRF-2003-037- B00102) and by NSF grants #REC-0087788 and #REC-0126070.

REFERENCES

1. G. J. Posner, K. A. Strike, P. W. Hewson and W. A. Gertzog, *Science Education*, **66**, 221-227 (1982).
2. C. Chan, J. Burtis and C. Bereiter, *Cognition and Instruction*, **15**, 1-40 (1997).
3. C. A. Chinn and W. F. Brewer, *Journal of Research in Science Teaching*, **35**, 623-654 (1998).
4. M. Lim6n, *Learning and Instruction,* **11**, 357–380 (2001).
5. G. Lee, J. Kwon, S. Park, J. Kim, H. Kwon and H. Park, *Journal of Research in Science Teaching*, **40**, 585-603 (2003).
6. Y. Kim and J. Kwon, *Journal of the Korean Association for Research in Science Education*, **24**, 687-708 (2004). (http://www.modelanalysis.net/ICARE/)
7. L. C. McDermott and the Physics Education Group at the University of Washington, *Physics by Inquiry*, Vol. Ⅱ. Wiley & Sons, New York, 1995.
8. Y. Kim, Characteristics of Students' Conceptual Change in Physics by Anxiety Types in Cognitive Conflict and Motivation Psychological Factors of Attributions. Unpublished doctoral dissertation, Korea National University of Education (2002).
9. Y. Cho, Y. Kim and J. Kwon, *Journal of the Korean Association for Research in Science Education*, **24**, 574-591 (2004).
10. S. A. Druyan, *International Journal of Behavioral Development*, **25**, 226-236 (2001).

Measuring Conceptual Change in College Students' Understanding of Lunar Phases

Rebecca S. Lindell

Physics Astronomy Chemistry Biology Education Research Group
Southern Illinois University Edwardsville

Abstract. To determine the overall effectiveness of instruction at producing a scientifically correct understanding, researchers need to be able to assess conceptual change. This paper details how using Model Analysis Theory (MAT) in conjunction with the Lunar Phases Concept Inventory (LPCI), provides researchers a more detailed picture of college students conceptual change with regards to lunar phases than traditional methods alone. A review of MAT is provided along with a detailed example of its application before and after instruction to determine conceptual change.

INTRODUCTION

Researchers now know that students do not enter the classroom as blank slates, but rather with pre-existing mental models of how they believe the world works. If rooted deeply enough alternate mental models may actually impair an individual's ability to learn a particular concept. To determine the effectiveness of instruction at overcoming these alternative mental models, researchers need a method by which to assess conceptual change.

While this can be accomplished qualitatively, it is not practical for the average university instructor. Quantitatively, researchers often employ a pre-test/ post-test design using a research-based concept inventory. Traditional statistics, such as gain or effect size, if used for assessing conceptual change are quite limited and do not provide an in-depth information necessary to measure conceptual change.

A new alternative is to employ Model Analysis Theory (MAT) [1] in conjunction with an appropriate research-based instrument to determine what mental models students possess and how consistently students use said mental models.

This paper is designed to provide a brief overview of MAT and provide a detailed example of how MAT was used to probe conceptual change in college students' understanding of lunar phase. The concept inventory utilized in this study was the Lunar Phases Concept Inventory (LPCI) [2], the first concept inventory specifically designed to be used with MAT.

MODEL ANALYSIS THEORY

According to Model Analysis Theory (MAT), an individual's conceptual understanding can be represented as a series of concept dimensions [3] relating to the phenomenon, with each dimension having multiple possible models. In this theory a model is defined as a layer of knowledge that forms a stable functional package. This set of models represents the scientifically acceptable understanding, as well as alternative conceptions uncovered by detailed qualitative interviews.

If multiple-choice questions are carefully designed, the results from the responses can be analyzed using MAT. An appropriate set of questions consists of multiple items for each of the different dimensions, with each item addressing only one dimension and the item distracters correspond to the different models of the dimension. By studying the patterns of individuals' responses to these questions, the probabilities of an individual utilizing each of the alternative models can be mathematically calculated. [3]

CP790, *2004 Physics Education Research Conference*, edited by J. Marx, P. Heron, and S. Franklin

To help visualize the consistency of use of a particular model, researchers can plot the probability of students answering with the correct model versus the most common incorrect model. [4] Depending on where the concept dimension model points fall on the model plot, one can determine how deeply rooted a model is for a particular concept dimension.

If the concept dimension point lies in the lower right of the plot (probability of utilizing correct model <0.3), the majority of the class will utilize the incorrect model consistently. While if the class model point lies in the upper left of the plot (probability of utilizing correct model >0.7), the majority of the class will consistently utilize the correct model. If the class model point lies in the middle of the plot, the class will utilize neither model consistently.

By producing model plots for each of the concept dimensions both before and after instruction, researchers can determine the effectiveness of the instruction at producing conceptual change.

THE LPCI

The LPCI is a twenty-item multiple-choice instrument based on a detailed qualitative investigation. [5] This investigation yielded detailed information about the different dimensions and models of post-secondary students' understanding of lunar phases. Additional models were taken from the alternative understandings previously uncovered by the literature. Table I shows the LPCI's concept dimensions with the correct model listed as the first response under each dimension, and the most common alternate model listed as the second response. Table II provides an overview of the 20 items on the LPCI. Item reliability and validity was previously reported. [2]

EXAMPLE

The LPCI was administered to a class of 200 students enrolled in a primarily non-science major introductory astronomy course at a large public American university. Students took the LPCI at the beginning of the semester, completed a twenty-minute inquiry-based tutorial [6] on lunar phases several weeks later and then took the LPCI again six weeks after completing the tutorial. Lecture was kept to a minimum during this tutorial and very little additional class time was spent covering lunar phases.

Table 1: Lunar Phases Concept Domain

1. Period of the Moon's orbit around the Earth
   - Approximately one month
   - Less than one month
   - More than one month
2. Period of the Moon's cycle of phases
   - Approximately one month
   - Less than one month
   - More than one month
3. Direction of the Moon's orbit around the Earth as viewed from a point above the north pole
   - Counter-clockwise
   - Clockwise
   - Random
4. Motion of the Moon
   - Moves like the Sun (East to West)
   - Moves opposite the Sun (West to East)
   - Random
5. Phase and Sun-Earth-Moon positions
   - Correct relationship
   - Opposite relationship
   - Other relationship
   - No relationship
6. Phase - location in sky - time of observation relationship
   - Correct relationship
   - Other relationship
   - No relationship
7. Cause of Lunar Phases
   - Alignment of Earth-Moon-Sun
   - Obstruction by the Earth's shadow
   - Obstruction by the Sun's shadow
   - Obstruction by Object
   - Combination
8. Effect on lunar phase with change in location on Earth
   - No change in Moon's shape
   - Moon's shape appears larger
   - Moon's shape appears smaller

The total scores and standard deviations for the LPCI are shown in Table 3.

Data from the LPCI was analyzed using MAT. For each of the different concept dimensions, the probability of students answering with the correct model, as well as the most common incorrect model were calculated and plotted. Figures 1 and 2 show the model plots of the concept dimension model points pre- and post-instruction respectively.

DISCUSSION

By studying the plots, we observe several interesting results. Prior to instruction, the plot points for dimensions 1, 2 and 4 show that probabilistically the majority of the students will use neither the correct model nor the most common alternative model consistently, while for dimensions 3, 5, 6, 7 and 8 the plots show that the majority of the students will utilize the alternative model to answer questions concerning these concept dimensions. After instruction, the plot shows that the majority of students will answer questions using the correct model for all concept dimensions, except for dimensions 6 and 7.

Table 2: Summary of questions in current LPCI (Corresponding Dimension)

1	Where to see Waxing Crescent Moon at Sunset (6)
2	Time to complete one orbit (1)
3	Moon orbits in which direction (3)
4	Cause of New Moon (7)
5	Frequency of New Moons (2)
6	Phase of Moon for Solar Eclipse (5)
7	Appearance of Full Moon in Australia (8)
8	Orbital period vs. phase period*
9	Time to observe 3rd quarter Moon setting (6)
10	Direction of 3rd quarter Moon Rising (4)
11	Earth-Moon-Sun geometry for Full Moon (5)
12	Time until Moon appears same (2)
13	Moon's appearance half-way around world (8)
14	Direction of Moon rising at sunset (4)
15	Phase of Moon rising at sunset (6)
16	Alignment between Sun/Earth/ Moon to produce Waxing crescent Moon (5)
17	Time difference between different phases (2)
18	Cause of phases (7)
19	Orbital period – time difference between observations (1)
20	Direction of Moon's orbit (3)

*Question 8 not used for Model Analysis Theory

Table 3: Summary of Average LPCI scores and standard deviations

	N	Average Score	Standard Deviation
Pre	194	6.54	2.43
Post	223	13.74	3.24

As a comparison to traditional methods for determining conceptual change, we can calculate the normalized gain [7]:

$$<g> = \frac{<Post\%> - <Pre\%>}{100 - <Pre\%>} \quad (1)$$

The normalized gain for this data set is 0.54, which is fairly high compared to other gains previously reported. [2]

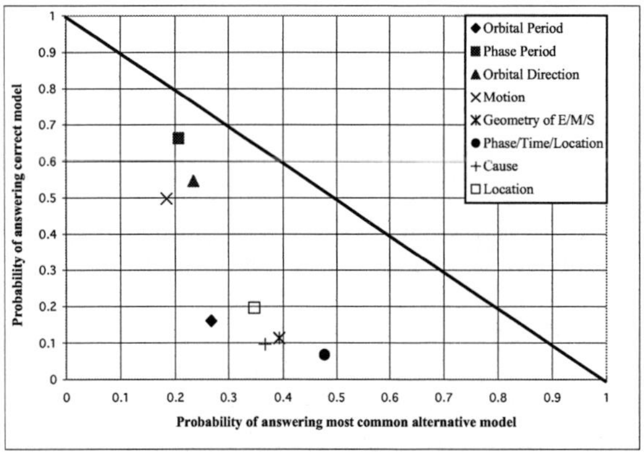

FIGURE 1. Pre-instructional probability of using different models for each concept dimension

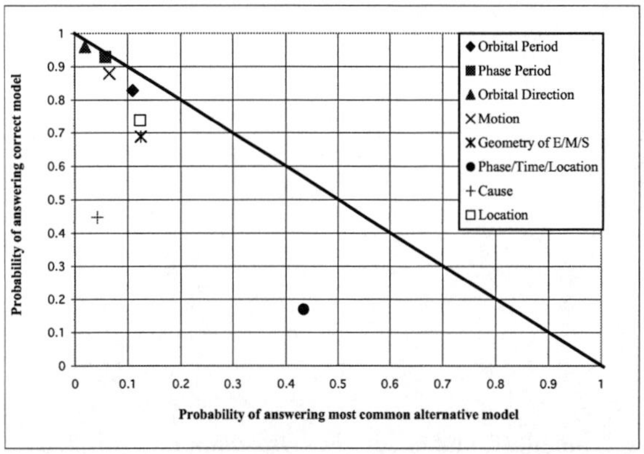

FIGURE 2. Post-instructional probability of using different models for each concept dimension

CONCLUSIONS

If we look at the normalized gain alone, we may have concluded that this instruction was fairly successful at promoting conceptual change. However, when we look at the results from MAT, we see that this may not be the case. The instruction on dimensions 1, 2, 3, 4, 5 and 8 was quite successful. We can conclude that the instruction on dimension 6 (cause of phases) was only moderately successful and on dimension 7 (relationship between time, phase and location) was a complete failure. It is interesting to note that these two concept dimensions' pre-instructional plots showed the most deeply rooted alternative models. In addition, these two concept dimensions are the only two that are typically assessed when evaluating instruction on lunar phases. This suggests that lunar phases instruction may indeed be more successful than was previously reported for the other dimensions. [8]

ACKNOWLEDGMENTS

This work supported by the Southern Illinois University Edwardsville's Funded University Research Program and the Office of Research and Projects. In addition, the author would to like to acknowledge the Physics, Astronomy, Chemistry and Biology Education Research (PAC_BER) at Southern Illinois University Edwardsville for the valuable insights they have contributed to this project. The author would also like to acknowledge Ed Prather and the rest of the Conceptual Astronomy and Physics Education Research (CAPER) Team at the University of Arizona for kindly collecting the data utilized in this paper.

REFERENCES

[1] L. Bao, Dynamics of student modeling: A theory, algorithms and application to quantum mechanics. Unpublished dissertation. University of Maryland (1999).

[2] Lindell, R. and Olsen, J., "Development of Lunar Phases Concept Inventory" Physics Education Research Conference Proceedings, ed. by S. Franklin, J. Marx and K. Cummings. PERC Publishing, NY, (2002)

[3] Also referred to as Physical Features in [1]

[4] Note that by making model plots of other probabilities, other research questions can be addressed

[5] R. Lindell, *Enhancing college students' understanding of lunar phases*. Unpublished Dissertation. University of Nebraska-Lincoln (2001)

[6] J. Adams, E. Prather & T. Slater, *Lecture tutorials for introductory astronomy.* Prentice Hall (2003).

[7] Hake, R. R., "Interactive-engagement versus traditional methods: A six-thousand-student survey of mechanics test data for introductory physics courses" American Journal of Physics, 66(1), 64-74. (1998).

[8] A. Lightman & P. Sadler, "Teacher predictions versus actual student gains" The Physics Teacher, 31, 162-167, (1993).

Probing Students' Epistemologies Using Split Tasks

Timothy L. McCaskey, Andrew Elby

Department of Physics, University of Maryland, College Park, MD 20742, USA

Abstract. Do students really believe the physical principles they learn in class? To explore this question, we gave an FCI "split" task in which students indicated the answers they think a scientist would give and also indicated the answers they really believe. To interpret the splits that students indicated between what they believe and what they were taught, we interviewed students about why they split. It turns out that a split does *not* indicate that the student disbelieves the scientist's answer. The splits actually arose for other reasons, one of which was students indicating a discrepancy between what they were taught and what makes sense to them. For this and other reasons, we devised a new split task focused on these discrepancies between "what makes sense" and what a scientist would say. The results of this new experiment, including validation interviews, will be discussed briefly. Evidence suggests that students are more willing to reconcile physics concepts with their everyday experience if epistemological development is an explicit goal of instruction.

INTRODUCTION

Do students really believe the answers they give on physics assessments? Students can (and often do) become facile at memorizing and repeating back information. However, most assessments don't gauge whether a student is giving answers that make sense to them. Since students try to tell instructors what they want to hear, they might answer questions in a way that contradicts their beliefs or intuitions.

Since the Force Concept Inventory (FCI) [1] is a popular assessment of physics concepts, we used it as the basis for a new probe called a split task, as explained in detail in previous work [2]. We'll briefly review the main points. At the beginning of the semester, our introductory algebra-based physics students, most of whom had taken high school physics, took the FCI with the following additional instructions:

Please ***circle*** *the answer* ***you really believe****. Please draw a* ***square*** *around the answer* ***you think scientists would give****.*

The instructions included a sample question marked up in this way and also made clear that students could circle and square the same answer, if appropriate. When the student's belief differed from the answer he thought a scientist would give — i.e., when he circled and squared a different answer — we called it a *split*. Our findings were intriguing. The average split rate was 24%, which means the typical student split about seven times out of thirty questions. Individual students' split rates ranged from 0% to 90%. Also, we found that women split more than men. Concerning the four FCI items dealing with Newton's third law, we found that even among the students who indicated the correct scientists' answers to *all* of those items, 80% of students (including all of the women) split on one or more of those items. These results led us to hypothesize that students —especially women — often do not believe the "accepted scientist answers" they give on assessments. Our previous work [2] ended there.

Since then, we have conducted validation interviews to get at the meaning of these splits. We wanted to know why students split and whether different students split for similar reasons. The interview data disconfirmed our initial interpretations of the splits. A split never indicated that the student disbelieves the scientists' answer. Their reasons for splitting were more nuanced and varied from student to student, as discussed in more detail below. Many times, students said that their circled ("belief") answers correspond to their intuition.

CP790, *2004 Physics Education Research Conference*, edited by J. Marx, P. Heron, and S. Franklin

For mainly this reason, we have subsequently changed the "circle" instruction to "circle the answer that makes the most intuitive sense to you." We hoped that examining when and why students split on this new task, and how "worrisome" they find these splits to be, will help us probe students' epistemological views about the role of common-sense ideas in learning physics. Our validation interviews confirmed that focusing on these "intuition splits" does indeed provide insight into students' propensity and ability to reconcile intuitive ideas with scientific concepts.

Analyzing FCI post-test results for traditionally taught Maryland students vs. students in our epistemologically focused reformed course, we found evidence that epistemologically focused instruction helps students reconcile. By looking only at students who got the FCI Newton's 3rd law questions correct (on their "scientists' answers"), we found that the reform-taught students split a far lower percentage of the time. This is true even though a much smaller — and therefore more selected — percentage of the traditionally-taught students got the scientists' answers right on those items.

VALIDATION INTERVIEWS

We now discuss our interviews associated with the "old" split task, where students indicated the answer they "really believe" vs. the answer they think a scientist would give. As discussed below, we designed an interview to determine why students split and their more general epistemological views about accepting knowledge from authority. For example, it is possible that students who split frequently are epistemological relativists, people willing to accept the "truth" of their own ideas *and* of the scientists' ideas, even when those ideas conflict. Such students would have no motivation to reconcile the conflicting sets of ideas. Along these same lines, it is possible that students who split rarely are either (i) great learners who have fully reconciled their own ideas with classroom physics concepts, or (ii) people who accept scientific authority and quickly adopt those beliefs as their own without trying to make sense of it all. It turns out, however, that none of our interview subjects fit *any* of these profiles.

Interview Protocol

We interviewed nine volunteers, eight of them women, for a half hour each. The sample included people who split frequently ("high splitters") and infrequently ("low splitters"). We asked our subjects the following introductory questions:

- What's your major/year in school?
- Why are you taking this course?
- Had you had physics in high school?
- If so, do you feel you got a lot out of the class?
- How do you know someone is an expert?
- What do you do when experts disagree?
- How do you know when you understand something?[1]

The next part of the interview was tied to each student's survey responses. We started by asking if the task made sense to the student. Then, for individual splits, we asked:

- Why do you believe the answer you circled?
- Why do you think a scientist would give the square answer?
- Is it worrisome that there's a difference?
- Do you think the scientist could be convinced to see your point of view? If so, how?
- Do you think you could be convinced of the scientist's view? If so, how?

Results And Discussion

Several patterns emerged from the interview transcripts. Low splitters typically did not think our task made sense. They often wondered why anyone would split on the FCI questions. Sarah[2] gave a typical version of this sentiment:

> "The only thing I didn't like was that every time I thought an answer would be right, I figured that's what a scientist would say too, so a lot of them ended up being a circle and square around the same one."

We had hoped that splits would indicate real differences between a student's belief and the answers she thought to be scientifically "correct." However, splits often corresponded to instances where the student did not have confidence in her belief. "Emily" clearly expresses this:

> "...for the ones where my circle and square were on different choices, I thought maybe it was just my intuition-based answer, so then a scientist might say something different because it wouldn't... they would be basing their answer on... facts, I guess."

[1] The last three questions are adapted from Belenky, et al [3].
[2] All student names are pseudonyms.

Not only does this response indicate a lack of self-confidence in Emily's "belief" thinking, it also contains an epistemological statement. Emily seems to believe here that scientist answers are based on facts and intuitions are not. In other students, this distinction between scientific and everyday reasoning is more pronounced. Christine's statement epitomizes a viewpoint expressed by several subjects:

> "It is true, if you listen to a scientist talk, a lot of times, you don't understand everything, and so if I didn't understand everything in the answer, it seemed like a plausible scientist explanation."

Relativism seemed to be absent from our interview data. No matter how many splits people indicated, they often believed the squared and circled answers *could* be reconciled; almost all the subjects said that only one answer could be right in the real world. (In addition, differences between a student's belief and scientist's answer often worried the student because they know that only one answer is acceptable on a test.) The strongest hint of relativism we spotted came from Sarah, responding to the question about splits being bothersome:

> "Well, just because then I wouldn't know which was correct, or anything, and I'd wanna know which one was, you know? There can't be two different answers. Well, I know that... I know that there's two different ways... if you can explain things and support your answers, I know there can be two different answers, but I would prefer to be saying the same thing."

For a moment, Sarah seems to say that there can be two different "right" answers to a question; but she quickly clarifies that the two "different" answers should actually be two ways of saying the same thing. Like our other subjects, Sarah thinks there should be one right answer to a given introductory physics question. But these subjects were not authority-driven absolutists. When discussing the nature of expertise and understanding in the context of politics, or cutting-edge science, they assert that there's no one right answer and that knowledgeable people can disagree. (We lack space here to give quotations supporting this conclusion.) So, students' epistemological views about the source and certainty of knowledge are more nuanced and context-dependent than are the categories we outlined above.

The interview questions about splits focused on subject matter students had not yet covered in class. In cases where the students had received a tutorial on the relevant topic between the initial survey and the interview, they often nullified their split. Questioning revealed that those students had, indeed, refined their intuitive ideas to be consistent with the physics concepts. More important from our perspective, however, was the expectation most students expressed that they eventually *would* be able to achieve a similar reconciliation regarding topics they hadn't yet covered.

A NEW SPLIT TASK

The above results suggest that (i) there are not "real" splits between students' beliefs and the answers they think scientists would give; (ii) even when instructed to indicate a "belief split," students gravitate toward expressing splits between their intuitive ideas and the answers they think scientists would give; and (iii) examining these "intuition splits" could provide insight into students' epistemological views concerning the reconcilability of informal and formal knowledge. For these reasons, we then decided to study intuition splits rather than belief splits.

The new survey task is the same as before, except the circle instruction now says,

> *Please* ***circle*** *the answer* ***that makes the most intuitive sense*** *to you.*

We gave the survey to a group of second-semester introductory physics students at the University of Maryland. A total of 153 students took the survey. Most (108) had taken a reform-oriented course during their first semester (the term where FCI material is covered), while the other 45 students came from a traditional first-semester class. The reform course incorporated tutorials and interactive lecture demonstrations designed to foster both conceptual *and epistemological* development.

The two groups behaved quite differently. Table 1 shows the average (and standard deviation) for each category.

TABLE 1. New split task data for UMD class

1st semester background:	**Reformed**	**Traditional**
N	108	45
Circle score	17.9 (6.2)	11.4 (5.5)
Square score	19.0 (5.4)	11.0 (6.4)
Number of splits	4.6 (5.6)	6.1 (3.8)

The difference in splits is marginally significant statistically (p=.06), while the score distributions are more significantly different, with $p << .0001$ for both circles and squares.

New Validation Interviews

New validation interviews were conducted for this intuition-split task. Seven students went through a similar interview protocol as our previous interview subjects. Most of these agreed to be interviewed for an hour. The differences in the protocol consist of the wording change in the new split task and additional background questions (for subjects willing to stay the full hour). Also, the Belenky questions were removed because they do not explicitly probe attitudes about reconciliation.

The results indicate that intuition-splits are much easier to interpret than belief-splits; the intuition splits really do indicate a discrepancy between a student's common-sense ideas and the answer he thinks a scientist would give.

Newton's Third Law

As instructors, we are most interested in the splits that arise when the student gets the scientist's answer correct, since those splits indicate a lack of reconciliation between the student's common-sense ideas and what she has *successfully* learned in class. To get at this issue, we focused on the FCI's four questions on Newton's Third Law (numbers 4, 15, 16, and 28). As expected, the reformed class performed better on the N3 questions, getting the scientist's answer right 85% of the time vs. 42% for the traditionally taught students. This result is nothing new. Previous studies, however, have not addressed this more subtle question: Is there a conceptual and/or epistemological difference between the traditional students who got the N3 scientist's answers correct and the reform students who got the N3 scientist's answers correct? Of the correct N3 responses among the traditional students, 41% had splits. By contrast, among the correct N3 responses given by the reform students, only 12% had splits. Low split rates indicate that the students believe they have reconciled Newton's 3rd law with the intuitive ideas. The interviews discussed above, along with students' written work on exams, suggest that the reform students don't just *think* they've reconciled; most of them have actually done so. Compared to the traditional students who answered the questions correctly, the reform students are more inclined and/or able to reconcile their intuitive ideas with Newton's 3rd law — an intended result of our N3 tutorial. In this sense, the traditional students who "learned" Newton's 3rd law didn't learn it as deeply as the reform students did. To confirm this interpretation, future work will analyze videotapes of students working through the N3 tutorial. For now, though, it's reasonable to hypothesize that even when traditional instruction "works," it works differently from epistemologically focused, reform-oriented instruction.

SUMMARY

Our old split task, which asked students to differentiate between their belief and a scientist's answer, yielded results that were difficult to interpret. A new split task asking for "the answer that makes the most intuitive sense" rather than a "belief" turns out to be more focused. With this new task, comparing traditional to epistemologically focused, reform-oriented instruction, we found that the reform students not only learn Newton's 3rd law better in the sense of learning the correct answer, but also reconcile that law better with their intuitive ideas.

ACKNOWLEDGMENTS

We would like to thank Melissa Dancy for administering our new split task to her students (although the results aren't discussed here) as well as for her support in preparing other papers and talks on this topic. This work is supported by a grant from the National Science Foundation (#REC-0087519).

REFERENCES

1. Hestenes, D., Wills, M., and Swackhamer, G., *Phys. Teach.* **30**, 141-158 (1992).

2. McCaskey, T., Dancy, M., and Elby, A. "Effects on assessment caused by splits between belief and understanding." Proceedings of the Physics Education Research Conference, Madison, WI (2003).

3. Belenky, M., Clinchy, B., Goldberger, N., Tarule, J. *Women's Ways of Knowing: The Development of Self, Voice, and Mind*, Harper Collins, New York, 1996.

Correlating Student Beliefs With Student Learning Using The Colorado Learning Attitudes about Science Survey

K. K. Perkins, W. K. Adams, S. J. Pollock, N. D. Finkelstein and C. E. Wieman

Department of Physics, University of Colorado, Boulder, CO 80309

Abstract. A number of instruments have been designed to probe the variety of attitudes, beliefs, expectations, and epistemological frames taught in our introductory physics courses. Using a newly developed instrument – the Colorado Learning Attitudes about Science Survey (CLASS)[1] – we examine the relationship between students' beliefs about physics and other educational outcomes, such as conceptual learning and student retention. We report results from surveys of over 750 students in a variety of courses, including several courses modified to promote favorable beliefs about physics. We find positive correlations between particular student beliefs and conceptual learning gains, and between student retention and favorable beliefs in select categories. We also note the influence of teaching practices on student beliefs.

INTRODUCTION

In addition to the traditional content within any course, there are extensive sets of attitudes and beliefs about science that are taught to our students. How we conduct our class sends messages about how, why, and by whom science is learned. Such messages are being studied with the goal of developing more expert-like views on the nature and practice of science in our students. [2][3][4]

Over the last decade, physics education researchers have used several survey instruments to measure these attitudes and beliefs and to distinguish the beliefs of *experts* from the beliefs of *novices*.[5][6][7][8][9] Experts think about physics like a physicist. For instance, they see physics as a coherent framework of concepts which describe nature and are established by experiment. Novices see physics as isolated pieces of information that are handed down by authority (e.g. teacher) and have no connection to the real world, but must be memorized.

Data have shown that, traditionally, student beliefs become more novice-like over the course of a semester.[5] Even in courses using reformed classroom practices that are successful at improving student conceptual learning of physics, student beliefs tend not to improve.[4] Some success has been achieved, however, in courses specifically designed to attend to student attitudes and beliefs.[2]

With these new measures of student beliefs about physics and about learning physics, the question emerges as to how these factors impact and are impacted by students' pursuit of physics study and their mastery of the content.[10] In this paper, we begin to examine the relationships among these different aspects of student learning. We look at: 1) the influence of teaching practices on student beliefs; 2) the relationship between students' beliefs about physics and their decisions about which physics course to take and whether to continue on in physics; and 3) the relationship between student beliefs about physics and their conceptual learning in the physics course.

DATA

The Colorado Learning Attitudes about Science Survey (CLASS)[1] was used to measure student beliefs at the start (pre) and end (post) of several introductory physics courses. The newly-developed CLASS survey builds on the existing attitude surveys (MPEX[5], VASS[6], EBABS[7]). The details of the design and validation of the CLASS are reported by Adams et al.[1] The survey consists of 38 statements to which students respond using a 5-point Likert scale. The 'Overall' favorable score is measured as the average percentage of statements to which the students

CP790, *2004 Physics Education Research Conference*, edited by J. Marx, P. Heron, and S. Franklin

answer in the favorable sense, e.g. as an expert-physicist would. The 'Overall' unfavorable score is similarly determined. The survey is used to measure specific belief categories by looking at subsets of statements. Here, we include measurements of the following facets: 'Conceptual Understanding' (physics is based on a conceptual framework), 'Math Physics Connection' (equations represent concepts), 'Sense Making / Effort' (I put in the effort to make sense of physics ideas), 'Real World Connection' (physics describes the world), and 'Personal Interest' (I think about physics in my life).

We look at the influence of teaching and at student's course selection and retention using data from six courses. These courses range in size (less than 40 to over 600 students), student population (non-science majors; pre-meds; physics, chemistry, and engineering majors), and school setting (from a large state research university, LSRU, to a mid-size multipurpose state university, MMSU). Table 1 lists the courses as well as the other data available for each course.

We look at the correlation with conceptual learning using data from LSRU's large introductory, calculus-based course in Spring 2004. The number of the students allows us to examine sub-groups and retain good statistics. The structure of this course included multiple reforms designed to improve student learning and student beliefs, including interactive engagement in lecture, tutorial-style recitations, and an emphasis on conceptual understanding. In addition, the development and application of expert-like beliefs and approaches to problem solving were emphasized across the course components. For a detailed description of the reforms and an analysis of the contribution of various reforms to learning see Pollock.[11]

RESULTS AND DISCUSSION

Influence of teaching practices. In Table 1, we show the average pre- and post- 'Overall' percent favorable score for the six introductory physics courses. In bold, we see a decline of 9.8% for the Alg-I and 8.2% for the Calc-I courses taught at MMSU. While the Alg-I course was traditionally taught, the Calc-I course was taught using interactive engagement methods; however, neither course specifically attended to improving student attitudes and beliefs about physics. These declines are consistent with those observed in similar courses.[5] In contrast, we do not see these declines in the LSRU courses. All of these LSRU courses incorporated teaching practices specifically aimed at improving student beliefs. Despite being constrained to a large lecture format, these courses resulted in increases of 1-2% in the 'Overall' score.

Course selection and retention. The courses listed in Table 1 represent a range of commitments to the study of physics. We see that the students' *incoming* favorable beliefs on the 'Personal Interest' category increases with the level of the physics course in which students enrolled. The non-science majors average only a 54% favorable belief while the average for the course for physics majors is 74%. Thus, students who make larger commitments to studying physics tend to be those who identify physics as being relevant to their own lives – as measured by 'Personal Interest'.

In addition, we see that the non-science majors who chose to continue and take the second term (Non-Sci II) had significantly more favorable 'Overall' and 'Personal Interest' beliefs than those who did not – their scores being 14% and 15% more expert-like, respectively. For the MMSU Alg-I course, a significant number of students (41) dropped the course. The students who completed the course had a substantially higher favorable 'Personal Interest' score initially (64%) than those who dropped (49%). These two pieces of data suggest a link between retention of students (both within and across courses) and students' favorable beliefs.

Student Beliefs and Conceptual Learning. The LSRU's large calculus-based courses were highly successful at achieving their goal to improve student learning.[11] On two standard exams for measuring conceptual learning, the students achieved median normalized gains of 0.67 on the FCI[12] (Fall 2003)

TABLE 1. Evident correlations between favorable 'Personal Interest' and physics course selection

Course Type	School Type/Term	Dominant student population	# of students w/ CLASS	Normalized learning gains	'Overall' %favorable$ Pre	Post	Shift (Std. Err)	'Personal Interest' %favorable on Pre-test (Std. Error of Mean)
Non-Sci-I	LSRU/Fa03	non-sci	76		57%	58%	+1.0% (1.5%)	54% (3%)
Non-Sci-II	LSRU/Sp04	non-sci	36		71%	72%	+1.4% (2.2%)	69% (5%)
Alg-I	MMSU/Fa03	pre-meds	35	g_FCI=0.13	**63%**	**53%**	-9.8% (2.8%)	64% (3%)
Calc-I	LSRU/Fa03	engineers	168	g_FCI=0.67	65%	67%	+1.5% (1.2%)	70% (2%)
Calc-I	LSRU/Sp04	engineers	398	g_FMCE=0.76	68%	70%	+1.5% (0.7%)	72% (1%)
Calc-I	MMSU/Fa03	physics maj	38	g_FCI=0.35	**65%**	**57%**	-8.2% (2.7%)	74% (4%)

I=1st semester, II=2nd semester; $ typical standard deviation for 'Overall' is ~16%

TABLE 2. Correlations between beliefs and learning

Belief category	Correlations of normalized FMCE gain with*	
	Pre-beliefs (p-value)	Post-beliefs (p-value)
Overall	0.21 (0.0008)	0.26 (0.00002)
Conceptual Understanding	0.22 (0.0005)	0.30 (0.00001)
Math Physics Connection	0.20 (0.001)	0.20 (0.001)
Sense Making / Effort	0.11 (0.09)	0.17 (0.007)
Personal Interest	0.03 (0.63)	0.15 (0.01)
Real World Connection	0.02 (0.79)	0.08 (0.19)

* for students in LSRU Calc-I Spring 2004 with FMCE pre-test<60.

and 0.76 on the FMCE[13] (Spring 2004). While the median normalized gain was quite high, some students had significantly smaller learning gains and a large number of students had much higher learning gains. With pre/post CLASS and FMCE data on 307 students from Spring 2004, we are able to explore the relationship between students' beliefs and their learning gains.

In Table 2, we show the correlations between students' beliefs and their normalized learning gain. We limit these to the 90% of the students who had FMCE pre-test scores < 60 (Thornton's level of conceptual mastery) because we are interested in students who have not already mastered the material (N=256).[13] The correlations in the various belief categories range from 0.02 to 0.22 for pre and 0.08 to 0.30 for post. We see the categories of 'Conceptual Understanding' and 'Math Physics Connection' have correlations with learning gains that are statistically different from zero (two-tailed p-values of 0.0005 and 0.001); where as, 'Real World Connection' and 'Personal Interest' do not. It makes sense that expert beliefs in these former two categories are more important to the form of learning measured by the FMCE. We also see larger correlations with post-beliefs than pre-beliefs in almost all categories. While these correlations are small, it is notable that they are larger than the correlations of learning gain with homework, attendance, and a math pre-test, which are all 0.22 or lower.[11]

In Figure 1, we show students average pre- and post- favorable beliefs as a function of normalized learning gain for 'Overall' and 'Conceptual Understanding'. Two trends emerge: (1) students with higher conceptual gains tend to have more favorable beliefs in these categories and (2) students in the lowest gain category tend to regress in beliefs, while higher performing students tend to post gains in beliefs.

It is important to note that the correlations between beliefs and learning gain listed in Table 2 apply to this course with its teaching practices, student population, and curriculum and with its choice of instrument for measuring conceptual learning. Changes to any of these elements can effect the correlations with beliefs.

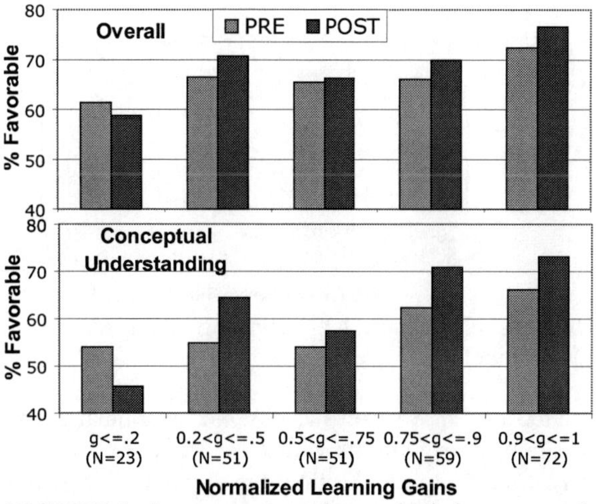

FIGURE 1. Average pre- and post- beliefs vs. normalized conceptual learning gain.

Analysis of the observed affect can then help interpret the meaning of the change and provide some more insight into the relationship of student beliefs and learning. For example, the Fall 2003 LSRU course had the same instructor (Pollock) using the same teaching practices and materials, but using a different measure of conceptual learning gain (the FCI). While the data *do* show the same general trends as observed in Figure 1, they show somewhat stronger correlations between normalized learning gain and students' beliefs. In addition, some categories which show relatively weak correlations here show stronger correlations with the FCI learning gain, and vise versa. These differences likely reflect the differences in the learning measured by the two instruments.[13]

The correlations reported here are between *individual* belief categories and normalized learning gain. If multiple categories contribute to improved learning, it is likely that a *weighted combination* of categories would result in higher correlation coefficients and is a topic of future analysis. Higher correlations may also result from the ongoing refinement of the CLASS statements and categories to better probe beliefs.[1]

As a start to understanding possible cause and effect, we have looked further at the relationship between pre-belief and normalized learning gain and the meaning of the observed correlation of 0.21. In Figure 2, we have binned the students by their incoming 'Overall' belief and plotted the percent of students within each belief bin who achieve learning gains of greater than 0.8, 0.8 to 0.3, and less than 0.3. We see that over 50% of the students with favorable incoming beliefs of 80-100% achieve normalized learning gains in excess of 0.8 and that this percentage decreases for each successive bin while the percentage of students with learning gains less than 0.3 increases. A chi-

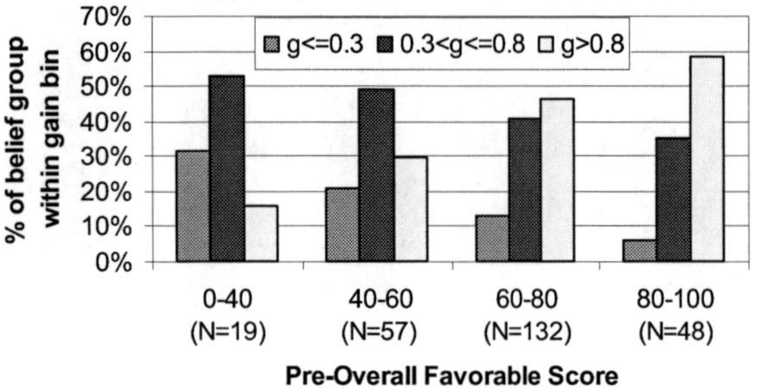

FIGURE 2. Conceptual learning gain vs. Pre-'Overall' belief.

squared test shows that the observed combination of distributions of students within each bin are statisticcally different (p=0.006) from the class-average distribution: 14% g<0.3, 43% 0.3<g<=0.8, 43% g>0.8. These data are consistent with the idea that students' beliefs about science when they enter a course influence their conceptual learning. Of course, we cannot rule out the alternative conclusion that other factors are simultaneously influencing both students' incoming beliefs and their conceptual learning.

It is also interesting to consider how learning is impacting or impacted by changes in students' beliefs over the course of the term. One approach is to select a subgroup of students with nearly identical incoming beliefs and look at the connection between their *shift* in beliefs and their learning gain. We preformed an analysis of this type for the data presented here. The results are suggestive of a positive relationship between occurrence of favorable shifts in beliefs over the course of the term and higher learning gains, but additional data are necessary to investigate this further.

CONCLUSION

Our analysis of students' attitudes and beliefs measured using the CLASS suggests that beliefs play a role both in the physics courses students choose to take and in their retention within those courses and in the discipline. In addition, we observe positive correlations between student beliefs and normalized conceptual learning gains. We see that students who come into a course with more favorable beliefs are more likely to achieve high learning gains. These data are consistent with the idea that beliefs are a factor in student learning; however, we must consider the possibility that other factors are simultaneously influencing students' beliefs, their learning, and their choices in pursuing physics. In our work, we found that it is possible to create environments, even in large lecture courses, that support improved student beliefs about physics and that this specific attention can avoid the common decline in beliefs commonly observed in introductory physics courses.

ACKNOWLEDGMENTS

The authors thank NSF for providing the support for this work. We are also very grateful to all of the instructors who generously agreed to administer the CLASS in their courses. We also give thanks to all the members of the Physics Education Research at Colorado group for many valuable discussions.

REFERENCES

1 W. Adams et al., "*Design and Validation of the Colorado Learning About Science Survey*", this volume, http://cosmos.colorado.edu/phet/survey/CLASS/

2 D. Hammer, *Am. J. Phys.* **68**, S52-S59 (2000).

3 A. Elby, *Am. J. Phys.* **69**, S54-S64 (2001).

4 Redish, *Teaching Physics with Physics Suite*, Wiley 2003

5 E.F. Redish, J.M. Saul, and R.N. Steinberg, *Am. J. Phys.* **66**, 212-224 (1998).

6 I.A. Halloun, "Views About Science and Physics Achievement: The VASS Story." In The Changing Role of Physics Departments in Modern Universities: Proceedings of the ICUPE.

7 A. Elby, Epistemological Beliefs Assessment for Physical Science http://www2.physics.umd.edu/~elby/EBAPS/home.htm

8 D. Hammer, "Misconceptions or p-prims: How may alternative perspectives of cognitive structure influence instructional perceptions and intentions?" *Journal of the Learning Sciences*, **5**, 97-127, (1996).
D. Hammer, "More than misconceptions: multiple perspectives on student knowledge and reasoning, and an appropriate role for education research," *Am. J. Phys.*, **64**, 1316-1325, (1996).
D. Hammer, "Discovery learning and discovery teaching," *Cognition and Instruction*, **15**, 485-529, (1997).

9 Chi, M.T.H, Feltovich, P.J., and Glaser, R. "Categorization and representation of physics problems by experts and novices," *Cognitive Science*, **5**, 121-152, (1981).

10 A. Schoenfeld, "What's all the fuss about metacognition," in A. Schoenfeld, *Cognitive science and mathematical education*. Philadelphia: Lawrence Erlbaum. 1987.

11 S. Pollock, "No Single Cause: Learning Gains, Student Attitudes, and the Impacts of Multiple Effective Reforms", this volume.

12 R. Hake, *Am. J. Phys.* **66**, 64-74, (1998)

13 R. Thornton and D. Sokoloff, *Am. J. Phys.* **66**, 338-352, (1998);
R. Thornton, D. Kuhl, K. Cummings, and J. Marx, "*Comparing the FMCE and the FCI*", unpublished.

Student understanding of Symmetry and Gauss's law

Chandralekha Singh

Department of Physics and astronomy, University of Pittsburgh, Pittsburgh, PA, 15213

Abstract. Helping students learn why Gauss's law can or cannot be easily applied to determine the strength of the electric field at various points for a particular charge distribution, and then helping them learn to determine the shape of the Gaussian surfaces if sufficient symmetry exists can develop their reasoning and problem solving skills. We investigate the difficulties that students in calculus-based introductory physics courses have with the concepts of symmetry, electric field and electric flux that are pivotal to Gauss's law of electricity. Determination of the electric field using Gauss's law requires discerning the symmetry of a particular charge distribution and being able to predict the direction of the electric field everywhere if a high symmetry exists. It requires a good grasp of how to add the electric field vectors using the principle of superposition, and the concepts of area vector and electric flux. We administered free response and multiple-choice questions and conducted interviews with individual students using a think-aloud protocol to elucidate the difficulties students have with the concepts of symmetry, electric field and electric flux. Here we discuss student responses to some questions on a multiple-choice test administered to them. The test can be used both as a teaching and assessment tool.

INTRODUCTION

A major goal of most calculus-based introductory physics courses is to help students develop the thinking skills of a physicist [1]. Physicists comfortably exploit symmetries to turn complicated problems into simpler ones. Gauss's law of electricity is an important topic in the second semester of most calculus-based introductory physics courses. Learning to reason whether Gauss's law can be exploited in a particular situation to calculate the electric field without complicated integrals can provide an excellent context for helping students develop a good grasp of symmetry considerations. Unfortunately, students often memorize a collection of formulas for the magnitude of the electric field obtained using Gauss's law for spherical, cylindrical, and planar geometries without paying attention to symmetry considerations involved in obtaining those results or even understanding the difference between the electric field and electric flux. Therefore, a majority of students have difficulty identifying situations where Gauss's law may be useful and overgeneralize results obtained for a highly symmetric charge distribution to situations where they are not applicable. Courses often do not sufficiently emphasize symmetry considerations or the chain of reasoning required to determine if Gauss's law is useful for calculating the electric field.

We administered free response and multiple-choice questions and conducted interviews with individual students using a think-aloud protocol [2] to understand the difficulties students have with the concepts of symmetry, electric field and electric flux. We subsequently developed a conceptual multiple-choice test that addresses these issues and administered it to 253 students in four different sections. The tests and interviews assess the extent to which students have become proficient in exploiting symmetry and in making conceptual predictions about the magnitude and direction of the electric field for a given charge distribution using Coulomb's or Gauss's laws. They also assess whether students can distinguish between electric field and electric flux, identify situations in which Gauss's law can easily be used to calculate the electric field strength from the information about the electric flux and the shapes of the Gaussian surfaces that would be appropriate in those cases.

DISCUSSION

Investigation of student difficulties related to a particular physics concept is important for designing instructional strategies to reduce them [3]. We investigated student difficulties with discerning symmetry and applying Gauss's law using two different methods: (1) design and administration of free-response and multiple-choice questions to elicit difficulties in a particular context, and (2) in-depth audio-taped interviews with individual students using a think-aloud protocol [2] while they solved those problems. The major advantage of written tests is that they can be administered to large student populations. When administered in conjunction with in-depth interviews with a subset of students, written tests can be very helpful in pinpointing student difficulties.

While a student can always calculate the electric field using Coulomb's law and their knowledge of the princi-

CP790, *2004 Physics Education Research Conference*, edited by J. Marx, P. Heron, and S. Franklin

ple of superposition without regard to symmetry, determining electric field using Gauss's law requires an explicit focus on the symmetry of the charge distribution. Although there are only three types of symmetry (spherical, cylindrical, and planar) for which Gauss's law can easily be exploited to determine the electric field at various points from the information about the electric flux, students need help in identifying when these symmetries are present. The principle of superposition is also a prerequisite for employing Gauss's law successfully to determine the electric field and helps determine if sufficient charge symmetry exists in a particular situation.

In addition to the consideration of symmetry, the area vector and the electric flux are often new concepts that are introduced in the context of Gauss's law. Students must be able to distinguish the electric flux from the electric field, a task that turns out to be extremely difficult. Students must learn that the electric flux, a measure of the total number of electric field lines passing through an *area*, is a scalar (which can have either a positive or a negative sign depending upon the relative directions of the electric field and area vector) while the electric field at a point is a vector. They must understand that the electric flux and the field even have different dimensions.

Students must also understand that Gauss's law only applies to closed surfaces and that for any closed surface, information about the net charge enclosed is sufficient to determine the net electric flux through it. On the other hand, determination of the electric field at various points due to a charge distribution is not simple and depends on the exact manner in which charges are distributed and may depend upon both the charges inside and outside the closed surface. For example, the electric field at various points on a Gaussian surface may be non-zero and vary from point to point even though the net electric flux through it is zero if the net charge enclosed is zero.

During the design of the multiple-choice test on symmetry and Gauss's law, we paid particular attention to the important issues of reliability and validity [4]. Reliability refers to the relative degree of consistency between testing if the test procedures were repeated for an individual or group. Validity refers to the appropriateness of test score interpretation. The test design began with the development of a test blueprint which provided a comprehensive framework for planning decisions about the desired test attributes. The degree of specificity in the test plan was useful for creating questions. We tabulated the scope and extent of content covered and the level of cognitive complexity desired. Using previous free-response and multiple-choice questions administered to students as a guide, we identified the desired performance and a description of conditions/contexts under which the performance was expected to occur. The rest of the test design proceeded in a manner similar to that described for the Energy and Momentum Conceptual Survey [5], and will be described in detail elsewhere.

The final version of the test contains 25 items. It was administered to a total of 253 students in four different introductory calculus-based physics courses at Pitt. One of these was an honors course, and the students in the other three courses had used some preliminary tutorials on symmetry and Gauss's law. The average score on the test is approximately 56%. The reliability coefficient α [4], which is a measure of the internal consistency of the test, is 0.8, which is considered good by the standards of test design. The point biserial discimination is related to the ability of a question to discriminate between students who overall did well on the test vs. those whose overall performance on the test was poor. This discrimination index for 16 questions on the test was more than 0.4. Only 4 questions had discrimination index less than 0.3 (only one question with less than 0.2) which is also very good by the standards of test design [4]. Below, we discuss some of the questions on the test:

• **Setup for the next two questions**

You perform two experiments (see figure) in which you distribute charge $+6Q$ differently on the surface of an isolated hollow insulating (non-conducting) sphere:

Exp(I): You put identical charges, $+Q$ each, on the spherical surface in six localized blobs (you can consider them point charges) such that the adjacent blobs are equidistant from each other.

Exp(II): You distribute charge $+6Q$ uniformly on the surface of the sphere.

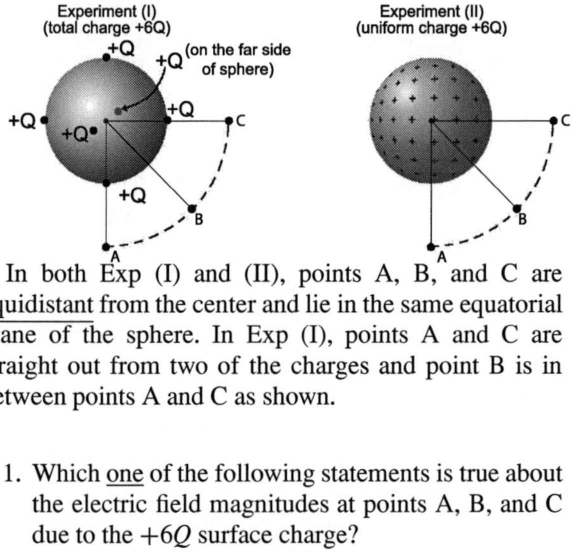

In both Exp (I) and (II), points A, B, and C are equidistant from the center and lie in the same equatorial plane of the sphere. In Exp (I), points A and C are straight out from two of the charges and point B is in between points A and C as shown.

1. Which one of the following statements is true about the electric field magnitudes at points A, B, and C due to the $+6Q$ surface charge?

   (A) In each experiment, the field magnitude is the same at points A, B, and C, but the magnitudes differ in the two experiments.

   (B) In each experiment, the field magnitude is the same at points A, B, and C, and the magnitudes are equal in the two experiments.

   (C) In experiment (I), the field magnitude is the

same at points A, B, and C, but not in experiment (II).
(D) In experiment (II), the field magnitude is the same at points A, B, and C, but not in experiment (I).
(E) None of the above.

2. Which one of the following is a true statement about the electric field magnitude inside the hollow insulating sphere due to the $+6Q$ surface charge (see figure above)?

(A) It is zero everywhere inside the sphere in both experiments.
(B) It is non-zero everywhere inside the sphere in both experiments.
(C) In experiment (I), it has a magnitude that varies from point to point inside the sphere, but it is zero everywhere inside the sphere in experiment (II).
(D) In experiment (I), it has the same non-zero magnitude everywhere inside the sphere, but it is zero everywhere inside the sphere in experiment (II).
(E) None of the above.

• Setup for the next two questions

A cubic Gaussian surface with 1 meter on a side is oriented with two horizontal and four vertical faces, as shown. It is in a uniform electric field of 20 N/C which is directed vertically upward. Point A is on the top surface and point B on a side surface of the cubic Gaussian surface.

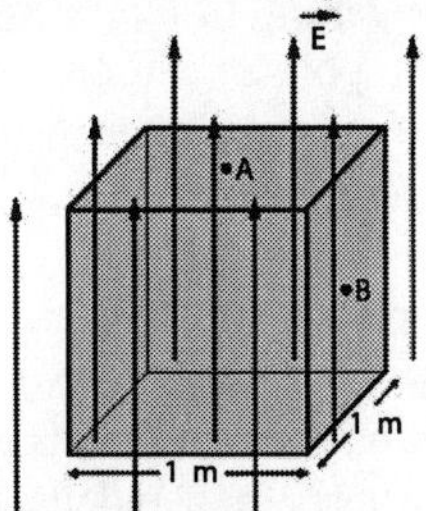

Cubic Gaussian Surface in a Uniform Electric Field

3. Which one of the following statements is true about the electric field at points A and B?
(A) The field is zero at both points A and B.
(B) The field is zero at point A but not at point B.
(C) The field is zero at point B but not at point A.
(D) The field is non-zero at both points A and B and its direction is the same at the two points.
(E) The field is non-zero at both points A and B but its direction is different at the two points.

4. Choose all of the following statements that are true about the electric flux.
(I) The net flux through the whole cubic surface is zero.
(II) The magnitude of the flux through the top face of the cubic surface is $20Nm^2/C$.
(III) The magnitude of the net flux through the whole cubic surface is $20Nm^2/C$.
(A) (I) only
(B) (II) only
(C) (III) only
(D) (I) and (II) only
(E) (II) and (III) only

• Setup for the next two questions

Shown below are four imaginary surfaces coaxial with an isolated infinitely long line of charge (with uniform linear charge density $\lambda\ C/m$):
(I) a closed cylinder of length L.
(II) a sphere of diameter L.
(III) a closed cubic box with side L.
(IV) a two dimensional square sheet with side L. The plane of the sheet is perpendicular to the line of charge.

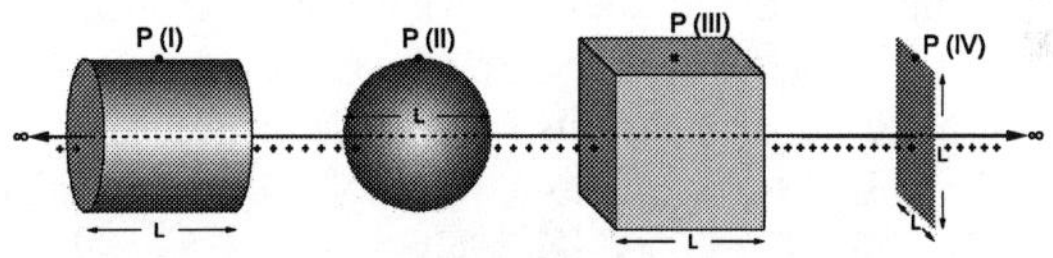

5. Choose all of the above surfaces through which the net electric flux is $\Phi_S = \lambda L/\varepsilon_0$:

(A) (I) only
(B) (I) and (II) only
(C) (I) and (III) only
(D) (I), (II), and (III) only
(E) (I), (II), (III), and (IV)

6. Choose all of the above surfaces which can be used as Gaussian surfaces to easily find the electric field magnitude (due to the infinite line of charge) at a point P shown on the surface using Gauss's law:

(A) (I) only
(B) (I) and (II) only
(C) (I) and (III) only
(D) (I), (II), and (III) only
(E) (I), (II), (III), and (IV)

The most common difficulty with question (1) was assuming that the magnitude of the electric field in experiment (I) will also be the same at all of the three points shown. In interviews and free response questions, some students explained this by claiming that since the six point charges and the three points are symmetrically situated, the field magnitude must work out to be the same at

TABLE 1. Percentage of students (out of a total of 253) who selected choices (A)-(E) on items (1)-(6). The correct response for each question has been italicized.

ItemNumber	A	B	C	D	E
1	18	13	0	*67*	2
2	32	4	*55*	4	5
3	2	0	35	*58*	5
4	25	12	7	*51*	5
5	28	3	15	*37*	17
6	*43*	15	15	15	12

all the three points shown while others explained their response by claiming that for a point outside, the six point charges on the sphere can be thought to be point charges at the center of the sphere. The most common difficulty with question (2) was assuming that the electric field inside the sphere in experiment (I) is also zero everywhere. In interviews and free response questions, some students explained their response by claiming that the electric field inside a sphere (or inside an object of any shape with charges on its surface) is always zero. When probed further, some students mentioned that this had to do with electrical shielding which guaranteed that the electric field will be zero inside. Even when interviewers drew students' attention to the fact that the sphere was made of insulating material, students adhered to the idea that the field inside must be zero everywhere.

The most common difficulty with question (3) was assuming that the electric field is zero at point B on the side surface of the cube although the problem statement explicitly mentioned that the cubic Gaussian surface is in a uniform electric field of $20N/C$. In interviews and free response questions, some students reasoned that since the area vector of the side surface is perpendicular to the direction of the electric field lines, the electric field must be zero at point B. This kind of confusion between the electric field at a point and the contribution to the electric flux from a certain area was found in other questions as well. The most common difficulty with question (4) was that students did not realize that both statements (I) and (II) are correct. The table above shows that many students only chose option (I) or (II) but not both.

Questions (5) and (6) were extremely difficult for students. Students were not comfortable with the statement of Gauss's law that relates the net flux through a closed surface to the net charge enclosed. They also could not differentiate between electric flux through a closed surface and electric field at a point on the surface. For example, in question (5), many students chose (I) (or (I) and (III)) and claimed in interviews and free response questions that only those surfaces can be used to determine the net electric flux through them because the other surfaces did not have the correct symmetry. Some students were also unsure about the distinction between open and closed surfaces and the fact that Gauss's law only applies to closed surfaces. For example, in interviews and free response questions, many students claimed that the net flux should be the same through all surfaces (including the square sheet) using Gauss's law. In Question (6), students had to choose the Gaussian surfaces that would help them determine the electric field at point P easily due to the infinite line of charge. All of the alternative choices were selected with an almost even frequency. Students were often unsure about the symmetry concepts relevant for making appropriate decisions and those who chose option (C) were often quite confident that the magnitude of the electric field due to the infinite line must be the same at every point on the cubic box as well.

WORK IN PROGRESS

Gauss's law provides an excellent opportunity to help students learn symmetry ideas and to develop their reasoning and problem solving skills. We are currently implementing and refining several tutorials that we have developed to help students learn to discern the symmetry of a charge distribution, determine whether Gauss's law can easily be applied to calculate the electric field strength at various points due to a given charge distribution, and predict the shape of the corresponding Gaussian surface if a high symmetry exists. The tutorials also help students learn to distinguish between the electric field and the electric flux, learn the superposition principle and determine if the electric field at two points is the same in magnitude and/or direction due to a given charge distribution. The tutorials in this series also help students calculate the electric field at various points due to highly symmetric charge distributions and revisit superposition principle after learning Gauss's law, for example, to determine the net electric field due to two spheres of charge. All tutorials have separate pre/post-tests that students take before and after working on the tutorials. The preliminary results are encouraging.

REFERENCES

1. A. Van Heuvelen, *Learning to think like a physicist: A review of research based instructional strategies*, Am. J. Phys. **59**, 891 (1991).
2. M. Chi, in *Thinking Aloud* Chapter 1, Eds. Van Someren, Barnard, and Sandberg, 1, Erlbaum, 1994.
3. L. McDermott, *Millikan Lecture: What we teach and what is learned-Closing the gap*, Am. J. Phys. **59**, 301 (1991).
4. G. Aubrecht and J. Aubrecht, *Constructing objective tests*, Am. J. Phys. **51**, 613-620 (1983); A. J. Nitko, *Educational assessments of students (2nd ed.)*, Englewood Cliffs, NJ: Prentice Hall/Merrill, (1996).
5. C. Singh and D. Rosengrant, *Multiple-choice test of energy and momentum concepts*, Am. J. Phys, Physics Education Research Supplement Section, **71(6)**, 607, (2003).

Rate of Change and Electric Potential

Rhett Allain & Robert Beichner

Southeastern Louisiana University & North Carolina State University

Abstract. This project aims to investigate a possible underlying cause of student difficulties relating change in electric potential to electric fields. A likely source of difficulties is the lack of students' understanding of the general concept of rate of change (both rate of change in time and in distance). To investigate this link, a diagnostic instrument was created that probed students' understanding of rate of change concepts and electric potential concepts. This paper will report on the creation of the diagnostic instrument and results from student responses.

INTRODUCTION

More and more colleges and universities are beginning to see the need for reforming introductory physics courses. Evidence for this can be seen in the development and implementation of innovative curricula such as SCALE-UP[1], Studio Physics[2], and Workshop Physics[3]. As developers work to modify the curriculum, they often apply a knowledge of which concepts are poorly understood after traditional instruction. But it may be possible to avoid some of this effort altogether if the underlying causes of the students' difficulties can be identified. If these can be addressed in prior courses, we might see students entering the introductory physics courses without some of the difficulties they now exhibit.

This study focuses on student understanding of the concept of "rate of change" as a possible source for their difficulties with electric potential. The electric field is the gradient of the electric potential. The term "rate of change" has traditionally referred to a derivative of a quantity with respect to time. We have chosen a more general definition that includes both time rate of change and the gradient or "position rate of change". Previous research[4] has shown that students do have difficulties with the concept of rate of change, so it seemed to be a logical place to start because of its underlying importance for understanding electric potential. The topic of electric potential was chosen for several reasons. First, electric potential is an important idea that later concepts build upon. Knowledge of DC circuits, electromagnetic induction, the photoelectric effect, and quantum mechanics depends upon the students' understanding of electric potential. Because of its fundamental importance, electric potential is one of the earliest topics discussed in the second course in introductory physics. There have already been many studies[5] of student difficulties in kinematics, forces and the other concepts traditionally covered in the first introductory physics course, but research on second-semester topics is limited.

DESCRIPTION OF RESEARCH

The main approach of this research project was to create a multiple-choice instrument on both rate of change and electric potential concepts (Rate and Potential Test – RAPT) and administer it to a variety of students. The purpose was not to measure the effect of a particular pedagogy or treatment, but rather to investigate the connection between two ideas that a student already holds. There was no need to administer both a pre and a post-test as the study was looking for a link in student understanding rather than looking for a change due to some curriculum. Think-aloud[6] interviews were used to validate the test items and to allow for investigation of any unexpected student responses. The combination of interviews and the

CP790, *2004 Physics Education Research Conference*, edited by J. Marx, P. Heron, and S. Franklin

distribution of the multiple-choice instrument allowed us to get a closer look at a few students as well as a more general look at a larger number of students.

The rate of change questions in the instrument ask about the rate of change of a quantity given information about that quantity. For example, there were items that asked about velocity when information about the position of an object was provided. There were not, however, items that focused on the position of an object given its velocity. In mathematical terms, there were questions that called for students to find derivatives, but not carry out integrations. This was done to make the instrument more manageable. If both "derivative" and "integrating" questions were asked, the instrument would be significantly longer. With a long instrument, it is less likely that cooperating instructors would be found. It might also have been less likely for students to take a very lengthy instrument seriously or to complete it without fatigue. In any case, if a connection would be found for "derivative" items, it is likely that there would be a similar relationship for "integration" items as well.

Development of the Instrument

The questions in the instrument were designed to address several possible modes of thinking students may have about rate of change. These modes were discovered in preliminary interviews. They are: 1) A quantity and its rate of change (like electric potential and electric field) look the same and behave the same. That is, if the quantity increases, the rate of change of that quantity increases. A student using this mode of thinking would say that a graph of the magnitude of the electric field versus position would look the same as a graph of the electric potential versus position. 2) If a quantity is zero, then its rate of change must also be zero. 3) If a quantity is not zero, then the rate of change of that quantity cannot be zero. 4) If a quantity is at its greatest value, then the rate of change of that quantity must also be at its greatest value.

It is important to realize the limitations imposed in the electric potential questions. The goal of these items was to probe students' understanding of electric potential as it related to the "rate of change" of a quantity. This being the case, the test did not include items that deal with the vector nature of the electric field. The test also was not concerned with the difficulties students have with the differences between positive and negative charges in an electric field.

Think-Aloud Interviews

As discussed earlier, part of the data collection process included think-aloud interviews of students taking the RAPT. In a think-aloud interview, the student answers the test items while verbalizing the thoughts in his or her mind. Ericsson and Simon[7] describe the general goal of the think-aloud interview as *"procedures to capture what actually goes on in a subject's mind."* Of course, the best we can do is approximate what the students are thinking. For this study, we wanted to make sure the students were reading and interpreting the questions as we thought they would. That is, we used the think-aloud interviews partly as a means of validating the questions.

The think-aloud interviews were conducted with eight volunteers from an introductory physics course. The students were paid for their time and the interviews took from 30 to 40 minutes. A multiple-choice question from the FCI was used as a practice question so that students could become familiar with the think-aloud protocol before they answered questions from the RAPT. The practice question also gave the interviewer a chance to comment to the student on how he or she performed with the think-aloud protocols.

Rate and Potential Scores

The most direct way to examine the relationship between answers on electric potential and rate of change items is to compare scores on these questions. To do this, several sub-scores were calculated. The *rate-score* is the number of rate of change items a student answered correctly, out of a possible 14. The *potential-score* is the number of electric potential items a student answered correctly out of a possible 11. The correlation between these two scores can then be calculated.

Students' wrong answers also reveal useful information. Do students choose similar incorrect answers for rate of change and electric potential items? To answer this question, two other sub-scores were calculated, the *rate-quantity* score and the *potential-field* score. The *rate-quantity score* is the number of times the students chooses the answer that corresponds to the mental model that says a quantity and its rate of change are directly related. Note that the *rate-quanity* score corresponds to particular responses, thus it is possible for a student to get a 0 rate-score as well as a zero *rate-quantity* score. As an example of the *rate-quantity* score, one item in the RAPT asks about the acceleration of the coin when it reaches its highest point. The correct answer states that the acceleration is in the negative direction and constant. The *rate-quantity* response for this item is that the acceleration is zero, implying that the acceleration (which is the rate of change of velocity) is zero because the velocity itself happens to be zero when the coin is at its highest point. There are, of course, other responses possible, but these are not associated with the mental model of, "if the quantity is zero (velocity), its rate of change (acceleration) must also be zero". Similarly, a *potential-field* score is calculated from responses on the electric potential items for student choices that are wrong but using the mental model that electric potential is proportional to the electric field.

RESULTS AND ANALYSIS

This section will examine the quantitative results from the RAPT. Qualitative data was collected from think-aloud interviews, but that will not be included in this paper.

Summary Statistics for the RAPT

The first goal of the study was to determine if the RAPT is an appropriate instrument for the task. The most common measure of reliability of an assessment instrument is the use of a reliability coefficient indicating internal consistency. This coefficient was calculated for the RAPT using the Kuder-Richardson[8] formula 20. The KR-20 calculation reported a reliability coefficient of 0.83 with the sample of 340 students. According to Doran's[9] guidelines for interpreting these coefficients, the reliability of the instrument is *"fairly high, possible for measurement of individuals"*.

Correlations Between Rate and Potential Scores

The correlation between the *rate-score* and the *potential-score* was found to be 0.45 using the Pearson product-moment coefficient. Fisher's *Z*-transformation[10] was then used to determine that this correlation coefficient is significantly different than zero at the 95% confidence level, indicating that there is some correlation between the two variables. The correlation between the *rate-quantity* score and the *potential-field* score was found to be 0.38 which is also significantly different from zero at the 95% confidence level.

CONCLUSIONS

It seems clear that students have difficulty with the concept of rate of change as well as the concept of electric potential. The aim of this research was to look for a link between these two difficulties. We have, in fact, uncovered a correlation. Why is evidence of such link important? It is important mainly for curriculum development in the realm of electric potential. In order to write curricula to address student difficulties, developers need to look not only at the concept at hand, but also at supporting concepts.

From the interviews and the multiple-choice data, it seems certain that students make similar mistakes in questions involving rate of change and those involving electric potential. This does not necessarily mean that one causes the other, it is possible there is some other conceptual difficulty that affects a student's understanding of both the rate of change concept and electric potential. If there is such a underlying cause, it could not be found from the data gathered in this study.

An important "next step" is to create a separate diagnostic instrument solely on the concept of rate of change. With such an instrument two things could be done. First, it could be used to determine what mathematical tools students carry as they enter the introductory physics course. Second, it could be used to investigate the possibility that rate of change difficulties are the source of problems in other areas of physics, such as electromagnetic induction.

This finding also suggests a path for curriculum development. It would seem the best way to approach the concept of electric potential would be to take time to first review the concept of rate of change. Situations where students already understand rates of change, such as the gradient of a hill, can be used as a "bridge" to the concepts the students don't understand. This is similar to the technique suggested by Clement[11] in which concepts students understand are used to bring students to an understanding of difficult new ideas.

ACKNOWLEDGMENTS

This work was supported in part by the National Science Foundation through Grant No. DGE-9554526. Suggestions and input were also obtained from the

Physics Education Research and Development Group at North Carolina State University.

REFERENCES

1. More information about the SCALE-UP project can be found online at <http://scaleup.ncsu.edu>

2. J.M. Wilson, "The CUPLE Physics Studio," The Physics Teacher, **32**, 518-523 (1994).

3. P.W. Laws (1997) Workshop Physics Activity Guide. New York, NY: Wiley <http://physics.dickinson.edu/~wp_web/WP_homepage.html>

4. G.S. Hauger, "Rate of change knowledge in high school and college students," A paper presented at the annual meeting of the American Education Research Association, San Francisco, CA 1995; see also R. Nemirovsky and A. Rubin, "Students' tendency to assume resemblances between a function and its derivative," MA TERC (ERIC Document Reproduction Service No. ED351193) 1992.

5. L.C. McDermott and E.F. Redish, "RL- PER1: Resource letter on physics education research," Am. J. Phys. **67**, 755-767 (1999). Available online at <http://www.physics.umd.edu/ripe/perg/papers/redish/rl.pdf>

6. K. Ericsson and H. Simon, "Verbal protocols as data," Psychological Review, **87**, 215-251 (1980).

7. Ericsson K. A., and Simon, H. A. (1984) Protocol Analysis: Verbal reports as data, Cambridge, MA, MIT Press.

8. K. Hopkins, (1998). Educational and psychological measurement and evaluation (8th ed). Boston, MA: Allyn & Bacon.

9. Doran, R.L. (1980). Basic Measurement and Evaluation of Science Instruction. Washington DC: National Science Teachers Association.

10. Fisher's Z transformation and other useful statistics can be found online at <http://icp.giss.nasa.gov/education/statistics/>

11. J. Clement; D. Brown and A. Zeitsman, "Not all preconceptions are misconceptions: Finding 'anchoring conceptions' for grounding instruction on students' intuitions," International Journal of Sci. Ed. **11**, 554-565 (1989).

Diminishing Forces – Implications for Contextual Dependence of a Misconception

Alicia R. Allbaugh

Rochester Institute of Technology, Department of Physics, Rochester, NY 14623, USA

Abstract. Evidence is presented to suggest a misconception concerning motion of an object when acted upon by a force that decreases with distance such as Coulomb's Law. This evidence was collected during interviews of several above average calculus-based physics students. The students stated that the motion of an object would slow, even stop, if the force on it decreased based upon its distance. This may not be surprising until viewed in the light that many of these students didn't reveal this impetus or Aristotelian notion except with diminishing forces.

INTRODUCTION

Over the last 25 years, a large portion of research has focused on the pre-existing mental models that a student brings to the classroom [1]. These have been called preconceptions, alternative conceptions, common sense concepts and misconceptions [2]. Studies have shown that people of all ages hold some type of misconception [3]. And students at all achievement levels have these misconceptions as well [4,5]. Students' misconceptions in mechanics and more specifically Newton's Laws have been of particular interest because Newton's Laws are so often referenced during other topics of instruction in an introductory physics course [6]. One particular misconception, motion implies force, is attributed to Aristotle and widely accepted to exist among introductory physics students [7]. This study has discovered implications that students who use Newtonian reasoning in one context have this Aristotelian misconception triggered when dealing with diminishing forces.

RESEARCH METHOD AND DATA

While conducting a research study on the problem-context dependence of Newton's Second Law [8], I believe that I happened upon a previously undocumented use of a misconception, or at least a complication, regarding forces which decrease in magnitude with distance such as Coulomb's Law.

The longitudinal study allowed me to interview 13 students on the topic of electric charges and fields (as well as a variety of other topics). These students were above average performers in a calculus-based introductory physics course at Kansas State University. The interviews were conducted after instruction. The particular interview discussed here occurred one week after the course exam covering the topics in the interview.

Students were first shown Fig. 1 and were told that the charged spheres were identical. The left one was fixed in place and the right one was set in place and released. There were no other instructions or caveats given such as to ignore friction. The student was then asked to describe what happened. A correct response would include reference to the Coulomb force making the released sphere accelerate to the right. The sphere would continue to accelerate. As it gets farther and farther away its increase in velocity would be smaller and smaller.

FIGURE 1. Electric Charge Question – Charged Sphere Released from Rest

CP790, *2004 Physics Education Research Conference*, edited by J. Marx, P. Heron, and S. Franklin

A majority of student responses included the statement that because the force diminishes as $1/r^2$, the particle would slow as it moves away from the other charge. Most students indicated that a point in space exists where the particle's speed would reach a maximum and then start to return to either a constant velocity or zero.

> "It would accelerate at the beginning until it reached a certain point I suppose where the field isn't so strong on it." (Student 7)

> "So it'll, at first it will probably accelerate and then get to a point when it starts slowing down again..." (Student 18)

These types of responses were given by several students. Some of whom responded to the next question in the interview (which had a larger, more massive sphere in the same scenario) in a Newtonian manner and some who did not. This idea of a maximum speed and then slowing down was quite prevalent. I was only able to eliminate it as a possible reasoning path for one of the 13 students. Yet, a response of this nature might not be indicative of an Aristotelian misconception, but rather the student just neglecting to mention friction or some other intuition from experience in the given response.

FIGURE 2. Electric Charge Question – Charged Sphere with Initial Velocity

However, the concept came up again when the scenario was changed slightly. The students were shown Fig. 2. Again, the charged spheres are identical and the left one fixed in place. In this scenario, a film has been capturing the sphere's motion and has been stopped when the sphere was moving at a velocity v as shown. The students were then asked to describe what would happen when the film was started again. The correct response would include the moving sphere would turning towards the right due to the Coulomb force between the spheres. Eventually, the sphere would stop turning because as it turns, the force line between the two spheres also changes until the force is in the direction of the moving sphere's motion. This latter detail was beyond the scope of the interviews and was not pursued. Seven of the students drew something similar to Fig. 3.

FIGURE 3. Written Response to Electric Charge Question (Student 2)

> "It'll look like a trajectory plot. It'll still accelerate this way identical to the first one. [Drawing arrow and labeling it a] In the x-direction it's going to do the same thing. However in the y direction it's going to continue to stay at this velocity. So for the most part it's going to be like that [Draws arc]." (Student 2 – while drawing Fig. 3)

One of the students drew a straight line with no change in trajectory. Two of the students drew diagonal lines angled to the right. Three of the 13 students drew responses similar to that drawn in Fig. 4 below:

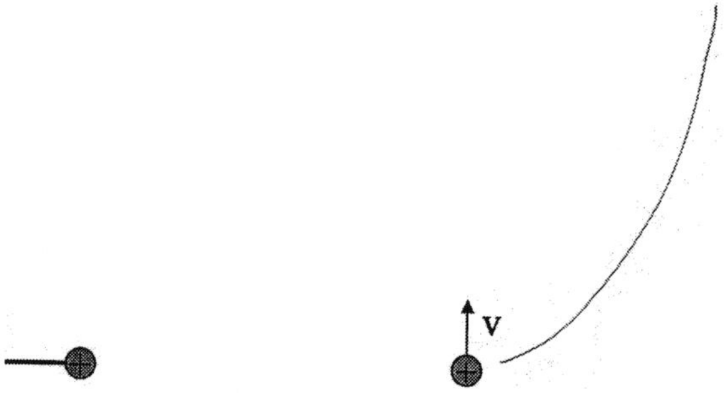

FIGURE 4. Written Response to Electric Charge Question (Student 17)

> "I would say that it should be accelerated in that direction but it will slow down as it gets farther away. So it should move [draws concave up arc]" (Student 17 while drawing Fig. 4)

Neither friction nor previous experience can account for such a trajectory shape. In fact, two of these students drew more intuitive trajectories initially then changed their minds as they discussed how they would include the $1/r^2$ factor of the force. An example is shown in Fig. 5.

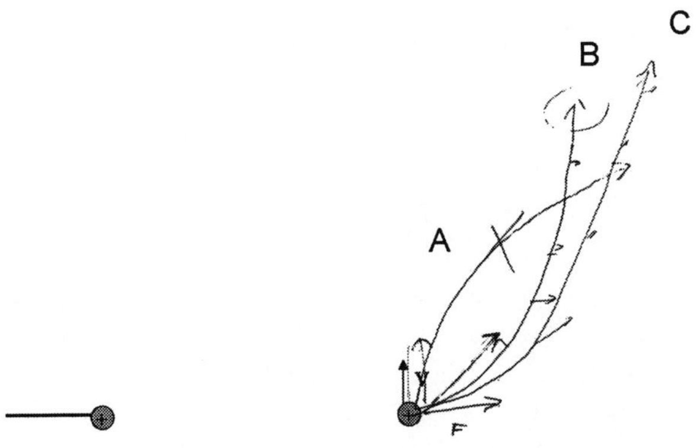

FIGURE 5. Written Response to Electric Charge Question (Student 15)

> "Maybe like that [drawing concave down arc A]...I'm thinking....since...only I'm going to change my mind...because as you get farther away this force is less so maybe it'll go like that instead [draws concave up arc B] since that is going to dominate the direction it goes since as you get farther away this force is going to be less. It's just a guess though. [Interviewer: So which are you going with?] This one, not that one [Crosses out arc A] It would be in this direction since as it's getting farther away the force vector is becoming less and less [draws arrows on arc B] but you still have this original velocity..." (Student 15 while drawing Fig. 5)

The student-drawn arrows pointing to the right represent the electrical force diminishing as the sphere gets farther away. This truly supports a diminishing forces causing diminishing velocity misconception regarding Newton's laws of motion.

Both the above students earned an A grade in BOTH semesters of the course. Student 17 received the highest score on the exam covering electrical charges which had been given in the week prior to the interview. In addition, both students used Newtonian reasoning in answering the next interview question when the mass of the moving sphere was doubled. By traditional measures and from previous and subsequent interview data, these students were exceptional. So just what is going on here? To find out, I had time available in one interview to request a comparison of nearly equivalent situations.

Later in the same interview the students were asked a similar question with a particle entering an electric field at a velocity v – See Fig. 6.

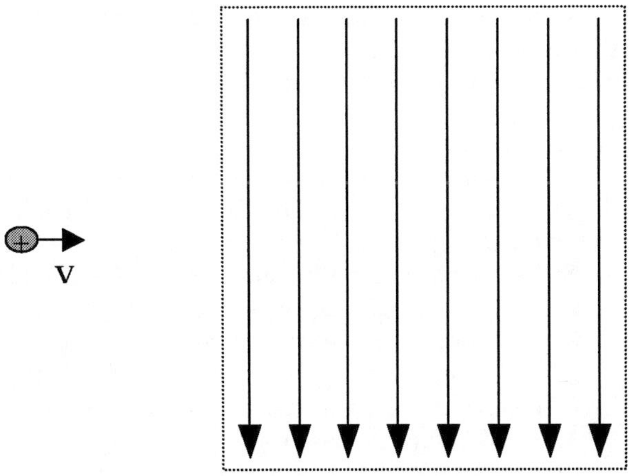

FIGURE 6. Electric Field Question – Charge Sphere with Initial Velocity

The students were asked to describe and draw what would happen. All three students who drew concave up arcs for the two charged spheres question, drew concave down arcs in response to the electric field question as shown in Fig. 7.

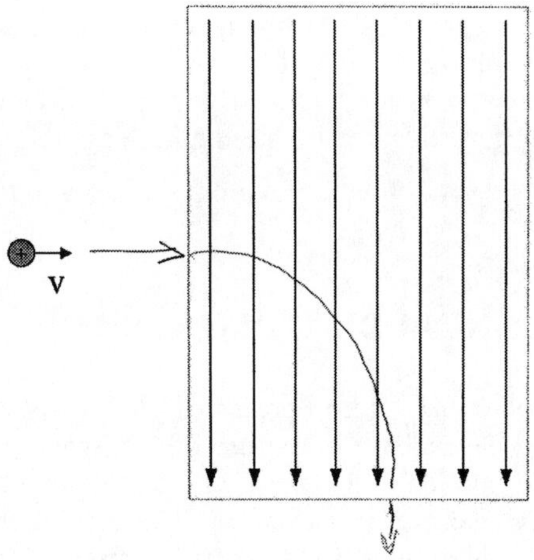

FIGURE 7. Electric Field Question – Charge Sphere with Initial Velocity

> "Well it's going to look the same as, like you know, a projectile. Since it's accelerating in one direction and its velocity is in the other. Like when you throw something horizontally and it drops. So it'd pull one of these numbers right there [draws arc]" (Student 17 while drawing Fig. 7)

Student 17 was asked to compare the two responses shown in Fig. 4 and Fig. 7. The question posed was whether the arcs should be the same.

> "Yeah…this is a really good question. Yes. Yes they should. I have no idea why it's drawn that way. [Interviewer: Really. Ok.] Huh. Cause it would seem like the force is going this way in both things. Accelerates. But this one [Fig. 4 – charge at v] acclerates it's DEcelerating as it goes this way. …So maybe it should be that way. Cause in that one [Fig. 7 - charge at v entering E-Field] it's ACCelerating as it goes that way should …should….should be …sure…I'm going to say yes they should be that way. Because this one [Fig. 4 - charge at v] is DEcelerating and this one [Fig. 7 - charge at v entering E-field] is ACCelerating in this direction. Hence it would go [gestures along the arc in Fig. 4 - charge at v] instead of going [gestures a different arc shape] you know…" [Student 17, with student's emphasis included as capitalization]

To summarize, the student stated the two paths should be dissimilar because one is accelerated by a constant field and the other is decelerating as it moves away. This student is associating the $1/r^2$ nature of the force with a sphere's deceleration and according to the student, while in the E-field the sphere continues to accelerate. This is clearly an Aristotelian misconception triggered by the diminishing nature of the force

CONCLUDING DISCUSSION

Evidence has been presented to suggest a contextual dependence of the use of a misconception concerning motion of an object when acted upon by a force which changes with distance. This evidence was collected while pursuing a different research goal which limits its thoroughness and amount. It is also important to note that forces of the $1/r^2$ form were first introduced in the course in addressing electrical charges. Gravitation was not covered although some students admitted to having learned it in high school

Despite these caveats, I believe the evidence suggests a contextual dependence or triggering of the use of an Aristotelian misconception when regarding diminishing forces that is possibly shared by many of the students I interviewed. These students were above average and quite possibly over thought the problem to the point of neglecting parameters they know to be true – like the continuity of the trajectory. This path to using misconceptions is likely triggered by some contextual feature, as in this case with diminishing forces.

FUTURE RESEARCH

As mentioned previously, this was a discovery made on the way to answering a different research question. The student interviews were not structured to investigate this particular situation and as such, useful follow-on questions were not pursued at the time. A future research investigation should be more thorough and include forces that diminish over time as well as distance and include as many differing physical contexts as possible. This would tease out the particulars of the issue and possibly help to discover what triggers the Aristotelian concept in students who otherwise have shown they understand Newtonian concepts.

ACKNOWLEDGMENTS

This work has been supported in part by NSF ROLE Grant #0087788. I would like to acknowledge of Dean Zollman and Paula Engelhardt for their insight and aid in preparing this document.

REFERENCES

1. L. C. McDermott and E. F. Redish,. *Amer. J. Phys.* **67**, 755-767 (1999).
2. A. Eryilmaz, *J. Res. Sci. Teach.* **39,** 1001-1015 (2002).
3. M. McCloskey, *Sci. Amer.* **248,** 122-130 (1983).
4. P. C. Peters, *Amer. J. Phys.* **50,** 501-508 (1982).
5. M. Steinberg, D. E. Brown, and J. Clement, *Inter. J. Sci. Ed.* **12,** 265 (1990).
6. I. Galili, *Inter. J. Sci. Ed.* **17,** 371-378 (1995);
7. I. Galili and V. Bar, *Inter. J. Sci. Ed.* **14,** 63-81 (1992);
8. A. R. Allbaugh, Unpublished Ph. D. Dissertation, Kansas State University, Manhattan, Kansas, (2003).

Student Understanding Of Gravitational Potential Energy And The Motion Of Bodies In A Gravitational Field

M.E. Loverude

Department of Physics MH-611,
California State University Fullerton, P.O. Box 6866, Fullerton, CA 92834

We have been investigating student understanding of energy concepts in the context of introductory courses for non-science majors as well as those for science and engineering majors. We have found that many students develop incomplete and incorrect understandings of the concept of gravitational potential energy. Moreover, students often have incorrect notions about the motion of bodies under the influence of gravity. These incorrect beliefs may prevent the development of a coherent understanding of energy as a conserved quantity.

INTRODUCTION

Energy is a fundamental concept in physics and physical science courses at all levels of the curriculum. In particular, the concept of gravitational potential energy (GPE) and descriptions of the motion of bodies in the presence of a gravitational field are taught in courses at every level from graduate level physics down to elementary school.[1] As part of an ongoing study of student conceptions of energy, we have sought to characterize student understanding of specific situations that are often covered in physics and physical science courses. There is considerable evidence to suggest that students conceptualize energy quite differently than do physicists.[2,3] [4] Informal discussions with students in class suggested difficulties with the standard physical explanation of the motion of a body in a gravitational field and even a failure to understand the variables on which GPE depends. We sought to document and analyze student difficulties and characterize student thinking.

Context for research

This study was performed in the context of courses at a public comprehensive university serving a diverse student population. Two sample courses enroll primarily non-science majors. The Course for Teachers is a physical science course for pre-service elementary school teachers. This course is taught in a lab-based, guided inquiry format, meeting six hours each week. All students are planning to teach at the K-8 level. The Survey Course is taken by students in order to satisfy a general education requirement. It includes three hours per week of traditional lecture and has an optional three-hour lab component. A third course, which we will call Introductory Mechanics, is a fairly typical introductory mechanics lecture and lab course for science and engineering majors, with three hours of lecture and an traditional three-hour lab.

In all courses, students were taught a model for energy conservation including the following features: definition of a system, potential and kinetic energy, and transfer of energy into or out of a system, including work. All students had been taught explicitly the formulas for GPE and kinetic energy (KE), with particular emphasis that GPE depends on mass, height above a reference point, and the strength of the gravitational field (which was assumed to be constant unless specified otherwise). The treatment in the course for science majors was considerably more quantitative and included the solution of standard numerical problems, and students had completed two labs on energy topics, one on collisions and the second on a ballistic pendulum. Unless otherwise noted, students had completed all instruction on energy at the time they completed the research tasks.

CP790, *2004 Physics Education Research Conference*, edited by J. Marx, P. Heron, and S. Franklin
 0-7354-0281-7/05/$22.50

RESEARCH METHODS

All data in this paper were collected by analysis of student responses to free-response written questions posed on course examinations and ungraded quizzes. The questions were constructed so that incorrect answers would correspond to the incorrect ideas we had observed in informal interactions with students. Students were asked to provide an answer as well as an explanation. Student answers were recorded and explanations were classified and assigned codes. The major categories of student answers and explanations are described below.

GPE AND MOTION OF A BODY

The initial focus of this study was student understanding of the periodic motion of bodies in the presence of a gravitational field, as in a simple pendulum. Therefore we posed problems in which students were asked to make predictions about this motion in various situations. For example, students are asked to predict the final height of an object released from rest, either a simple pendulum or a ball or block sliding in a frictionless bowl. Students are told that the object is released from rest and asked whether the level at which the object comes to rest (on the opposite side of its first cycle) is *higher than, lower than, or the same height as* the point of release. Students are explicitly asked to ignore the effects of friction and air resistance. Table I shows responses from several height comparison problems.

Belief that final height will be greater

In each set of responses, at least one in eight students predicted that the final height would be *greater than* the initial height. These responses raise a question whether students are in a position to develop a good model for the behavior of this system. The energy description of this system is based on observations of the motion of bodies like a pendulum, but the belief that the final height is greater explicitly contradicts conservation of energy.

Students giving this response frequently appeared to use a colloquial association of energy with activity. For example, one student making such a prediction explained: "The height will be greater because the ball gains energy in the swinging process." Other responses included the term momentum: "As the pendulum continues in its motion, I believe the maximum height will be slightly above point A due to the momentum built up as it swings back and forth." Another student wrote: "It will be higher because it is getting more momentum (energy is being transformed to kinetic energy) so it will be pushed upward more."

Final height will be less

At least 20% of the students predicted that the final height would be lower, even when explicitly asked to neglect the effects of friction and/or air resistance. The explanations given by a few of these students suggest that they may have misunderstood the problem statement and considered friction. However, many explanations suggest that these responses reflect more than a mere misunderstanding of the assumptions in the problem. For example, one wrote, "Below the release point because even though friction and resistance is disregarded, there is still gravitational energy that is pulling down on the ball as it rolls to the opposite side." Another wrote, "It will be below due to the fact that it is using energy as it swings so when it swings back it is below the beginning place."

These responses raise some questions about development of the energy model. Can these students distinguish the actual motion of a pendulum with resistance from the motion of an idealized system without resistance? The first answer above seems to indicate confusion between energy and force, and indicate that the gravitational force, rather than friction, eventually causes a pendulum or ball to slow down and stop. This notion led to the additional questions described in the next section.

TABLE 1. Student responses to written problems. All numbers rounded to the nearest 1%.

	Pendulum	**Ball in Bowl**	**Ball in Bowl**	**Ball in Bowl**
Max.height after first half-cycle will be:	Course for Teachers ($N = 48$)	Survey of Physics ($N = 40$)	Course for Teachers ($N = 25$)	Calc-based Mechanics ($N = 36$)
Same height	52%	18%	64%	67%
Lower	35%	70%	20%	22%
Higher	15%	13%	16%	11%

WHY DOES IT STOP?

Based on the results described above, we asked additional questions that stated explicitly that the object eventually stopped moving. Students were then asked to give an explanation of this behavior in terms of the concept of energy. Similar problems have been posed by other researchers.[5]

A physicist would say that the swinging pendulum comes to rest because of a decrease in the mechanical energy of the pendulum-earth system; the forces of interaction between the ball and the surrounding air act to transfer energy from the ball to the air and to increase the systems' internal energy. Student responses often differed considerably from this ideal.

Belief that energy is used up or consumed

Many students in their explanations referred to loss, consumption, or use of energy. When asked to explain why the pendulum eventually came to rest, one student wrote, "...the energy gets weaker and weaker. When the pendulum finally stops, it mean there is no energy at the end." Another wrote: "The pendulum uses all of its energy... After release its gravitational potential energy and its kinetic energy are both being used. Since the pendulum is being pulled down by gravity both energies decrease after each swing."

Belief that 'gravity' or GPE stops objects

In both the pendulum height and 'why does it stop' questions, many students identified either 'gravity' or GPE, rather than friction or air resistance, as the agent that eventually causes bodies to stop moving. The language used by students is imprecise, so it is difficult to know whether the term 'gravity' refers to a force, a field, or to GPE. In one section of the Survey of Physics ($N = 69$), 45% of students identified one of these in a question asking why a pendulum came to rest. Only 39% gave explanations classified as correct. One student wrote, "The gravitational PE energy will eventually pull the pendulum to rest," suggesting confusion between energy and force. Another response suggested that GPE would be greater after the pendulum had stopped, "The pendulum loses kinetic energy and gains potential energy as it slows down to its resting point."

In some cases, these students also identified friction and/ or air resistance. They certainly identify gravity, rather than any resistive force, as the ultimate cause of the stopping. Responses to these problems are consistent with previous research results.

HEIGHT DEPENDENCE OF GPE

The responses described above led us to pose additional questions. In particular, some responses suggested confusion between GPE, gravitational force, and the more general notion of 'gravity.' Thus we asked questions involving comparisons of GPE for objects of the same mass. Results from final exam questions suggest that, even at the end of a course, many students do not understand that GPE depends only on the mass and height of an object (given a constant gravitational field). Many students seemed to be distracted by the motion of bodies or by the presence of additional forces on the bodies. Therefore we developed questions to probe whether students could correctly apply the definition of GPE despite these distractions.

Students in the Survey of Physics course (N = 51) were asked to compare the GPE of the five balls in Figure 1. Only 21% gave correct rankings. Students in the Course For Teachers ($N = 24$) were asked to compare the GPE of the two identical balls in Figure 2, one of which is at rest on a ledge and one of which is moving downward. One third answered correctly. Below we examine common incorrect answers and explanations.

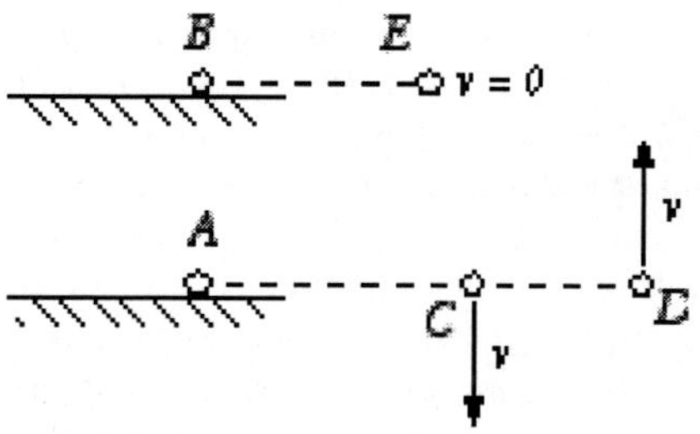

FIGURE 1. GPE comparison for five identical balls with different heights and velocities. Students were asked to rank the GPE of the ball-earth system for each of the five balls.

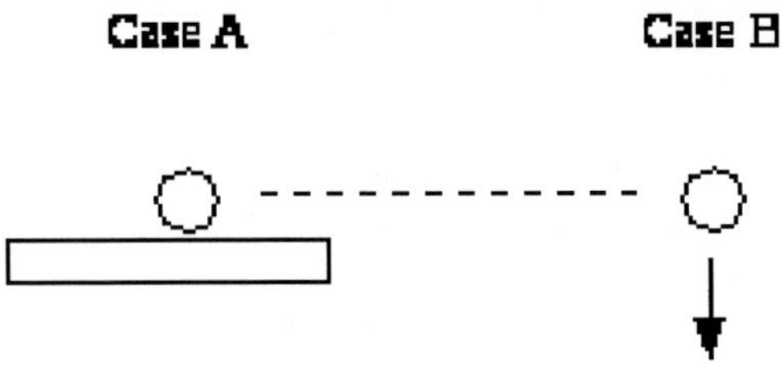

FIGURE 2. GPE comparison for two balls. Students were asked to state whether the GPE in the ball-earth system was greater in case A, greater in case B, or equal in the two cases.

Belief that GPE is influenced by motion

Only 25% of the students in the Survey of Physics course ranked balls A, C, and D correctly. Approximately 40% of students stated that ball D, moving upward, would have a greater GPE than ball C, moving downward, despite their equal height. Some students placed the GPE of ball A between the two, and others found it to be the least of the three.

In reponses to the two balls question, a number of students in the Course for Teachers stated that ball A would have more GPE and referred to the motion of ball B in their explanation. One such student wrote, "A>B, because the ball is at rest at the top of the ledge at A, while the ball is moving downward in case B." Some explanations may reflect a misapplication of energy conservation: "B is actually moving so its energy has transferred to kinetic energy, not just GPE like A."

Belief that GPE is influenced by support:

Student responses to the problem shown in Fig. 2 suggest that many students believe that GPE is influenced by the presence of a supporting object. For example, one student wrote: "In case A the ball is on a ledge, therefore doesn't have any [GPE]. In case B it has the ability to drop and is above the ground so therefore has a greater [GPE]." Students gave this response even while noting that the balls are at the same height: "Case A has the ball securely on a stable setting, it GPE is much less than in case B, where it is sure to fall. Although they are at the same height, B > A with gravitational potential energy." Responses of this sort echo a common student statement in class, that GPE and other forms of potential enegy describe the "potential" for an object to move.

In responses to the problem shown in Fig. 1, approximately 60% of students stated that the GPE of balls B and E were different, with most stating that ball E would have the greater GPE. These responses are consistent with the explanation above and the notion that the supporting ledge influences the GPE of ball B. However, explanations were not very detailed.

SUMMARY AND FURTHER STUDY

Even the simplest questions on GPE proved difficult for many of the students in this study. Many students fail to recognize the variables on which GPE depends, and appear to be confused by the effects of external forces and the speed of a body. Many students make incorrect predictions about the motion of bodies under the influence of gravity. Finally, students fail to apply correctly the energy model in accounting for the cessation of motion under resistive forces. The current report is admittedly incomplete, but suggests that there is fertile ground for further examination of these questions. We intend to pursue these questions further, and examine student thinking on GPE in more detail. We have already performed preliminary interviews and posed additional written questions and will report additional results soon.

ACKNOWLEDGMENTS

This work supported in part by NSF grant DUE 0341560.

REFERENCES

[1] California State Board of Education, *Science Content Standards for California Public Schools: Kindergarten through Grade Twelve*, Sacramento: California Department of Education (2002).

[2] See, for example, Singh, C., and D. Rosengrant,. *Am. J. Phys.* 71, 607-617 (2003), Driver, R., and L. Warrington, *Physics Education* 20, 171-176 (1985). Duit, R., *Physics Education* 19, 59-64 (1984), and the resource letter by McDermott & Redish, *Am. J. Phys.* 67, 755-767 (1999) and the references within.

[3] For discussions of innovative approaches to teaching energy, see, for example, Van Heuvelen and Zhou, *Am. J. Phys.* 69, 184-194 (2001), Chabay R.W.& Sherwood B. A., *Am. J. Phys.* 67, 1045-1050, (1999), and Arons A. B. *Am. J. Phys.* 67, 1063-1069 (1999).

[4] See, for example, D. Maloney, *JRST* 22, 261-278 (1985), and.Grimellini-Tomasini, et al., *Sci. Ed.* 77, 169-189 (1993).

[5] Kesidou and Duit, *JRST* 30, 85-106 (1993).

Student Understanding Of Sound Propagation: Research And Curriculum Development

Katherine V.P. Menchen[1] and John R. Thompson[1,2]

[1]*Center for Science and Mathematics Education Research and* [2]*Department of Physics and Astronomy*
The University of Maine, Orono, ME; katherine.menchen@umit.maine.edu

Abstract. We describe recent results in an ongoing investigation of student understanding of sound, especially of sound propagation in solid media. We have recently administered written questions and conducted interviews to probe student reasoning about this topic, specifically the effect on the pitch or frequency of a sound that is propagated from one object to another. Our findings show that the concepts of propagation and resonance are not functionally distinguished by these students; many students seem to be distracted by the resonant properties of an object that is propagating sound. Furthermore, students have more difficulty when considering propagation of a sound through objects with sound generating properties than with objects not associated with sound generation. This work forms the basis for instructional materials to improve student learning, especially among preservice teachers. The curriculum developed to date is reasonably successful at addressing these student difficulties.

BACKGROUND AND CONTEXT

Overall, little research has been done on how well sound is understood by students at either the college or precollege level.[1-5] Sound is typically taught at the elementary level. Because elementary teachers need to have a thorough understanding of the topic in order to teach it, our efforts toward research and curriculum development are targeted primarily toward preservice teachers. We have been identifying student difficulties as part of the process of developing a guided-inquiry-based curriculum on sound; this curriculum is currently used for primary and secondary teacher preparation at the college level. In this paper we will describe some difficulties that K-12 teachers have with specific concepts within sound, as well as discuss some details of the curriculum.

The main population discussed in this paper is a class of primarily elementary and secondary education majors taking a hands-on, inquiry-based physics course at The University of Maine.

The questions discussed here were asked as a pretest for the second section of the curriculum. In the first section students associate sound with vibrations and the properties of sound with those of vibrations. We note that the students are led to understand that pitch is an indicator of frequency. Students listening to sounds that had identical frequencies in different materials reported hearing the same pitch.

Previous research included analysis of written responses to questions asked of several classes including both preservice and in-service teachers.[6] In this paper we discuss new results that build on our earlier work. We have asked pretest, homework, and examination questions as well as interview tasks that are variations on the previously asked questions. Our data come from 14 students in the class who took both pre- and post-tests. Additional data are provided by individual interviews of two students from a different semester of the same course.

THE STUDY

Previous examination of written pretest responses left us with several questions to explore in greater detail. We are particularly interested in student

CP790, *2004 Physics Education Research Conference*, edited by J. Marx, P. Heron, and S. Franklin

understanding of propagation of sound through various materials. In particular, we study student conflation of transmitted pitch with the resonant properties of the material propagating the sound. In the basic scenario, a tuning fork is struck and then placed on a wooden meter stick, at the other end of which is a second tuning fork, already in contact. The students are asked if a sound will be heard from the second tuning fork, and (if so) what frequency the sound will have. Variations on this question (illustrated in Figure 1) include changing the connector material or the size of the second tuning fork. One additional variation involves the use of a light bulb as the second object.

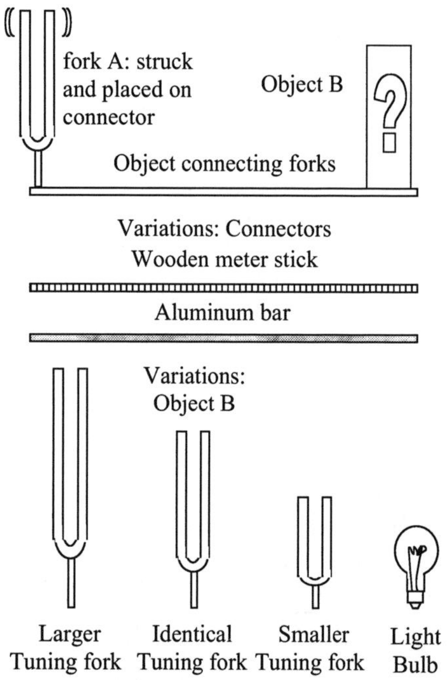

Figure 1. Format of question posed to students, including variations. The theme of this question is that fork A is struck and placed on the "connector"; students are asked to predict what sound will be heard from Object B. Variations concern the material of the connectors and the type of object at the end of the connector.

The responses students provide on written questions may lack sufficient detail to allow us to understand the students' thinking. Thus, an interview protocol was developed based on the written questions to clarify vague or ambiguous student explanations, rather than resorting to unwarranted interpretation of written responses. Interview subjects were encouraged to explain where their answers came from, and were not given any answers until the interview was completed. The subjects were probed to determine their understanding of (1) how the length of a tuning fork affects the pitch produced when the fork is struck, and (2) what causes the tuning fork to make a sound. Both subjects demonstrated adequate understanding of these two concepts. Subjects were then asked about the scenarios shown in Figure 1 that involve a second tuning fork as Object B. In addition, a variation was asked in which a string is used as the connector and Object B is a plastic cup.

RESULTS AND DISCUSSION

The data obtained from these questions point to three ideas that various students seem to hold about what characteristics of this system affect the frequency of the propagating sound: the material of the connector, the size of the second tuning fork, and whether or not the second object is typically associated with a specific frequency. We discuss these three ideas below, along with relevant information about the curriculum and post-test data.

Material of Connector

Earlier data suggested that students may think the material of the connecting bar alters the pitch of the sound heard at the far (right) end of the bar. In order to explore this issue, we asked a version of the pretest using identical forks, first with a wooden meter stick, and second with an aluminum bar of the same length. The students were aware that the tuning forks were also made from aluminum. Using identical forks allows us to filter out responses influenced by the size of the final tuning fork (discussed in the next section). Students were asked to compare the pitch heard at the second fork in the two situations. Responses to this pretest are summarized in columns 1 and 3 of Table 1. This question yielded an interesting result: the majority of students discussed the volume of the sound heard with the aluminum relative to the wood, rather than the pitch; the students who did mention frequency in this part of the question gave no reasoning for their response.

TABLE 1. Results for question varying connector but with an identical tuning fork as Object B. N = 14.

Connector	Wooden Meter Stick		Aluminum Bar	
Response	**Pre-test**	**Post-test**	**Pre-test**	**Post-test**
f_A	7	13	2	12
Lower than f_A	1	0	1	0
Higher than f_A	1	0	1	1
Frequency not specified	2	0	8	1
No sound	3	1	2	0

After working through the first section of the curriculum, students easily grasped that the size of a tuning fork affected the pitch it produces when struck. Some students seemed to be extending that effect to the connector. This interpretation is supported by Mark (pseudonym), who gave this response in an interview. Mark explained that the pitch heard in the meter stick "would be lower, because it's bigger than the tuning fork, ... and it'll have a lower sound."

After taking the pretest, students explore a variety of different arrangements. Students listen as sound propagates from a tuning fork to each of the following: a short aluminum rod, a meter stick, and two meter sticks clamped together at a 90° angle. A "telephone" made of cups and string provides another means for exploring propagation; a tuning fork then replaces one of the cups (as in the interviews). Students are then asked to generalize from these experiments. Students do not set up the pretest scenarios until later in the curriculum.

The pretest question described above was later given as homework. As Table 1 shows, there was marked improvement among students: for the scenario using the wooden meter stick as a connector, the number of correct responses improved from 7 to 13; for the aluminum bar, the gains were more dramatic, with 2 correct responses on the pretest and 12 on the homework.

Size of Object B

In addition to altering the connector material, different versions of the pretest varied the size of the tuning fork at the far end of the wooden meter stick. For example, fork B was *larger* than fork A in pretests administered to 5 separate classes at three different universities (N = 69; Column 1 of Table 2). Twenty-six of these students responded that the pitch heard at fork B would be the pitch produced by fork B, f_B, which is lower than that produced by fork A, f_A. We also asked the question in which fork B was *smaller* than fork A. Responses stating that the pitch was raised could be attributed to the size of the second fork, while a lower pitch could be assumed to be caused by the longer connector.

As seen in Table 2, student performance on both versions of the pretest was similar: a far larger percentage stated that the sound would have the frequency f_B, whether fork B was larger or smaller, rather than f_A. In fact, for this group, none of the students gave a correct response on the pretest (this was also true of one class previously). It is worth noting that both interview subjects also chose f_B as their response to this question.

Students were given the same question as part of the same homework assignment described in the previous section. This was the only question on the set of four that did not show any pre/post improvement, indicating that the difficulty elicited by the presence of a different tuning fork, a distinct "sound-making" object with its own (resonant) frequency, is persistent. Three students changed their response to a correct one, while three others who gave a partially correct response on the pretest gave completely wrong responses on the post-test.

TABLE 2. Results for question varying the size of tuning fork B, with a wooden meter stick as the connector.

Size of Tuning Fork B	Larger $N = 69$	Smaller $N = 13$		
Response	*Pretest*	*Pretest*	*Post-test: HW*	*Post-test: Exam*
f_A	17	0	3	10
f_B	26	7	7	1
Frequency not specified	7	2	0	0
No sound	12	0	3	2
Other	7	4	0	0

A second post-test question of this type was given on the final exam. The scenario involved two students who each have a (different) tuning fork. The students in the scenario need to propagate the sound with the pitch of the first student's fork down a radiator pipe and, more importantly, across a gap between two radiator pipes; the second student's fork is used to fill the gap. On the exam, the students are asked what pitch will be heard *after* the gap if the first fork is struck and placed on the first pipe.

We have data for 13 students for all three questions (pretest, homework, exam). Of these "matched" students, 10 provided responses on the exam question similar to the following: "[the second] fork will produce the same frequency as [the first], even though they are different sizes. [The second] tuning fork is just acting as another surface to vibrate." Of the 3 students who answered incorrectly, one stated that the pitch of the second fork would be heard, since something can only vibrate at its own "unique frequency." The other two said that because the second fork is different, it cannot vibrate at the frequency of the first one, and so no sound will get through at all. Most students understood the dynamics of propagation by the end of the course, but this example demonstrates the persistence of the idea that things only vibrate at "their own" frequencies.

Classification of Object B

A final situation was presented to students on the pretest and later in homework. In this scenario, Object B was a light bulb, chosen as an object that is not associated with sound at all to see whether students would still expect the pitch to be altered by an object that has no obvious resonant sound properties.

On the pretest, only three students gave correct responses (see Table 3). Results on this question improved dramatically after instruction, when 11 students were correct. All of the incorrect responses referred to a *higher* frequency being heard at the light bulb. Furthermore, the reasoning behind these responses deals with the size of the bulb, which was drawn as smaller than the tuning fork. Clearly these students are associating the size of the object with a specific pitch, despite the lack of an obvious association with sound.

TABLE 3. Results from question using a wooden meter stick as a connector and a light bulb as Object B. N = 14.

Response	Pretest	Post-test
f_A	3	11
Higher than f_A	0	3
Frequency not specified	5	0
No sound	2	0
Other	4	0

In interviews, students were asked similar questions, again starting with the tuning-fork–meter-stick–tuning-fork (tf-ms-tf) scenario, and comparing that to a tuning fork tied with a string to a plastic cup.

When considering the string and cup situation, Melissa (pseudonym) stated correctly that the pitch heard from the cup would be the same as that from the tuning fork, and that the sound vibrations "travel" through the string. When comparing this situation to the tf-ms-tf scenario, Melissa said that the string and meter stick had similar roles – "sound travels through" each one. However, when it came to the receiving tuning fork and the cup she said, "I think the pitch through the cup is going to be same as [the tuning fork]. I thought that the pitch would be different… [in] the meter stick [situation]." Melissa restated and confirmed that she had two distinctly different ideas about how the pitch from the tf-ms-tf system compared to the tuning-fork–string–cup system. When asked if a bigger or smaller cup affects the pitch (again, in an analogy to the larger or smaller tuning fork in the tf-ms-tf system), she said that the pitch would still be the same.

Like Melissa, Mark originally stated that he would hear the pitch produced by the tuning fork coming from the cup. However, in contrast to Melissa, upon comparison with his earlier reasoning about the tf-ms-tf situation, Mark stated that a bigger cup *would* change the pitch heard from the cup somehow. Mark chose incorrectly to be consistent between these objects, whereas Melissa chose to distinguish between the properties of the two systems.

We see in the interviews a difference in students' ideas about how sound works with a sound related object, compared to a non-sound related object. This is consistent with the data from the light bulb question: associating an object with sound *generation* causes confusion when considering sound *propagation.*

CONCLUSIONS

Our findings show that the concepts of propagation and resonance are not functionally distinguished by these students; many students seem to be distracted by the resonant properties of an object that is propagating a sound. This distraction is not alleviated when considering objects not associated with sound generation.

Simple experiments that provide students with empirical observations seem to be effective at addressing some of these difficulties; however, the tendency to associate the size of an object with any sound, regardless of origin, is persistent.

REFERENCES

1. See R. Driver, A. Squires, P. Rushworth, and V. Wood-Robinson, *Making sense of secondary science* (London: Routledge, 1993), p. 133.
2. C.J. Linder and G.L. Erickson, *Int. J. Sci. Educ.* **11**, 491-501 (1989).
3. C.J. Linder, *Int. J. Sci. Educ.* **15**, 655-662 (1993).
4. M.C. Wittmann, *Int. J. Sci. Educ.* **24**(1), 97-118 (2002).
5. Z. Hrepic, D.A. Zollmann, and N.S. Rebello, *2002 Proc. Phys. Educ. Res. Conf.*, pp. 59-62 (2003).
6. K.VP. Menchen and J.R. Thompson, *2003 Phys. Educ. Res. Conf. Proc.*, K. Cummings, S. Franklin, J. Marx, Eds. (2004).

Contrasts in Student Understanding of Simple E&M Questions in Two Countries

Gordon J. Aubrecht, II*† and Cristian Raduta†

* *Department of Physics, Ohio State University, Marion, OH 43302 USA*
†Department of Physics, Ohio State University, Columbus, OH 43210 USA

Abstract. We administered a survey on electricity and magnetism to two populations of undergraduate students: one from Ohio State University, the other from Bucharest University (Romania). The survey had two multiple part questions. One question invited use of Gauss's Law in several different region. A bare majority of students could solve the simplest problem, that of the electric field inside a conductor. The other question asked about the force on and trajectory of charged particles in regions of magnetic field. These latter questions rely on understanding the Lorentz force and on transfer of general knowledge from classical mechanics studied earlier. Our results show that mechanics knowledge learned earlier does not transfer to electricity and magnetism. Transfer of learning about electricity and magnetism in both countries as measured by our instrument is less successful than we, as teachers, would have wished.

Keywords: student understanding, electricity and magnetism
PACS: 01.40.Fk, 01.40.Gm

I. INTRODUCTION

Much work has been done in the past decade in considering the way students approach electricity and magnetism (E&M) problems. Much of this research took place in other countries. For example, Törnkvist et al. [1] found that Swedish students have great difficulties with the concept of field. Electric field lines in the neighborhood of charges and materials constitute a representation that is not generally understood by more advanced university students. Viennot and coworkers found that students have difficulty understanding the principle of superposition [2] and applying it to electric fields. Eylon and Bagno [3,4], working with Israeli teachers and students, found three critical areas of deficiencies in students concerning E&M: (1) qualitative understanding the importance of central ideas; (2) conceptual understanding of the relationships between the electric field and its sources; and (3) ability to apply central relationships in problem solving.

Raduta [5] identified seven general classes of E&M misconception in an extensive literature survey. In addition, Raduta further identified four areas of student "misconceptions" that had been missed or understudied in current research. These are:

A. mathematics-related misconceptions—e.g., misuse of mathematical tools,
B. tempting analogies between electric and magnetic fields—e.g., electric field : charge :: magnetic field : "charge",
C. lack of ability to see the connection between Maxwell's equations and the laws—e.g., Gauss's Law for electric fields; Ampère's law for magnetic fields, the Biot-Savart law for magnetic fields, Faraday's law connecting changing electric and magnetic fields, and Coulomb's law for electric forces, and
D. geometry of the Lorentz force law—e.g., belief that **F** is always perpendicular to **v**.

Despite the body of work, much further work remains to be done on this subject.

We are interested in the similarities and differences between American students and those in other countries. Our null hypothesis is that, overall, students would exhibit no differences between countries. We here take advantage of Raduta's commuting between the U.S. and Romania to begin to perform some tests of possible similarities and differences between students studying physics in these two countries while investigating student understanding of E&M.

We decided to determine whether university physics students who had studied E&M really were able to apply and understand the Gauss and Lorentz force laws, identified by Raduta as underresearched areas. We asked one relatively straightforward (multiple part) question about

CP790, *2004 Physics Education Research Conference*, edited by J. Marx, P. Heron, and S. Franklin

each topic on a survey administered to over 50 students from each country.

American students attend Ohio State University and had all completed the first quarter of the engineering sequence, which focuses on classical mechanics, and had just finished E&M in the second quarter. The Romanian students were second-year physics students at the University of Bucharest. Generally, the material learned in the last years of Romanian high school physics resemble that learned in the first two years of American university physics courses, so these students should be considered somewhat more advanced than the U.S. students.

In Sec. II, we consider the Gauss's Law question, presented to 74 American and 52 Romanian students (some data are still being evaluated; we report on 8 of these responses here). In Sec. III, we discuss the Lorentz force question using results from all 74 Americans and 52 Romanians. Three students from each country were interviewed in detail about their ideas. In Sec. IV, we summarize and discuss our results.

II. THE GAUSS'S LAW PROBLEM

We first discuss how students approached a Gauss's Law problem (given below). Gauss's Law relates the flux of a field (the surface integral of the field lines penetrating a surface) to the sources of the field enclosed within the surface and is found in student textbooks in both countries. The problem was presented as follows:

A small solid insulating sphere of radius a whose surface is uniformly charged with positive charge +q is surrounded by a larger hollow sphere whose inner surface (radius b) is uniformly charged with charge -q.

a. Determine the electric field at distance r from the center of the sphere when $r < a$.

b. Determine the electric field at distance r from the center of the sphere when $a < r < b$.

c. Determine the electric field at distance r from the center of the sphere when $b < r < c$.

d. Determine the electric field at distance r from the center of the sphere when $r > c$.

e. If you know that the electric potential at r = c is V(c), what is the electric potential for $r > c$?

It is easy to show that **E** is zero for (a) and (c). For (b) and (d), charge is contained within the Gaussian surface, so $E = +kq/r^2$ at any point. For case (e), we integrate to find

$$V(r) = \int_c^r -E_r\, dr = \int_c^r \frac{-kq}{r^2}\, dr = \frac{+kq}{r} - \frac{+kq}{c} + V(c).$$

We have not yet completed the analysis of our results from this question, so we present the preliminary analysis. An overview of our results from the five parts is given in Table 1, which gives details of student answers, and Fig. 1, which shows the numbers of students correctly answering the parts of the Gauss's Law question. It is not apparent from Table 1, but no student correctly answered all parts. Table 2 provides more detail on student answers (all students together).

TABLE 1 Preliminary results from the Gauss's Law question, Americans (A): 74, Romanians (R): 8

Characterization of answers	A	R
Mostly correct, explanations given	12	2
Some right, some explanation provided, misuse of formulas	3	1
At least one answer correct (usually a or c), explanations and formulas provided	14	0
At least one answer correct (usually a or c), no explanations but formulas provided	19	0
Formulas used, all of them incorrect	8	0
Formulas and explanations, incorrect	5	0
Some explanations, but incorrect	2	0
Blank, or wrong answers; no formulas or explanation provided	11	5

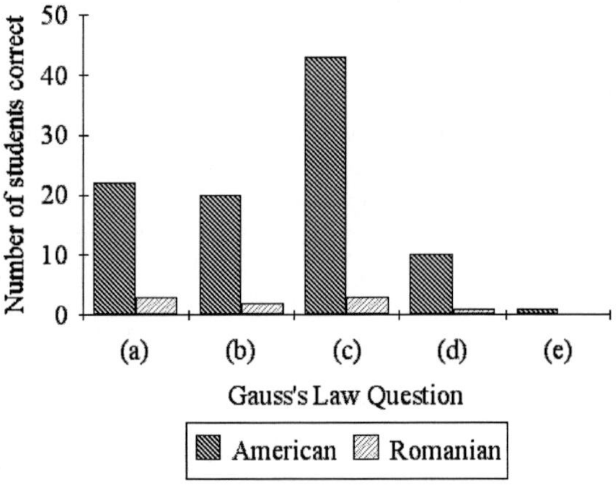

FIGURE 1. Differences in student answers for the Gauss's Law question; Americans (N = 74), Romanians (N = 8).

The "correct" designations were assigned generously: for example, the statement "the electric field is zero because it is within the object" was counted correct in part (c). It should also be noted that students could have simply memorized the maxim "there is no electric field inside a conductor" and produced the correct answer by applying it. Given the lack of detail in student answers (discussed further below), we were not able to identify which students might have found the answer by memorization. Clearly instruction at both universities had left students ill-prepared to find the potential from the field (e, Table 2).

TABLE 2. Numbers of students answering each part correctly.

Task	Americans Correct	(%)	Romanians Correct	(%)
a	22	29%	3	38%
b	20	27%	2	25%
c	43	58%	3	38%
d	10	14%	1	13%
e	1	1%	0	0%

III. THE TRAJECTORY PROBLEM

We turn to the second question:

There is a charged particle inside a region containing a constant uniform magnetic field.

a. What is the magnetic force (magnitude and direction. acting on the charged particle if the initial velocity is zero? What is the trajectory of this particle?

b. What is the magnetic force acting on the charged particle if the initial speed of the charge is v (known, but unspecified here) and the direction is parallel to **B**? What is the trajectory of this particle?

c. What is the magnetic force acting on the charged particle if the initial speed of the charge is v (known, but unspecified here) and the direction is perpendicular to **B**? What is the trajectory of this particle?

d. What is the magnetic force acting on the charged particle if the initial speed of the charge is v (known, but unspecified here) and the angle between **v** and **B** is α? What is the trajectory of this particle?

Expected solutions are **F** = 0 in a and b, so the velocity remains constant. For case c, F = qvB, and the trajectory is a circle about the axis along which the magnetic field lies. For case d, $\mathbf{v}_{\|}$ remains unchanged (as in b), while $\mathbf{v}_{\perp}$ makes the particle move in a circle (as in c), so the particle's trajectory is a spiral.

Students do not distinguish in many cases between a scalar and a vector—students often have a scalar on one side of the equal sign and a vector (or a vector product) on the other side. Even if a student made this mistake, we decided to include them as having a correct answer (but only if everything else was correct).

The results on this set of questions are summarized in Fig. 2. Romanian students are less knowledgeable about magnetic forces than American students, but slightly more likely to be able to identify the trajectories caused by the force when the velocity is perpendicular to the force. Romanian students who answered correctly were more likely to use a greater number of words to express their answers than American students. American students preferred to "let equations do the talking" (see Table 1). As a further example of the differences, we present contrasting excerpts from a Romanian "good" answer to thc trajcctory problcm,

(a) The magnetic force for a charge in an uniform field is: **f** = qv**xB**. If v = 0, than f = 0, and it will not be accelerated in the field, hence we can't speak of direction of the force, but we can say that the magnitude is always zero.

(b) α(**v**,**B**), v ≠ 0, **f** = qv**xB** = qvB sin α; α(**v**,**B**) = 0 ... **f** = 0; Hence the trajectory is a straight line parallel the lines of magnetic field. The equation of the motion will be: x = x(0) + vt, where v = ct.

and an American "good" answer,

(a) **F**(L) = qv**xB**; if v = 0, then F(L) = 0; ⊥ to magnetic field.

(b) **F**(L) = qv**xB** = qvB sin α = qvB sin 0 = 0; across magnetic field.

Overall, Fig. 2 shows how little students from both groups understand charged particle trajectories, even after instruction.

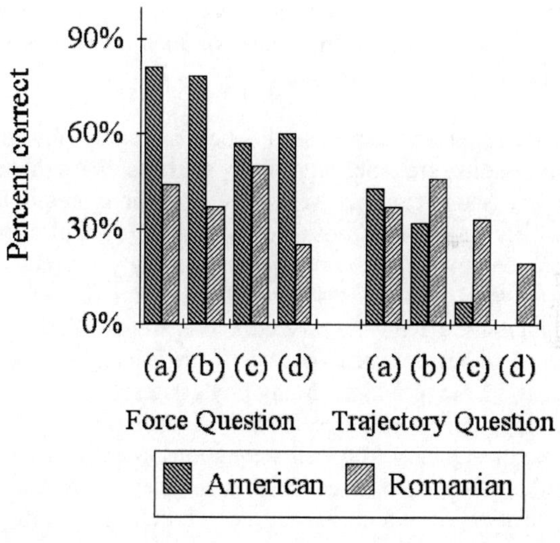

FIGURE 2. Number of American (N = 74) and Romanian (N = 52) students with correct answers to the force and trajectory parts of the question.

To probe this question further, three American and three Romanian students were interviewed and videotaped while answering the Lorentz force questions. They were asked to explain out loud their solutions for each part of the trajectory problem as they thought about them, and we attempted to observe differences between these two small (we hope representative) samples of the larger student populations. From the transcripts, Americans were more likely to speak "telegraphically," while Romanians were more likely to expand their answers (that is, to use longer phrases, which contained a greater number of words) and to refer to matters off the topic. Our results for the analysis of part (c) of this problem [for which $\mathbf{v} \perp \mathbf{B}$] are presented in Table 3 and do not contradict the results obtained in the written survey.

TABLE 3. Characteristics of students' interview answers to the trajectory problem. Names were changed; A: American; R: Romanian.

Student		Phrases used	Words used, part c answer	Times referred to other subjects
Mike	A	short	36	0
Barney	A	short	40	2
Timothy	A	long	68	1
Artenie	R	long	89	4
Ilie	R	long	58	1
Cristi	R	long	67	3

IV. SUMMARY AND DISCUSSION

McDermott, Shaffer and coworkers have attempted to address some student difficulties with field representations and Gauss's Law through Part II of their *Tutorials in Introductory Physics* [6]. Much of their work has been presented at AAPT meetings.

These questions we asked appear simple to physicists. Our results are sobering. Few students were able to answer satisfactorily, even though we interpreted their answers in the best possible light. Few of the answers were complete and none were completely correct. The best results for Gauss's Law occurred for answers that might have been "memorized." We are especially concerned about the lack of student proficiency in solving Gauss's Law problems. Many physics teachers avoid this by skipping Gauss's Law entirely, but that makes it very difficult to justify Maxwell's equations. Most students are unlikely to be exposed to this wonderful example of unification elsewhere in their careers. Given its practical and theoretical importance, it would be near tragic for them to lose that opportunity. The results drawn from the second problem show students' inability to find the trajectory corresponding to a given magnetic force.

These questions—two easy standard problems—were given to the American students two weeks before the end of the quarter, when one would expect students to be comfortable with the main concepts of E&M and students had just studied Gauss's Law. While Romanian students (as second-year students) were more advanced in terms of their coursework than the American students, they exhibited similar limitations. One would have expected them to have been better able to connect their mechanics and E&M knowledge than we found to be the case. The null hypothesis cannot be ruled out except for the better ability of American students to identify the force in the Lorentz force question, where the majority of American students were correctly able to identify the forces, while the majority of Romanian students were not. Overall, the results showed that both groups lacked E&M knowledge.

We are left with many questions about these topics.

•What can we do to teach more effectively? How might we continue to teach Gauss's Law but increase our students' ability to understand the deep connection to Maxwell's equations? Are those who would drop Gauss's Law from the elementary course altogether correct that the gains of doing so are outweighed by the difficulties?

•Why aren't students able to retain simple ideas about kinematics for even one quarter? This compartmentalization has often been observed in other, broader, contexts (for example, between mathematics and student application of their mathematics knowledge in physics, or between knowledge from chemistry courses and application of that same knowledge in physics courses). How can we better communicate the unity of the physical approach?

•Are the (conceivably major) stylistic differences we believe we have observed between these two samples of students doing just two problems representative of the outcomes in other countries? What is the connection between culture, if any, and student approach?

•Are these results characteristic of all international groups? Clearly, there are a lot of points of similarity; both groups of students can be said to belong to the Western tradition of thought, and perhaps groups differing more in culture would perform differently.

REFERENCES

1. S. Törnkvist K.-A. Pettersson, and G. Tranströmer "Confusion by representation: On students' comprehension of the electric field concept," *Am J. Phys.* **61**, 335-338 (1993).
2. L. Viennot and S. Rainson "Students' reasoning about the superposition of electric fields," *Int. J. Sci. Ed.* **14**, 475-487 (1992). S. Rainson, G. Tranströmer, and L. Viennot, "Students' understanding of superposition of electric fields," *Am J. Phys.* **62**, 1026-1032 (1994).
3. E. Bagno and B.-S. Eylon "From problem solving to a knowledge structure: An example from the domain of electromagnetism," *Am J. Phys.* **65**, 726-736 (1997).
4. E. Bagno, B.-S. Eylon, and U. Ganiel, "From fragmented knowledge to a knowledge structure: Linking the domains of mechanics and electromagnetism," *Am J. Phys.* **68**, S16-S26 (2000).
5. C. Raduta, *Students' Misconceptions Related to Electricity and Magnetism*, Ohio State University General Examination paper, April 2001. available from the authors.
6. L. C. McDermott, P. Shaffer et al., *Tutorials in Introductory Physics,* Prentice Hall, Upper Saddle River, NJ 1998.

Grounding Inquiry-based Teaching and Learning Methods in Physics Experiences for Prospective Elementary Teachers

Gordon J. Aubrecht, II

Department of Physics, Ohio State University, Marion, OH 43302 and Columbus, OH 43210, USA

Abstract. In teaching inquiry classes in physics, I ask students to reflect on their learning in journals. One of the journal questions deals with student expectations of transfer of the inquiry techniques used in our class into their own classrooms when they become teachers themselves. I report on students' answers to this question over a five-year period, which gives insight into how much or how little the students think the techniques are worth to themselves as both students and prospective teachers.

Keywords: physics inquiry, preservice teachers.
PACS: 01.40.Fk, 01.40.Gm, 01.40.Jp

I. INTRODUCTION

Children have natural curiosity. But after a few years of school, children seem to become less curious. Elementary and middle school teachers are required to know so many things about so many subject areas, they have little likelihood of becoming science experts. Perhaps it is because of this that many teachers discourage their students from asking questions of them and of nature. Teachers' lack of subject-matter confidence in science may lead them to interpret students' questions about content as a questioning of the teachers' authority.

I teach three inquiry courses in physics for undergraduates, each of which is adapted from and based on the research of the Physics Education Group at the University of Washington [1]. Many prospective teachers on my campus take these physics courses, and their prior experience is that *teaching* means *telling*. They are unaware that any other way of teaching exists. They have learned to, and therefore they expect their students to learn to, cram, memorize, and immediately forget.

Physics by inquiry attempts to break this cycle by providing an alternate experience for prospective teachers. The course emphasizes the scientific approach to nature—providing experiences from nature, discussing those experiences, proposing models that explain those experiences, evaluating the models by doing more experiments, etc. Because I rely on student experiences, and the students themselves set up the experiments, they have personal, kinesthetic knowledge of their subject.

I believe along with Zull [2] that the role of concrete experience must be supplemented by reflection for learning to occur. I build on the familiar learning cycle, Fig. 1, which Zull argues is effective as a result of its reflection of the structuring of connections among areas within the brain.

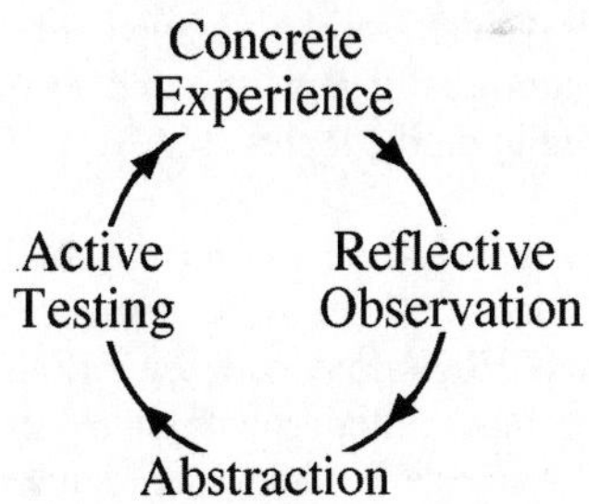

FIG.URE 1. The learning cycle

The rest of this paper is organized as follows. In Sec. II, I describe course conditions. In Sec. III, I exhibit a selection of typical answers to one journal question, in Sec. IV I discuss the results, and in Sec. V I summarize the results of our investigation.

II. COURSE ORGANIZATION

Ohio State University offers its own version of three physics by inquiry courses: Properties of Matter, dealing

CP790, *2004 Physics Education Research Conference*, edited by J. Marx, P. Heron, and S. Franklin

with mass, area, volume, density, and concentration; Electric Circuits, dealing with current and potential difference in the context of arrangements of batteries, bulbs, and wires; and Optics and Astronomy, dealing with single and multiple sources of light, shadows, and effects of the motion of the moon and sun. OSU uses the *Physics by Inquiry* books [3] as course textbooks. Typically, under 100 pages of the text will have been covered by the end of a one-quarter (10 week) course.

Physics by inquiry courses are taught in a way that encourages students to cooperate (students must work in groups.) It tries to assure student mastery of the topics. The syllabus they are given the first day of class shows students that only about half (53%) the final course grade comes from examinations. The remainder of the grade is based on attendance (which is important for both understanding and group success), homework, reworked pretests, and journals.

The courses are taught with concept mastery in mind (while complete mastery is not always gained, the average course grade is near a B+). The research basis of the materials is the elicit-confront-resolve model of learning. Most students come with ideas, which are brought forward by asking students to make a prediction (the question being determined by research on student thinking about the subject to elicit student answers inconsistent with reality), by doing an experiment that shows that the actual working of nature is discrepant from the expected student response, and by helping students build a model that encompasses the experimental results, resolving apparent discrepancies. Models are continually revisited and form the basis of student reasoning on the topics.

One important feature of the course is a complete lack of any lecture. Teachers interact with students mainly through discussions with the individual groups at appropriate points, called checkpoints, in the book. (I proffer help as needed, of course.) Students are led to construct their own (group) understanding of the phenomena being studied. The discussions follow from questions the instructor bases on the material covered since the previous checkpoint. Reflection is supported in several ways: group discussions, questioning at checkpoints, reworking of pretests, and the journal.

III. JOURNAL QUESTIONS ON INQUIRY

As noted above, the journal is an important component of the course. In it, I solicit answers to both content and course-related questions. The journal topic for week 6 provides information on students' likelihood of using inquiry themselves when they become teachers. The questions are posed as follows: "For the following four questions only, imagine you are an elementary or middle school teacher. (1) What physics concepts taught so far in this class could be taught in your classroom? Would you teach the concepts the same way you learned them or modify their presentation depending on the age of the students? If you modify, how would you do that? (2) How would you adapt the structure of the material to work in a time period of about 45 minutes each day? (3) Would you use the same inquiry method of teaching as the instructors are doing right now? Why or why not? Would it be necessary to modify the teaching method also? (4) How would you keep elementary or middle school students on track in a group effort? How would you determine whether the group is functioning, and how strict you should be with students who are disrupting the group and/or falling behind?"

In this paper, I consider the student responses only to the underlined parts (treated as one question) contained within the larger set of questions. I categorized the responses into the following six categories:

- yes
- yes, but ... [something more than the "inquiry" as used in our class, but without major changes]
- yes, but have some lecture
- have students do some experiments, but teach mostly by lecturing, giving demonstrations
- probably not, absolutely not
- don't know (what the student would do if a teacher).

In addition, I noted whether students said they would plan to do experiments or referred to experiments or demonstrations in their response.

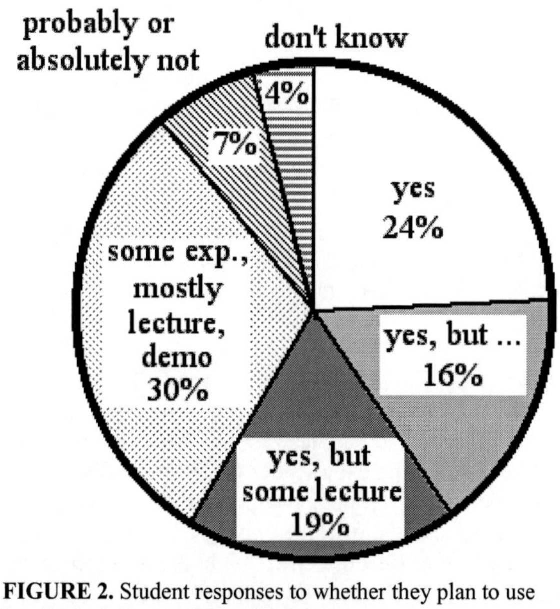

FIGURE 2. Student responses to whether they plan to use inquiry in their own future classes.

The responses of the 251 students included in this preliminary study fall into categories as shown in Fig. 2. About one-quarter plan to use the unmodified inquiry method, while one student in six plans to use the method, with slight modification, in teaching in the future. Only about one-tenth of students are adamantly opposed to using the method. To provide more context, selected student statements that fall into each category are provided below. The range of the answers gives a sense of student thinking that the summary table cannot.

Yes

"I can't think of a better way to teach science in general. Hands on experience is something I think would promote student interest as well as learning."
"I would use the same class structure as we are using now. It gets REALLY frustrating at times ... BUT I'm learning a lot because of the way that the class is set up."
"I realize now why it is so important to communicate in a simple, clear, and precise manner; it is easy to get lost, lose interest and understanding when the person explaining or demonstrating throws in everything and the kitchen sink and most of the babble ..."
"I think using the inquiry method could be just as effective on younger students as it is on college students, so I would probably use it."

Yes, But ...

"The way it is taught in class seems to be one of the most effective but I think I would step down on some of the more complex parts (such as math)."
"I would let the students learn on their own to a certain point but as previously stated i [sic] would try to help the students as much as they needed."
"I would use the inquiry method but I wouldn't allow students (groups of students) to work at their own pace."
"I think I would not lecture, but give some guidelines and something to think about before they perform the experiments."

Yes, but Have Some Lecture

"I think we would work as one big group instead of more than one since they will probably be less likely to get work done and understand concepts on their own or just working with each other."
"I would modify the material, I would leaev [sic] it to major concepts only, and I wouldn't put as much emphases on a complete understanding, as I would on just knowing the material."
"I would do it much the same way done in this class but probably demonstrate a few things in each section first before turning them loose."
"Students enjoy demos or lab but they need to be structured and expectation set from the beginning and the paper work to keep track would be tremendous."
"I would discuss the main points of the concept but still have them try it. I feel that experimentation is still a good method."

Have Students Do Some Experiments, but Teach Mostly by Lecturing, Giving Demonstrations

"Just teaching them the basic idea of how these things work and having them perform some exercises to see the ideas in motion would suffice."
"I would still use prediction but by raise of hand or group discussion. Then perform the experiment, check against predictions and discuss reasons for results."
"I don't think that I would require them to do as much thinking on their own, I would definitly [sic] help them out a lot more."
"... the teacher would have to just tell the students what certain things are ..."
"... less inqury [sic] and more demonstration, allowing the students to tell what happened ..."
"I need to give them information instead of let [sic] them think because they are too young for that method."
"I don't think that it would be necessary for me to use inquiry with my students because it might be too hard for them to think of things on their own, however if they understood the concept well enough then I would go into some inquiry"
"I just don't think it feasible, so it would become a demonstration in front of the class asking questions and having the students assist me as one large group to figure out the answers, having me perform a couple of the different ideas."

Probably Not, Absolutely Not

"I think that the inquary [sic] idea is a joke. I do not feel that is benificail [sic] at all."
"I would teach them the concepts through lectures and class demonstrations."
"I would not use the same inquiry method of teaching as the instructors are right now. Why, because it would be too compilicated [sic] for the students."
"I would not use the inquiry method. I would probably just instruct while they tried doing it themselves at their desks, and then asked them questions as a class, like why do you think that the candy had the same mass of the square nut?"
"I do not think that elementary kids could find these things out on their own."
"I wouldn't use the inquiry method......just because it takes too long for a class of 30 YOUNGER students to grasp. I would teach with experiments as well as lectures and explanations."

IV. DISCUSSION

The time spent in the course plays an important role in student responses. I have noted that the fourth or fifth week is the nadir—many students seem frustrated and ready to call it quits. After this, though not on a predictable schedule, many students change their minds and come to see things differently. A student writes "Some days I really hate going in for class. But lately I have felt better and have actually enjoyed it. I liked it more when I felt as if I had really found out something ahead of where we were. I don't know if it happened later in the quarter because I was more used to how it all works or if I am just understanding it more." Ideally, we would like to have two quarters to influence the students, but many students (those we don't reach before the end of the course) opt out of a second quarter. Time also plays a role in coming to terms with the experiments, as it should. It takes time to assimilate new ideas, which come rapidly in the physics by inquiry courses. An example of this effect is seen when one student writes "When I look back at some of the experiments we did, I think to myself 'why did I have such a difficult time grasping the concept?'."

About one-quarter of students say they would apply the inquiry method when they become teachers, and another 16% are interested in using it, but with minor modifications. On a final evaluation in the journal, one student writes "I feel inquiry based instruction is very open. We learn everything by doing it ourselves, and if there is something we have a question on, it is usually figured out by thinking up an experiment to try it out. There are no study and answer kind of sheets, or tell all formulas." I regard this student's answer as a great success for the course. She has taken responsibility for her learning. She appeals to experiment as the final arbiter of answers about how nature works. She recognizes she can design an experiment herself to address her particular questions about nature.

About one-half the students say they would partly or mostly lecture, and almost 10% were completely turned off by inquiry (at least as I have practiced it here). Even those who are convinced have mixed feelings, as exemplified by this student's comment in the final journal: "I know what the course is designed to do, and I think to an extent it works, but since it is so different from how we're traditionally taught, many people got frustrated and may not have gotten as much out of the class as possible. This type of curriculum stimulates those who learn in a different manner than by hearing or reading. Many people learn best with a hands-on process. This stretches those who learn in other ways. You are right in that we will always remember this course. I am still debating as to whether I liked it or not to be honest."

V. SUMMARY

It appears that many preservice teachers believe that elementary-school children are too young to be able to think or to solve problems. An awareness of barriers such as this can help science methods instructors identify ideas about teaching and learning that need to be addressed or confronted in their classes.

I find that 62% mentioned doing experiments or having students do experiments in their future classes. Since physics by inquiry classes so emphasize experimentation, it may not be a surprise that students would mention it in the journal.

During the past year as part of the first journal, I have asked students what they had thought teaching was before entering my course. Virtually all responses refer to lecture, as in, for example "One of the usual ways ... is being lectured. Most of the time I spend my time in class writing notes as fast as I can, and never really getting the time to process what I am being taught because its impossible to do both," or are vague references to ideals: "Teachers need to guide children." If I assume that previous classes would have written similar responses, the results presented here imply that a number of students have changed their minds about what constitutes teaching by week 6 of a single ten-week quarter.

At the very least, I have demonstrated to students that there are alternatives to lecture, that teaching needn't always be telling. The proof of the effectiveness of this pedagogy will come only after these students become teachers, and these teachers' own students begin to arrive already functioning as inquirers in the classes I teach.

REFERENCES

1. L. C. McDermott, "What we teach and what is learned—Closing the gap," *Am. J. Phys.* **59**, 301-315 (1991). L. C. McDermott, "How we teach and how students learn—A mismatch?," *Am. J. Phys.* **61**, 295–298 (1993).
2. J. Zull, *The art of changing the brain*, Stylus Publishing, Sterling, VA, 2002.
3. L. C. McDermott et al., *Physics by Inquiry*, Vol. I: Properties of matter; Heat and temperature; Light and Color; Magnets; and Astronomy by Sight: the sun, moon, and stars. New York: J. Wiley & Sons (1995). L. C. McDermott et al., *Physics by Inquiry*, Vol. II: Electric circuits, Electromagnets, Light and Optics, Kinematics; and Astronomy by Sight: the Earth and the Solar System, J. Wiley & Sons, New York, 1995.

Virtual reality in introductory physics laboratories

Dedra Demaree, Stephen Stonebraker, Wenhui Zhao, and Lei Bao

Department of Physics, The Ohio State University, Columbus, OH 43210, USA

Abstract. Physicists consider laboratories to be a vital part of any introductory course, yet The Ohio State University's existing labs are not meeting their educational goals and students consistently rate them as having low value. This paper explores some of the reasons that standard introductory physics laboratories are not having the expected impact, and describes the implementation of Virtual Reality based experiments to improve upon lab effectiveness. Student response to these experiments and preliminary results regarding their impact on student learning will be discussed.

INTRODUCTORY LAB PROBLEMS

Physics laboratory (lab) activities have the potential to achieve a variety of goals and provide a valuable enhancement to introductory courses [1]. The current labs at The Ohio State University (OSU) utilize Socratic [2], Physics Education Research (PER) – based manuals aimed primarily at improving student conceptual understanding. Although initial reports [3] indicate these labs improved student gain on the Force Concept Inventory (FCI) [4], currently observed FCI gains are lower, and comparable to traditional courses [5]. In addition, OSU students often rank these labs as the least helpful component of the introductory physics course [6].

Before implementing enhancements to the labs, it is necessary to gain some idea of problems with the existing labs. The lab manual contains exercises which can be quite good at improving conceptual understanding, but this depends on how students approach the labs. Although the following problems are speculation, they have been observed by the authors as well as several of the lab instructors. We feel it is likely these problems are common, and are not specific to OSU.

Although some lab programs explicitly address error analysis as part of the experimental process, the existing OSU labs do not. It is documented that students have difficulty carefully addressing error even when emphasized during the lab [7]. Student comments indicate they think error is "bad". This can lead to frustration, or worse: students ignoring inconsistencies with predictions by blaming them on experimental error. We often implicitly expect students to ignore effects such as friction, while observations under incorrect assumptions may lead to wrong conclusions. Due to the unfavorable instructor-student ratio in large introductory service courses it is often difficult to address these issues properly.

The OSU lab questions are aimed at addressing common conceptual difficulties and when not carefully considered can appear simplistic to our students. They are sometimes overconfident, thinking they already "know" the concepts from attending lecture and reading the text. Many students are also reluctant to do the labs carefully, believing the lab activities will not help them with their exams. In reality, low FCI gains show that students are missing fundamental physics concepts, which is not only disturbing for instructors, but should be important to students, since an unstable conceptual foundation can lead to mistakes on graded work.

Physics Education Research has shown that active engagement computer-based activities are more effective than passive programs [8-10]. FCI gains have been shown to improve significantly with the replacement of a single traditional lab with a PER-based activity [11]. However, even effective computer-based activities can sometimes lead students to view the computer as authoritative [12], with information output accepted as fact without critical consideration. Virtual Reality (VR) software developed at OSU has the potential to improve upon problems associated with OSU's existing labs. In addition, VR can be fun and engaging to students, and could even improve their overall attitudes towards physics [13].

CP790, *2004 Physics Education Research Conference*, edited by J. Marx, P. Heron, and S. Franklin

VIRTUAL REALITY TO THE RESCUE

The PER group at OSU has developed a VR platform that can run several VR-based experiments. These allow students to view processes in a much more detailed and controlled way than is possible with traditional equipment. Parameters can be easily adjusted for careful exploration of phenomena, providing more freedom to choose what conditions to test. Friction can be turned off and on to allow students to carefully observe how motion is affected. Processes that are normally too fast to observe can be run in slow motion, revealing details students might otherwise miss. VR can accommodate both very large and very small scales, allowing for a more detailed study of a process.

This virtual environment is manipulated using a touch-sensitive joystick; if the student pushes harder a larger force is applied to the object in the simulation. This allows students to apply a controlled force at a distance and instantly see how their force affects the motion of an object. Unlike typical computer-based activities, the VR output is not authoritative information; it is a direct, real-time response to student input. This provides a true complement to traditional lab experiments.

The VR platform is engaging. The viewpoint moves with the object of interest through scenery. Many students enjoy video games, and the VR scenery and joystick mimics them. Students often challenge each other to see who can best control the motion of an object, or see if they can cause some strange effect to occur. These voluntary explorations give them added experience with Newton's laws. Although we encourage this and enjoy the exploratory attitude of the students, all VR activities are written with a sequence of prediction, testing, and resolution of differences to maximize the educational impact of the VR features. Because this platform is seen as fun by many students, concepts can be repeated with lower student resistance than in a traditional activity.

THE THREE NEW VR LABS

Three VR programs were ready for full implementation at the start of this investigation; those dealing with linear motion, circular motion, and collisions. Each of these was incorporated into a new lab for OSU's introductory calculus-based mechanics course spring quarter.

The Linear Motion VR lab

The Linear Motion VR lab allows students to probe Newton's 2nd law using the Linear Motion VR software (Fig. 1). A box rests on a scenic track, and the students can apply an external force to the box using a joystick. All parameters are adjustable: the initial conditions, the mass of the block and the coefficient of friction. The force diagram for the box can be displayed in real time, in addition to vectors representing the velocity and acceleration of the box. Motion graphs can also be displayed on the screen.

The current OSU linear motion lab falls short of the full time period. Therefore, we implemented this software by integrating questions requiring its use into the existing lab. In this way, students made predictions and observations of the motion of a cart with external forces applied, and then further explored Newton's 2nd law using predictions and observations through the VR environment.

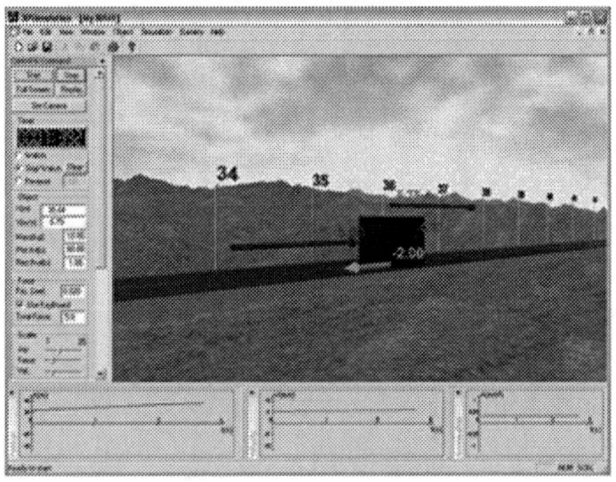

FIGURE 1. The Linear Motion VR program

The Circular Motion VR lab

Before now there has not been a circular motion lab in this course. The new Circular Motion VR lab (Fig. 2) contains a ball and a circular track. The initial conditions of the ball can be set and the external force, velocity, and acceleration vectors can be displayed. The object of the program is to apply external force with the joystick to keep the ball in uniform circular motion. This can be done with and without friction.

Even if students had studied central forces in lecture and predictions were made prior to using the VR, students often first tried to drag the ball around the track. This instinct is then better understood and corrected when the experiment is repeated with friction turned on. In addition to the VR software, students were given a small smoothly rolling ball and a rough puck to contrast with the motion observed using the computer.

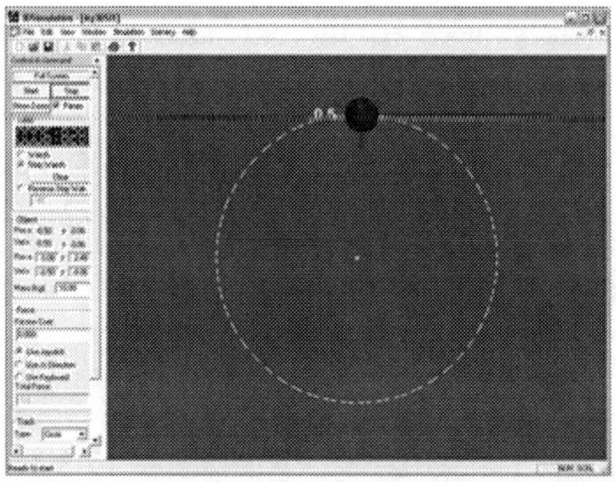

FIGURE 2. The Circular Motion VR program

The Collisions VR lab

The Collisions VR program (Fig. 3) contains two carts on a scenic track. All initial conditions can be adjusted for each cart. The user can also set both the coefficient of friction and the elasticity of the bumpers. As with the Linear Motion VR program, the force diagrams, velocity and acceleration vectors, and motion graphs can be displayed in real time.

Although there is normally a lab covering collisions at OSU, a completely new lab was written following the suggestions in [14]. The VR software permits a detailed exploration of the forces acting on each cart during the collision, permitting an approach to conservation of momentum directly from Newton's 3rd law. During one lab period, students were able to predict and observe the forces while a cart collides with a wall, a spring, another cart moving towards it, and another stationary cart. Many other interesting combinations are possible, including collisions between carts of very different masses, initial speeds, and/or elasticity. Students sometimes explored these on their own at the end of class.

FIGURE 3. The Collisions VR program

WERE THE VR LABS EFFECTIVE?

In order to test the effectiveness of the new VR labs, students were given the FCI during the first and last lab period. In addition, students were given a set of nine questions written to test their physical intuition about how force affects motion. In order to determine the effect of the VR labs on student attitudes, a qualitative questionnaire was also given.

The introductory calculus-based physics students were split into two groups, referred to as the Linear VR group and the Collisions VR group. The Linear VR group, containing 136 students, did the new Linear Motion VR lab and the existing collisions lab. The Collisions VR group, containing 118 students, did the new Collisions VR lab and the existing linear motion lab. Each group served as the control for the other group and both did the Circular Motion VR lab. The two groups were shown to be statistically similar based on their FCI pretest scores. None of the lab instructors were involved with developing the VR labs, nor are they involved in PER.

Looking at the normalized gains (as defined in [5]) for specific questions on the FCI, a significant difference was seen for questions that corresponded to the activities performed in the VR labs. The Linear VR group had a gain of 0.35 on FCI questions 25 and 26, which directly relate to activities done with the joystick, while the control group gain was only 0.23. It is significant to note that the FCI posttest was given a full 6 weeks after the Linear Motion lab. The Collisions VR group had a gain of 0.38 on FCI questions 4, 15, 16 and 28 pertaining to Newton's 3rd law, while the control group gain was 0.26. This second result is also partly due to the completely different structure of the new VR lab. Similar results were seen with the nine extra questions which will be addressed along with a more detailed quantitative analysis of our results in a longer publication to follow shortly.

In a multiple-part qualitative feedback question we asked the students if they preferred existing labs or the VR labs. We asked 4 sub-questions spanning from educational impact to level of enjoyment. The students had a slightly higher overall preference for the VR labs. This result is very encouraging when considered along with the fact that when asked which individual lab the students like best and why, students strongly prefer a subset of the existing labs called experiment problems. In these labs, students design their own experiment for determining an unknown quantity. In fact, all three experiment problems ranked in the top four (also included was the Collisions VR lab). This means that despite the students' strong preference for the experiment problems, half of them would still rather have VR labs than existing labs.

Students were asked to give specific feedback as to their preference between VR and existing labs. The majority of students who prefer the regular labs state that it is because they like having something hands-on (21%) or "real" (11%). The majority of students who prefer the VR labs state that it is because there is "no error," the environment is "exact" (19%). These students would best make use of a mix of physical experiments and VR experiments to directly contrast ideal with real. Some representative student comments on the VR labs are given in Fig. 4.

Some Student Comments Regarding the VR labs:

- They are a great complement (to regular labs)
- Allows us to see the inner workings of a system
- There is more control over the environment
- It gave instant data feedback

FIGURE 4. Student Comments

Students were asked if using the joystick helped them get a better feel for how force affects motion. 95% of the students doing the Linear Motion lab found the joystick more helpful than traditional lab equipment. Among the 19% of students who ranked the VR labs low, 31% of them attributed their ranking to technical reasons, such as "the joystick was hard to control" and "the software was difficult to use". These drawbacks will be improved before our next implementation.

Lab instructors felt that the new VR labs challenged the average student and were more interesting than existing labs. They observed a higher level of interest and excitement when the students used the VR as compared to the traditional lab equipment. We hope the students came away with a better attitude towards physics, which is a topic being further explored in our research.

CONCLUSIONS

This paper discussed the motivations for implementing VR labs in the introductory calculus-based mechanics course at The Ohio State University (OSU). Existing labs are not meeting their educational goals and are not seen as a valuable component of the course by students.

Three VR-based labs were implemented to help enhance the existing labs. Students doing the new labs scored higher on specific FCI questions which were most related to the VR activities. Students responded positively to the VR labs stating overwhelmingly that the joystick was more useful for understanding forces than the normal lab equipment.

Although students were mixed as to their preference of VR over existing labs, there is strong evidence that students would value having a mix of both types. Progress is being made towards combining the VR experiments with traditional lab equipment. This provides the students with both the hands-on and ideal environments they enjoy while possibly improving their conceptual understanding.

ACKNOWLEDGMENTS

The authors would like to thank Professor E. Sugarbaker and Dr. E. Adelson. This work is supported in part by NSF grant REC-0087788, and REC-0126070.

REFERENCES

1. American Association of Physics Teachers, *Phys. Teach.* **35** (1997).
2. R. R. Hake, *Phys. Teach.* **30**, 546-552 (1992).
3. D. J. VanDomelen and A. Van Heuvelen, *Am. J. Phys.* **70**, 7 (2002).
4. D. Hestenes, M. Wells and G. Swackhammer, *Phys. Teach.* **30** (1992).
5. R. R. Hake, *Am. J. Phys.* **66**, 4 (1996).
6. OSU student evaluation data from 1996 to 2002.
7. For example, see: S. Allie, A. Buffler, L. Kaunda, B. Campbell, and F. Lubben, *Int. J. Sci. Educ,* **20**, 4 (1998).
8. R. K. Thornton and D. R. Sokoloff, *Am. J. Phys.* **58**, 9 (1990).
9. E. F. Redish, J. M. Saul and R. N. Steinberg, *Am. J. Phys.* **65**, 1 (1997).
10. F. Reif and L. A. Scott, *Am. J. Phys. Suppl.* **67**, 7 (1999).
11. D. S. Abbott, J. M. Saul, G. W. Parker, and R. J. Beichner, *Phys. Educ. Res., Am. J. Phys. Suppl.* **68**, 7 (2000).
12. R. N. Steinberg, *Phys. Educ. Res., Am. J. Phys. Suppl.* **68**, 7 (2000).
13. E. F. Redish, J. M. Saul and R. N. Steinberg, *Am. J. Phys.* **66**, 3 (1998).
14. R. D. Knight, *Five Easy Lessons, Strategies for Successful Physics Teaching*, Addison Wesley, San Francisco, CA, 2004.

Helping Preservice Teachers Implement and Assess Research-based Instruction in K-12 Classrooms

Donna L. Messina, Lezlie S. DeWater, and MacKenzie R. Stetzer

Department of Physics,
University of Washington, Seattle, WA 98195

Abstract. The Physics Education Group at the University of Washington offers special physics courses for preservice teachers. The three-quarter sequence helps prospective teachers develop an in-depth understanding of some of the important basic concepts they will teach. The guided-inquiry pedagogical approach provides them with an opportunity to learn as they will be expected to teach. As a result of the course, preservice teachers also come to recognize some conceptual and reasoning difficulties commonly encountered by students. A culmination of their experience is a teaching *practicum* in which the preservice teachers apply what they have learned to instruction in middle or high school classrooms. Observations of the preservice teachers as they design, teach, and assess their lessons contribute to our understanding of the type of preparation needed for them to be able to teach physics and physical science by inquiry.

INTRODUCTION

The Physics Education Group (PEG) at the University of Washington offers special courses in physics and physical science for both preservice and inservice science teachers. A primary purpose is to provide teachers with an opportunity to develop an in-depth understanding of topics relevant to K-12 instruction and to do so in an inquiry-oriented manner consistent with how they are expected to teach. For inservice teachers, this process occurs during a six-week NSF-funded Summer Institute and an academic-year Continuation Course. During the academic year, the PEG also offers similar courses for preservice teachers. This paper focuses on the courses for undergraduate physics and mathematics majors who anticipate becoming middle or high school teachers.

PRESERVICE COURSE: OVERVIEW

During the academic year, students interested in teaching physics or mathematics enroll in a three quarter physics sequence at either the 200 or 400 level [1-3]. The sequence for which a student registers depends on whether the student will teach at the middle school or high school level and on his or her physics and mathematics background.

Throughout the academic year, the preservice teachers investigate various topics in physics and physical science. As they work through *Physics by Inquiry (PbI),* developed by the PEG, they complete carefully sequenced experiments and exercises designed to guide them in developing an in-depth understanding [4]. Common conceptual and reasoning difficulties are specifically addressed. Working in pairs, the students progress at their own rate, keeping careful notes of their observations, data, and answers to questions posed by the curriculum or that arise during their study. The notebooks reflect the inductive and deductive reasoning through which they developed basic concepts and learned to apply them. At pivotal points, the instructional staff probes student progress through semi-Socratic dialogues. Student understanding is also assessed via exams, homework, and papers in which students must provide thorough explanations of reasoning.

In addition to helping students develop the in-depth content understanding that is imperative for success in the classroom, the courses encourage students to reflect on instructional strategies they experience [1]. A special focus is on the inquiry approach that has

CP790, *2004 Physics Education Research Conference*, edited by J. Marx, P. Heron, and S. Franklin

proved effective in helping students develop a coherent conceptual framework and in eliciting and addressing common student difficulties [5,6]. The paper assignments, in particular, are designed to help the preservice teachers examine and articulate how their conceptions developed through the experiments and exercises they performed. During the course, the preservice teachers begin the process of translating their own learning experiences into effective classroom practice. The culmination, which takes place in the third quarter, is a teaching *practicum* in which the preservice teachers design, implement, and assess a lesson for middle or high school students.

Your partner is in a dark hallway. You are facing a door that enters into the hall. No other lights are on. Your partner aims a flashlight with a mask so that it projects a **very** narrow beam of light at the far end of the hall.

Would you be able to tell whether the flashlight is on? If so, how? What would you see? If not, why not?

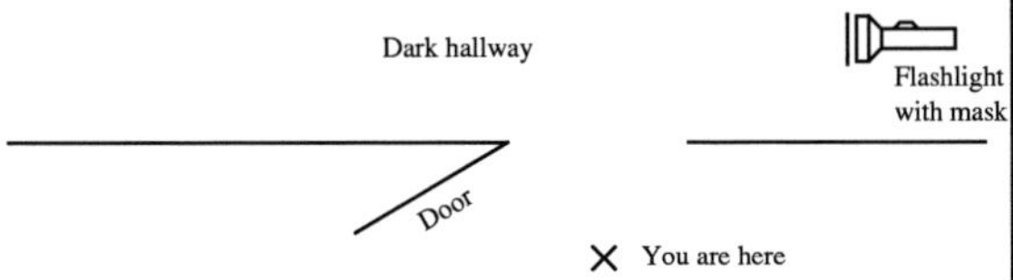

FIGURE 1. Pretest question used to assess student understanding of light.

TEACHING PRACTICUM

After developing their own understanding of several topics, the preservice teachers have an opportunity to teach a topic they have studied. They try to provide for middle or high school students learning experiences similar to those they have had themselves. The *practicum* is designed to serve as a context in which the preservice teachers can:

- recognize the value of identifying initial, current, and final levels of student understanding
- experience the iterative process of curriculum development
- gain practice in teaching by guided inquiry, and
- reflect on teaching and learning as processes of inquiry

These goals are accomplished through four phases of the *practicum:* preparation, adaptation, implementation, and assessment/reflection. The topic varies from year to year. For specificity, the discussion below is in the context of the 2004 *practicum* on geometrical optics.

Preparation phase: During the preparation phase, the preservice teachers continued to work through *Physics by Inquiry*. They also read relevant research literature on the teaching and learning of science [7,8]. Meanwhile, the students with whom they were going to interact were given a pretest by their teacher. (See Fig. 1.) The preservice teachers analyzed the responses to try to identify specific difficulties.

The following pretest responses were considered correct: "No [I could not see the light], because in order to be able to see something, light must first hit that object and then enter your eye." and "Yes, when there is dust in the air the light would show the dust. It would look like a beam of dust." Responses considered incorrect or incomplete include: "Yes. Even the smallest ray of light will pierce the darkness." and "Yes, you would see the slit of light in front of you almost like there was a rope or line." Many students failed to articulate (implicitly or explicitly) that, for the observer to see light from the flashlight, the light must first strike an object (*e.g.,* a dust particle) so that the (reflected) light has an unobstructed linear path to the observer's eye.

Only 35% of the secondary students who have taken the pretest (N=153) gave correct reasoning. (This result includes students given the same pretest the previous year.) It is interesting to note that these results are similar to those of preservice and inservice teachers who have taken the pretest. (See Table 1.)

Adaptation phase: Analysis of the pretest informed the preservice teachers of the need to help the students develop an understanding that light travels in straight lines and the ability to apply this idea. Therefore, the preservice teachers adapted a section of the Light and Color module from *PbI.* This section covers the linear propagation of light and also helps develop the idea that light is emitted in all directions from each point on an extended source. The preservice teachers, working in groups of three, formulated questions to use as verbal checkouts to evaluate understanding as the students worked through the materials. In addition, the preservice teachers designed post-test questions to assess student learning upon completion of the curriculum. Members of the university instructional staff assisted the preservice teachers throughout this phase of the *practicum.*

Implementation phase: After the adaptation phase, the preservice teachers were ready to venture into the secondary classroom. Each three-person group of preservice teachers was assigned 15 secondary students with whom they would work for 3

TABLE 1. Assessment of student learning. (The total numbers of pretest or post-test responses are given in parentheses.)

Population	Pretest		Post-test
	Fig. 1 % correct	**Fig. 2 (or similar) % correct**	**Fig. 2 (or similar) % correct**
Secondary students taught by preservice teachers	35% (153)	–	50% (42)
Secondary students taught by very experienced teacher*	–	–	85% (57)
Preservice math and physics teachers	55% (24)	35% (29)	85% (60)
Inservice K-8 teachers	30% (43)	30% (43)	90% (47)
Undergraduates in calculus-based introductory course with *Tutorials in Introductory Physics* [9]	–	20% (>2100)	80% (~360)

* The teacher had a Ph.D. in physics education research and had been involved in the design of *PbI* modules on light.

hours over 2 days. While in the classroom, the preservice teachers used their own learning experiences as a model for their instructional approach. Rather than answering students' questions directly, the preservice teachers guided students to use their developing understanding in constructing a model for light. The preservice teachers also moved around the classroom observing the students and listening to their conversations. Several preservice teachers commented that these observations were quite informative since they provided insight into the intellectual struggles of the students.

Assessment and reflection phase: A few days after completing the curriculum, the secondary students were given a post-test. The post-test consisted of several questions, some designed by the preservice teachers and others developed by the PEG. We have noticed that preservice teachers often find it difficult to design questions that effectively probe student understanding. By including well-tested questions, we ensure that all preservice teachers obtain results they can use to assess student learning (even if their own questions prove to be less effective). The questions also allow comparisons to be made between different populations (including various secondary student *practicum* groups). Below, we limit our discussion of post-test results to the PEG questions on geometrical optics.

The post-test question in Fig. 2 was given to the secondary students who participated in the 2004 *practicum*. A correct response requires understanding of the ideas that: light travels in straight lines and extended light sources may be treated as collections of point sources. The correct answer (see Fig. 3) can be obtained by treating the long-filament bulb as a continuum of point sources, each of which produces a crescent-shaped image on the screen.

Many of the preservice teachers were disappointed by the results from this portion of the post-test. Only 50% of the secondary students gave correct or nearly correct answers (that is, responses in which extended sources were treated as continuous or discrete collections of point sources). In contrast, after instruction that uses *PbI*, preservice teachers typically perform at the 85% level on similar questions. See Table 1 for results from comparable post-test and pretest questions from various populations.

A mask that contains a small crescent-shaped hole is placed between two long-filament bulbs and a screen as shown below. Sketch what you would see on the screen when the bulbs are lit. Explain.

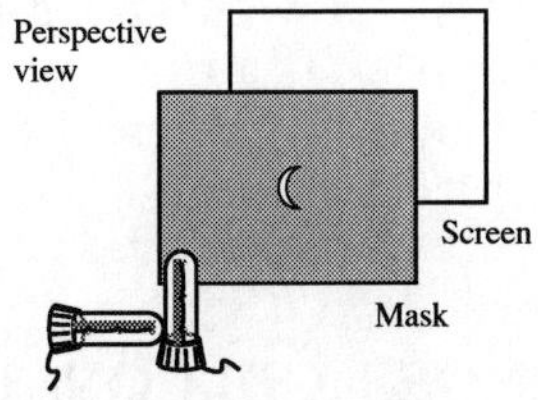

FIGURE 2. A post-test question used to assess student understanding of light after instruction.

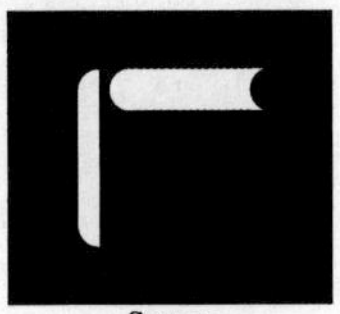

FIGURE 3. A correct response to the post-test question shown in Fig. 2.

The secondary students' post-test results provide an excellent opportunity for the preservice teachers to analyze remaining student difficulties as well as to reflect on their own insights into teaching and learning by inquiry. The reflection is formalized in a paper that the preservice teachers write. The paper includes:

- an analysis of the student post-test data

- an evaluation of the effectiveness of the adapted curriculum, including proposed revisions
- a reflection on pedagogical issues, and
- a description of interactions with students

The paper also provides insight into the usefulness of the *practicum* for the preservice teachers. In the 2004 *practicum* papers, many of the preservice teachers stated that the questions they had asked at pivotal points as students worked through the curriculum failed to probe student understanding adequately. They realized that the questions they had designed prior to working with the students had not been effective in guiding them to their own understanding of the material. Several preservice teachers also became aware that, during the Socratic dialogues, they often "heard from the students what they wanted to hear rather than what the students actually said." These insights helped the preservice teachers recognize that teaching by inquiry requires not only a deep understanding of the content, but also pedagogical skills to support inquiry learning. One preservice teacher commented:

> "When I spoke with [students] who really understood the material, I got to learn how other people think about physics. When I was talking with [students] who didn't understand, I had to reach inside of myself for different ways of leading them to their own understanding."

CHALLENGES AND CONCLUSION

The *practicum* is a relatively new component of the PEG courses for preservice teachers. We have evidence that this is an important experience for K-12 teachers. For example, we worked with an inservice high school teacher whose students were studying astronomy. Although she had worked through materials on astronomy in the preservice course (without the *practicum*) and could account for the phases of the moon, she did not recognize the careful consideration that must be given to curriculum adaptation and implementation. Failing to recognize the need for a coherent curriculum that explicitly addresses common difficulties, she eliminated some crucial experiments needed for a logical progression of ideas. Only 30% of her students (N=78) were able to identify the role of the sun/moon angle in the moon's phases, a disappointing result since 35% had correctly identified the role of the angle in a pretest. The *practicum* allows PEG members to guide and advise prospective teachers throughout the adaptation and implementation process. Thus, preservice teachers, in collaboration with staff, can examine the instructional materials critically, search for gaps in the logical progression of ideas, and identify common difficulties not explicitly addressed in the draft curricula.

Courses that focus on content, but that are taught in a manner consistent with how teachers are expected to teach, play an invaluable role in preparing prospective teachers to meet the challenges of the K-12 classroom [1]. The incorporation of activities such as the *practicum* into preservice physics courses enables prospective teachers to gain practice in adapting, implementing, and assessing guided inquiry lessons. Such opportunities provide preservice teachers with a chance to reflect on the inquiry approach, both as a learner and a teacher, in a way that helps them draw upon their own learning experiences when working with secondary students.

ACKNOWLEDGMENTS

The authors appreciate the substantive contributions by colleagues in the Physics Education Group and the support of the NSF.

REFERENCES

1. L.C. McDermott, *Am. J. Phys.* **58** (8) 734 (1990).
2. L.C. McDermott and L.S. DeWater, "The need for special science courses for teachers: Two perspectives," an invited chapter in *Inquiring into Inquiry Learning in Teaching and Science,* J. Minstrell and E.H. van Zee, eds., Washington, D.C.: AAAS (2000), pp. 241-257.
3. L.C. McDermott, P.S. Shaffer, and C.P. Constantinou, *Physics Education* **35** (6) 411 (November 2000).
4. L.C. McDermott and the Physics Education Group at the University of Washington, *Physics by Inquiry,* New York, John Wiley & Sons, Inc. (1996).
5. P.R.L. Heron and L.C. McDermott, *Opt. & Phot. News* **9** (9) 30 (1998).
6. K. Wosilait, P.R.L. Heron, P.S. Shaffer, and L.C. McDermott, *Am. J. Phys.* **66** (10) 906 (1998).
7. R. Driver and E. Guesne, "Chapter 2: Light," in *Children's Ideas in Science,* R. Driver, E. Guesne, and A. Tiberghien, eds., Milton Keynes, UK: Open University Press (1985), pp. 10-32.
8. M. Kennedy, "Defining Optimal Knowledge for Teaching Science and Mathematics," Research Monograph No. 10, National Institute for Science Education, University of Wisconsin-Madison (1997).
9. L.C. McDermott, P.S. Shaffer, and the Physics Education Group at the University of Washington, *Tutorials in Introductory Physics, First Edition,* New Jersey, Prentice Hall, Inc. (2002).

Can Computer Simulations Replace Real Equipment in Undergraduate Laboratories?

N. D. Finkelstein, K. K. Perkins, W. Adams, P. Kohl, and N. Podolefsky

Department of Physics, University of Colorado, Boulder, CO 80309-0390

Abstract. This paper examines the effects of substituting computer simulations in place of real laboratory equipment in the second semester of a large-scale introductory physics course. The direct current (DC) circuit laboratory was modified to compare the effects of using computer simulations with the effects of using real light bulbs, meters and wires. Three groups of students, those who used real equipment, those who used computer simulations, and those who had no lab experience, were compared in terms of their mastery of physics concepts and skills with real equipment. Students who used the simulated equipment outperformed their counterparts both on a conceptual survey of the domain and in the coordinated tasks of assembling a real circuit and describing how it worked.

INTRODUCTION

Since the introduction of the microcomputer several decades ago, researchers and educators have developed, explored, and studied mechanisms for employing computers in the classroom. Over this time, computers have made their way into nearly every element of university courses in physics. Microcomputers have been used to augment large-scale lectures [1] to promote individual student development [2] and seemingly everything in between [3]. One of the most common applications of computers has been to supplement recitations or laboratories in large-scale introductory physics courses [3,4,5]. These applications have included using computers to facilitate data acquisition, to provide real-time data display, to analyze these data, and to simulate complex phenomena. Such efforts have been shown to be as or more effective than their non-computer based counterparts, be they traditional or reformed, PER-based activities.

While computers have been used to supplement labs, we examine the effectiveness of completely replacing traditional equipment with computer-based simulations. Given the constraints (and large expense) of traditional labs, we explore whether it is possible to achieve the conceptual learning gains and mastery of mechanical skills with real equipment by working with computer simulations.

The following study took place in a traditional large-scale introductory algebra-based physics course at a large research university. The course was the second in a sequence of two, covering electricity, magnetism, optics and modern physics. Students, typically in their 2nd or 3rd year of study, received 5 credit hours for participating in three lectures and one integrated two hour laboratory / recitation section per week. Weekly homework assignments were the traditional end-of-chapter style questions, offered and graded by a computer system (CAPA) [6]. Two instructors and 7 TAs were assigned to the 363 students enrolled in this course that met for 15 weeks.

The laboratories where this study occurred were offered every other week, alternating with the recitation sections. Six laboratory sessions were offered over the course of the term. The laboratories themselves had recently been revised in order to emphasize an inquiry-based approach. The labs emphasized discovery rather than verification [7] and included some elements of "Scientific Community Labs" [8]. This study occurred in the second laboratory of the course: DC-circuits. Students engaged in a series of exercises including: examining resistors in series and parallel, building a simple circuit and then predicting, observing and reconciling its behavior as various elements (resistors or light bulbs) were added or rearranged, and finally developing methods to measure resistance in multiple ways in these circuits. The goals of the lab were for students

CP790, *2004 Physics Education Research Conference*, edited by J. Marx, P. Heron, and S. Franklin

to develop an understanding of simple circuits (the concepts of voltage, current, and series, parallel and equivalent resistance), to develop the skills associated with connecting light bulbs, resistors and wires in various combinations, and to collect data and make arguments about these circuits' behaviors. Each lab began with students turning in pre-lab work and asking TAs questions about the material.

Some of the students in this study worked with a computer simulation in lieu of real equipment. The simulation, the Circuit Construction Kit (CCK), is part of the Physics Education Technology (PhET) project at the University of Colorado. The PhET project has developed approximately forty research-based simulations which are available online [9]. These simulations are highly interactive, engaging, and open learning environments with animated visual feedback to the user. They emphasize students constructing conceptual models through a physically accurate, highly visual, dynamic representation of the physics. The PhET simulations are research-based, tested both in interviews of students and in-class use, and go through several design cycles. More on the PhET project and the research methods used to develop the simulations is available in ref [9]. The Circuit Construction Kit includes an open workspace where students can place resistors, light bulbs, wires and batteries. Each element has operating parameters (resistance, voltage, etc.) that may be varied by the user and measured by a simulated voltmeter and ammeter. The underlying algorithm uses Kirchoff's laws to calculate current and voltage. Moving electrons are explicitly shown to represent current flow (and conservation). A screen shot appears in Figure 1.

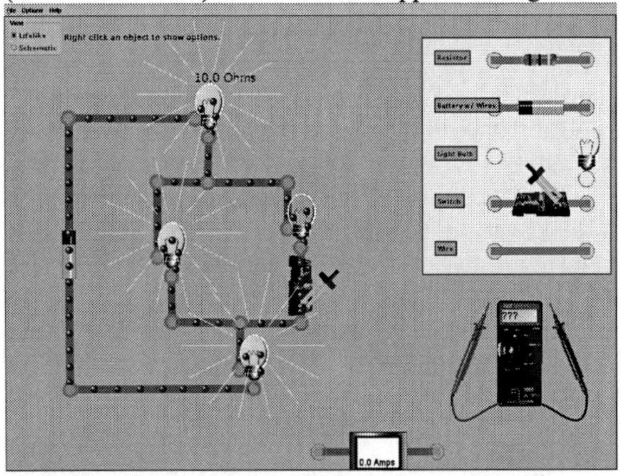

FIGURE 1. PhET simulation: Circuit Construction Kit.

STUDY DESIGN

Each of the fifteen sections of the algebra-based introductory physics course conducted the DC-Circuits lab. The sections were split into experimental conditions: CCK, (4 sections; N= 99 students) which conducted the laboratory with the simulations, and traditional conditions, TRAD, (6 sections; N = 132 students) which conducted the laboratory with real equipment. (The remaining 5 sections participated in the labs in slightly varied conditions, but are not included in this study for the sake of brevity.)

Each student was assigned a pre-lab activity, which varied by section (CCK or TRAD). Three of the four questions on the pre-lab were identical; the fourth question, which asked students to make a light bulb light with battery and a single wire, varied. The traditional group was asked to draw a circuit that would light the bulb; the CCK students were asked to build the circuit using the simulation and submit the results. All students turned in their assignments at the beginning of their lab section.

The DC-Circuits lab was matched for the two groups. The written introduction to the equipment was the same for both groups, each receiving instructions on reading resistors etc. However, additional instructions on operating the simulation were provided to the CCK group. The actual laboratory activities (challenges and wording) were the same except the traditional groups were instructed to manipulate real equipment, the CCK groups to manipulate simulated equipment. Students worked in groups of two to four.

The last 30 minutes of each laboratory section was set aside for students to engage in a challenge worksheet consisting of three questions. The main challenge for *all* students was to build a circuit (shown in Figure 1) using *real* equipment, and describe what happens and why when the circuit was broken where the switch is shown. Notably the CCK group had no formal exposure to this equipment before this challenge. As an added control, this same challenge sheet was given to students (N=107) in the calculus based introductory sequence (electricity and magnetism), during 4 of the recitation sections. The calculus based course has a detached laboratory course which students generally had not yet taken; however, these students had spent three lecture periods (including a demonstration of this very circuit), and a homework set covering DC-circuits before attempting the challenge worksheet.

DATA

The following data were collected and analyzed during the course of this study:

- Pre-lab worksheets (for each student).
- Observational notes of the sessions (both by TAs and by researchers in this study).
- Timing data. How long it took students to build the assigned circuit as a group and to complete the challenge sheet at the end of the lab individually.

- Lab challenge write-ups (written up and turned in by each student).
- Three questions on DC-circuits on the final exam issued 12 weeks later.

Here, we report only on the timing, the performance on the challenge worksheets, and the performance on the final exam questions.

Circuit challenge:

During the laboratories, all students completed a challenge worksheet where they were asked in their groups to build the circuit shown in Figure 1, show the TA, and then break the circuit at the point designated by the switch. Students answered the following question: "In 50 words or more, describe what happens and why the bulbs change brightness as they do. You may use words and formulas..." and turned in their answers individually. Meanwhile TAs were given observation sheets that included a section asking them to report on how long it took students to complete the circuit challenge, "For each group in your section (i.e. groups 1-10 or so), please note roughly how long it took them to complete the circuit challenge." The same challenge and observations sheets were provided to students and TAs for the calculus- based introductory physics course.

Timing:

The timing data -- how long it took groups of students to build the circuit, break the circuit and write-up the challenge -- are plotted in Figure 2. The averages for the experimental condition (CCK), the group using traditional equipment (Trad), and calculus-based course sections with no lab (No Lab) are plotted.

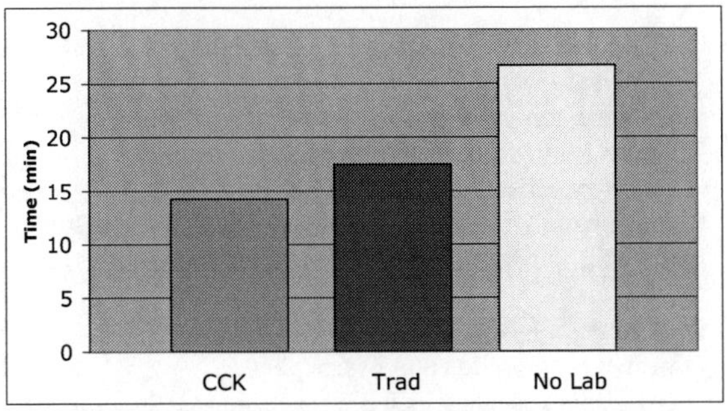

FIGURE 2. Time for students to build a circuit with real equipment and write about it. Mean times for all groups plotted. Student times for individual groups varied ±7 minutes for most all (90% or more) of the groups.

Evaluation of write-ups:

Each circuit challenge write-up was evaluated by the authors [10] as to overall correctness on a scale from 0 to 3. 0 represented no demonstrated knowledge of the domain, while 3 represented correct and complete reasoning. The fraction of students scoring 0, 1, 2,and 3 are reported in Figure 3 for each of three groups, the experimental group (CCK), the traditional group (Trad) and the calculus-based group (No Lab).

The average score for the experimental group (CCK) is 1.86 and the real-equipment group (Trad) is 1.64 – a statistically significant shift ($p<0.03$). The calculus-based no-lab group averaged 1.91 (significantly different than the traditional group on a two tailed z-test, $p< 0.02$).

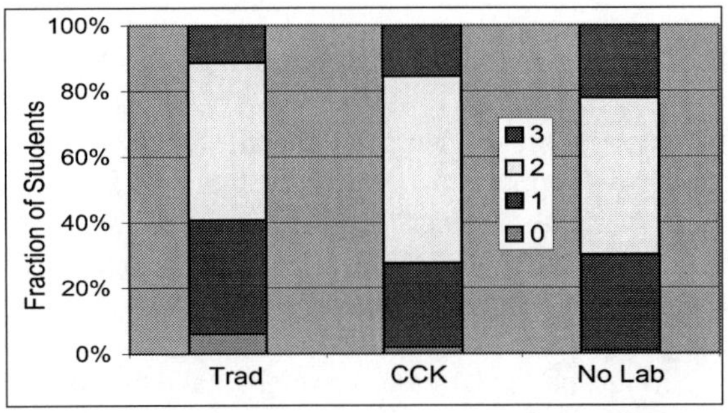

FIGURE 3. Performance on write-up of circuit challenge (0- no demonstrated knowledge to 3 – correct / complete)

Final Exam:

Referring to a schematic drawing of a series and parallel circuit, like the one shown in Figure 1 (but in schematic form and with a closed switch), students were asked to: (q1) rank the currents through each of the bulbs, (q2) rank the voltage drops across the bulbs in the same circuit, and (q3) predict whether the current through the first bulb increased, decreased or remained the same when the switch was opened [11]. Figure 4 plots student performance for the experimental (CCK) group and the traditional (TRAD) group for each of these three questions (q1, q2, q3), and their performance on the remaining 26 questions

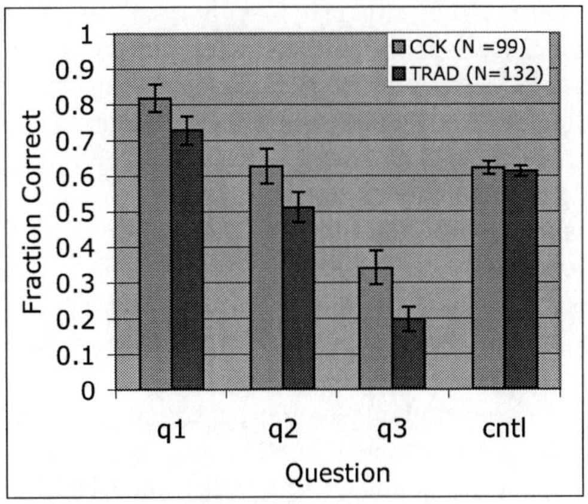

FIGURE 4. Student achievement on three conceptual circuits questions on final exam (q1-q3) and the remaining 26 questions on final (cntl).

on the final (labeled "cntl"), which covered other material in the course. The mean for all three questions is 0.593 (□=.27) for CCK and is 0.476 (□=.27) for TRAD groups; $p<0.001$. Notably, these results occur roughly more than two months after the lab was conducted, on the final (after many other forms of exposure to the same material – lectures, homework, exams, and studying).

DISCUSSION

While only three forms of data are presented here, the other data in this study (observations of the laboratories, pre-lab work, and detailed analysis of the challenges) appear to corroborate these findings. In short, we might answer the question posed in the title in the affirmative—in the right conditions, simulations can be substituted effectively for real laboratory equipment.

In this particular case, we might make the stronger claim: it is preferable for students to work with simulations over real light bulbs and resistors. We are not claiming that all circuits labs ought be replaced, but rather the conventional wisdom that students learn more and need the hands on experience is not borne out by measures of mastery of the content, nor their ability to construct circuits [12]. In a hands-on, inquiry-based lab, students using the simulations learned more content than did students using real equipment.

No less significant is student facility at actually constructing physical circuits. The data suggest that students working with simulations are indeed no slower at constructing and writing about circuits. In addition to more correctly and thoroughly writing about the circuits, the students take less time on average than their counterparts at building and describing these circuits.

Others have suggested a list of important characteristics of computer-based activities and simulations – from actively engaging the students to providing tools to reduce unnecessary drudgery [5,13]. To this list we add two more items which might begin to explain our findings. Computers can:

- make visible the models that are useful for forming concepts, and
- constrain the students in productive ways.

The first point is demonstrated by the CCK simulation where making the current flow (moving electrons) visible to the students emphasizes that current is conserved. As to the second point, we note that students' "messing about" can be productive in the simulation where explorations lead to investigation of circuit properties. (Notably, students' investigations are constrained to the concepts at-hand.) With the real equipment "messing about" can be unproductive (intellectually) and students end up building wire-bracelets. As Otero finds, students use computer simulations productively to produce conceptual models that are then effectively applied to physical (dare we suggest "real world') applications [14].

We do not suggest that simulations necessarily promote conceptual learning and facility with equipment, but rather they are useful tools for a variety of contexts which can promote student learning. Redish asks "Is the computer appropriate for teaching physics?" [15] – we suspect its time may have arrived. "To simulate or not to simulate?" asks Steinberg [2]. We answer yes… providing simulations are applied in the appropriate contexts.

ACKNOWLEDGEMENTS

This work was conducted with the support of the NSF, the Kavli Foundation, and the University of Colorado. The authors are grateful to Carl Wieman, the PhET team (particularly Sam Reid, author of CCK, and Ron LeMaster, lead software architect for the project) and members of the newly formed Physics Education Research Group at Colorado (PER@C).

REFERENCES

1. D. Sokoloff and R. Thornton, Phys.Tch,**35**,340 (1997).
2. R.N. Steinberg, G.E. Oberem and L.C. McDermott, *AJP.* **64** (11) 1370 (1996).
3. A useful starting point to find references: Z. Zacharia and O.R. Anderson, AJP 71(6), 618, 2003.
4. P. Laws, *Physics Today*, Vol. 44, No. 12, Dec. 1991
D. Sokoloff and R. Thornton, AJP **58**, 858 (1990)
5. E. F. Redish, J. M. Saul, and R. N. Steinberg, AJP. **65** (1997) 45-54.
6. http://www.lon-capa.org/
7. E. Lewis Multi-Initiative Dissemination Project http://www.cchem.berkeley.edu/midp/
8. R. Lippmann, Dissertation. University of Maryland 2003. http://www.physics.umd.edu/rgroups/ripe/perg/dissertations
9. C. Wieman, Physics Education Technology Project, http://www.colorado.edu/physics/phet
10. Five graders each evaluated these responses, grading at least one section of each type (CCK and TRAD).
11. These questions follow from those in the ECCE (Thornton& Sokoloff) and are part of Workshop Physics: http://physics.dickinson.edu/~wp_web/wp_homepage.html
12. We suspect there are some benefits to using real equipment, but that we did not measure them here.
13. R.N. Steinberg, Am. J. Phys. **68**, S37-S41 (2000).
14. V. Otero, *Proceedings of the International School of Physics "Enrico Fermi"*, Course CLVI, edited by M. Vicentini and E.F. Redish (IOS Press, Amsterdam, 2004).
15. E.F. Redish, *Computers in Physics* **7** (December 1993).

Learning Physics by Listening To Children

Danielle B. Harlow and Valerie K. Otero

University of Colorado, Boulder, CO 80309-0249
Danielle.Harlow@colorado.edu

Abstract. This paper examines the role that physics subject matter knowledge plays in one aspect of elementary teaching: listening to children's science ideas. Preliminary results of this study show that prospective teachers use their conceptual understanding of physics to analyze the "physics talk" of elementary students. This is demonstrated by their restatements of the children's ideas about physics phenomena.

INTRODUCTION

Physics faculty who educate prospective elementary teachers have the difficult challenge of helping teachers attain physics knowledge in a form which is useful in tasks of teaching. This responsibility is complicated by the fact that what constitutes useful content knowledge for teaching science to elementary children is not yet fully understood [1]. However, many researchers in teacher education agree that learning to listen to and analyze elementary students' ideas about science is one skill that is both vital and difficult for teachers to acquire [2][3][4].

Researchers who have studied the influence of scientific language on the learning of science have found that specialized language in the sciences in general and physics in particular causes difficulties for science learners of all ages [5][6]. In many cases, words have scientific meanings which differ from their everyday meanings. This disconnect between everyday meanings and scientific definitions of these terms may make learning to listen to and interpret children's ideas about science difficult. Children use everyday words and everyday meanings of words to talk about science and teachers must translate between the child's words and his science ideas. For example, a teacher must use her understanding of science to discern what a child means when he makes comments such as, "the force runs out." When a child says, "the force runs out," he may be thinking of *force* as defined by physicists, or he may be talking about *motion* or *energy*, terms with different meanings in physics but similar meanings in everyday speech. How a teacher interprets this child's statement may have implications for her teaching. Based on her interpretation, she may decide that the child needs to learn more about the idea of force or she may decide he needs only to use a new word (such as *motion*) for a concept he already understands.

In the study discussed in this paper, we investigated the question: How do prospective teachers use content knowledge when analyzing video tapes of elementary children learning physics? We found that prospective teachers used content knowledge in three different ways: (1) restating children's ideas, (2) reflecting on their own learning, and (3) discussing the content involved in the elementary student videos directly. In this paper, we present detailed examples of events included in the *restating children's ideas* category which illustrate prospective teachers' use of pedagogically-relevant content knowledge, a type of knowledge used by teachers which is often referred to as pedagogical content knowledge or PCK [7] in the teacher education literature.

THE PET CURRICULUM

The participants in this study were perspective teachers enrolled in a physics course that used the NSF-funded curriculum, Physics for Elementary Teachers (PET) [8], a physics curriculum targeted toward future teachers. PET differs from traditional physics curricula in its pedagogy, its content, its focus

CP790, *2004 Physics Education Research Conference*, edited by J. Marx, P. Heron, and S. Franklin

on the nature of science and its focus on elementary students' ideas. In PET activities, prospective teachers state their initial ideas in small groups and whole class discussions and then conduct hands-on and computer-simulation activities which challenge their initial ideas. They then discuss their observations and interpretations in their small groups and present their ideas in a whole class discussion. Through this discussion, they reach a consensus of ideas which they agree satisfactorily explain the observed phenomena. The physics content objectives of PET are based on the National Science Education Standards [9] and the Benchmarks for Science Literacy [10]. The PET curriculum addresses only those standards and benchmarks that have strands in K-5 grades, but at a level appropriate for university students preparing to teach elementary children.

Special activities focusing on Elementary Students' Ideas (ESI) are integrated into the PET curriculum. ESI activities guide prospective teachers through the analysis of video clips of elementary children discussing and investigating physics phenomena. The children participate in activities similar to the activities in the PET curriculum but at a level appropriate for elementary children. Prospective teachers are asked to identify and analyze the children's ideas soon after they have covered the same content in their physics class. In many cases the children in the video clips express ideas which are nearly identical to those ideas held by the prospective teachers only a few days before. These ESI activities are designed to help prospective teachers learn to recognize and value the ideas of elementary students, value their own prior knowledge, make connections between their ideas and the children's ideas and to apply their physics knowledge in a meaningful activity in the context of their chosen profession.

DATA

We have analyzed transcripts and video tapes of two PET classes. During these classes, prospective teachers participated in two Elementary Students' Ideas (ESI) activities. Here, we focus on only one activity, but analysis and claims are based on the entire data set. This ESI activity, *Applying Our Ideas about Friction*, follows activities in which prospective teachers develop ideas about friction forces.

ESI activity: Children discuss why things slow down

In *Applying Our Ideas about Friction* prospective teachers analyze a video of 2nd and 3rd grade children investigating the question: "Why does a toy car slow down and stop after having been given an initial push?" The video illustrates elementary children's initial, developing, and final ideas. The children's initial ideas about why a toy car slows down include that the car slows down because it does not have batteries, because of gravity, and because of little cracks in the sidewalk. The children in the movie then measure how far a toy car travels along different surfaces (different grades of sand paper) after leaving a ramp. During the experiment, the children talk about their interpretations of their observations (developing ideas) and present their final ideas. In the final ideas discussion, each child who presents his or her final ideas talks about the *surface* as playing a central role in the car's slowing motion.

Data: Prospective teachers discuss children's ideas

As homework, prospective teachers watched the video described above and responded to questions which ask them to make claims about the children's ideas and to provide evidence from the transcript and video to support their claims. During the following class meeting, prospective teachers discussed their responses with their small groups and then with the whole class. The data reported here consists of transcripts of segments of the discussion among the prospective teachers about the children's ideas.

ANALYSIS AND FINDINGS

We began analysis of prospective teachers' discussions about the video tapes with an a priori coding scheme which included four categories: 1) *talk using physics content, 2) talk using common children's ideas, 3) talk about how to use children's ideas,* and *4) talk about the nature of science.* This paper reports the analysis of data within the *talk using physics content* category. Data coded into this category included any utterance by a prospective teacher that could be directly mapped to the content learning goals of the PET curriculum and included ideas such as a *force is a push or a pull* and *a force is an interaction between two objects.*

We then further analyzed the data in the *talk using physics content* category to determine when talk about physics content occurred and how prospective teachers used content knowledge. Three patterns emerged. Prospective teachers used content knowledge when 1) restating children's ideas about science, 2) reflecting on their own learning and 3) discussing the content directly. In the remainder of this paper we present two detailed examples of prospective teachers using content knowledge when restating children's ideas. This category is of particular interest because it may represent pedagogically-relevant content knowledge.

Example 1: Kyle

After watching the second and third grade children conduct the friction experiment and analyzing the children's developing ideas, the prospective teachers talked in their small groups about what they thought the children understood about friction. Prospective teacher, Kyle, presented his group's thoughts during the whole class discussion. As illustrated in the transcript below, Kyle claimed that although the children never used the word *force*, they understood that there was an interaction between the car and the surface. He further claimed that the children attributed the car's slower motion to external factors (such as the surface) rather than an internal force or energy.

> Kyle: it's pretty surprising how much they do know. But they're not using the word force or things like that. But they clearly see that it goes fast on the ramp and that when it hits the ramp, the bumpy surface, they use the word bumps and things a lot, but they basically understand that when it hits a rougher surface, the car slows down. So although they're not understanding the fact that it is force acting on a car, they see the interaction taking place.

Kyle's evidence for the above claim was transcript excerpts from the ESI video featuring Maria and Allie talking about the car rolling down the ramp:

> Maria: I think it [the toy car] goes fast here [*pointing to smooth ramp*] but here it gets slower [*pointing to the rough sand paper*].
>
> Allie: Yeah, because there's bumpy things.

In restating two children's ideas in his own words Kyle made inferences about the children's thinking on the basis of his understanding of the physics content. In Kyle's claim above we identified two physics ideas: *friction is a force* and *force is an interaction.* Kyle used his understanding of *friction as a force* when he made the statement, "they're not using the word force…but they basically understand that when it hits a rougher surface, the car slows down." He also used his understanding that *a force is an interaction* when he stated, "they see the interaction taking place." These two physics concepts helped him to both restate the children's ideas in science terms and to make the claim that the children did understand some of the key concepts of the experiment even if they did not express the concepts in scientific language.

Example 2: Emily

Prospective teacher, Emily, discussed the conceptual development of the children over the entire sequence of video clips. Emily concentrated on one child, David, and inferred that he changed his ideas as a result of the lesson. She claimed that he initially attributed the slowing of the car to an internal mechanism (battery) during the initial ideas discussion and then changed his idea to one involving an interaction between the surface and the car when he stated his final ideas.

Emily restated David's ideas and in doing so made inferences about his thinking based on her own understanding of the content. She said, "He talks about that the cars stop because of the surface they're traveling on versus what the energy the car has…blocks or stops would be another word for pushes against." This statement was based on a comparison of David's initial ideas to his final ideas.

> David (initial): It's cause it's not like have batteries cause if it has batteries the tires move by themselves so you don't have to give them a push you just have to turn it on. But if you give it a push and it doesn't have batteries so it won't keep on going.
>
> David (final): The cars are stopping because the bumps stop the cars and it kind of blocks the cars, it blocks the cars from going really really um fast-and then it looses its speed.

Emily restated David's ideas in terms of the surface it was traveling on and the energy the car had. David, however, did not talk about an internal push or the interaction between the car and the surface. Rather, he talked about batteries and bumps. When David talked about batteries, Emily inferred that he attributed the slowing to an internal mechanism. When David talked about bumps and blocking motion, Emily inferred that he was thinking about the surface pushing against the car. To make the claim that David was no longer thinking about internal energy running out of the car, but rather about the surface slowing the car down, Emily used her understanding that *a force is a push or a pull*, and that *forces such as friction change the motion of objects* (rather than an internal push running out).

In the examples above, Kyle and Emily restated children's ideas in light of their own content understanding. That is, they used their physics knowledge to help them make sense of children's thinking. To formatively assess children's learning and determine which course of action will best facilitate students' further learning; teachers must first understand what a child is thinking. This understanding will inform teachers' pedagogical decisions. Thus, making sense of children's thinking is the first step in taking pedagogical action and therefore represents the use of pedagogically-relevant content knowledge.

DISCUSSION & IMPLICATIONS

Analyzing children's ideas requires applying physics ideas to words and thoughts not typically found in a physics course. Being able to think reasonably about what a child is saying is an important skill for elementary teachers and a meaningful application of their physics knowledge. Identifying children's ideas about physics requires more than memorized vocabulary words and phrases because children use their own informal talk to discuss their science ideas. Such analysis of children's science talk requires conceptual understanding. Thus, a necessary part of learning to teach elementary children is the acquisition of useful and relevant content knowledge.

The analysis of the prospective teachers' talk in this study provides evidence that these prospective teachers used their content knowledge to analyze the children's ideas. Specifically, the prospective teachers in this study demonstrated that they were able to use physics content knowledge to make reasonable inferences about the scientific meaning of children's informal science talk when participating in the ESI activities as part of the PET curriculum.

One implication of this study is that the integration of physics content instruction with activities which ask students to apply their developing content knowledge in activities which are relevant to teaching may facilitate the development of a form of content knowledge that is usable for teaching. Further research involving how teachers actually use their physics content knowledge in teaching situations would be useful toward substantiating this claim.

Based on the research conducted in this study, we inferred that the evaluation of children's ideas through videos can provide a meaningful context for applying conceptual physics knowledge in physics courses. This may be even more relevant to future teachers than physics application activities which are not directly related to teaching tasks. Activities that are embedded within a physics curriculum, such as the Elementary Students' Ideas (ESI) activities studied here may help prospective teachers learn to use physics knowledge in exactly the type of activity in which their content knowledge will be most useful: listening to and interpreting children's science ideas.

ACKNOWLEDGMENTS

The research in this paper is supported by grant NSF grant #ESI-0096856. We would like to thank Fred Goldberg and Benjamin Williams for their valuable assistance in the implementation of the curriculum and the collection of data used in this study.

REFERENCES

1. Wilson, S., Floden, R., & Ferrini-Mundy, J., *Teacher preparation research: current knowledge, gaps and recommendations*, Document R-01-3, Center for the Study of Teaching and Policy, University of Washington, 2001.

2. Ball, D., "What do students know? Facing challenges of distance, context, and desire in trying to hear children" in *International handbook of teachers and teaching*, edited by B.J. Biddle, T.L. Good, & I.F. Goodson, Doderich, the Netherlands, Klewer, 1997.

3. Paley, V., *Harvard Educational Review,* ***56(2),*** 1986, 122-131.

4. Gallas, K., *Talking their way into science,* New York: Teachers College Press, 1995.

5. Lemke, J.L., *Talking Science: Language, Learning, and Values.* Norwood, NJ: Ablex Publishing Corporation, 1990.

6. Moje, E., Collazo, T., Carrillo, R., & Marx, R., *Journal of Research in Science Teaching, 38*(4), 2001, 469-478.

7. Shulman, L., *Harvard Education Review, 57(1)*, 1987, 1-22.

8. For more information see http://cpucips.sdsu.edu/web/pet

9. NRC, *National Science Education Standards*, Washington DC, National Academies Press, 1996.

10. AAAS, Educating teachers of science, mathematics and technology: new practices for a new millennium, New York, Oxford University Press, 1993.

Teaching, Learning and Physics Education Research: Views of Mainstream Physics Professors

Charles Henderson* and Melissa H. Dancy†

*Department of Physics, Western Michigan University, Kalamazoo, MI, 49008, USA
†Department of Physics, University of North Carolina at Charlotte, Charlotte, NC, 28223, USA

Abstract. This paper presents the preliminary results of interviews with four thoughtful senior faculty who are not part of the Physics Education Research (PER) community. The interviews focused on their general beliefs about teaching and learning as well as their use of and attitudes towards PER and PER-based instructional strategies. We found that these instructors have beliefs about teaching and learning and instructional goals that are more PER-compatible than their self-described instructional practices. We hypothesize that one factor impeding more complete incorporation of PER is instructors' either misinterpreting or having a low opinion of the trustworthiness of educational research results.

INTRODUCTION

Substantial time and money has been spent developing research-based instructional strategies for introductory college physics courses. Despite wide availability, existing evidence suggests that these strategies and materials are only fully incorporated into a small number of introductory physics courses. Perhaps some physics instructors don't care about teaching or what their students learn, but our experience is that most physics faculty take their teaching responsibilities seriously. What then is preventing these instructors from more fully adopting the results of educational research? Common suggestions [1] include satisfaction with current practices, unawareness of alternatives, and incompatibility of goals and beliefs with alternatives. This paper suggests that these factors do not predominantly explain the slow adoption of research-based strategies by introductory physics instructors and hypothesizes about other important factors.

RESEARCH METHOD AND ANALYSIS

Semi-structured, exploratory interviews were conducted with four tenured physics faculty from three different institutions who had no formal connections with the Physics Education Research (PER) community and, in our opinion, were particularly thoughtful and reflective teachers. If, as is commonly stated, the goal of the physics education reform movement is to create a critical mass of instructors using reformed pedagogical approaches, this type of instructor can be expected to form the core of that critical mass. Each interview lasted over 1 hour and contained questions about instructional goals, current and past instructional practices, attempts to change practices, and familiarity with educational research. The analysis focused on identifying (1) expressed beliefs about teaching and learning, instructional goals, self-described instructional activities, and inconsistencies between these three aspects; (2) current and past efforts to change instruction and factors that facilitated or hindered such changes; and (3) knowledge about and attitudes towards PER-compatible instructional strategies.

Instructional Goals

Each instructor was consistent in describing his or her instructional goals throughout the interview. Similar to other studies [2], all described their primary instructional goals as developing students' understanding of physics principles and ability to solve problems related to these principles. None of the instructors felt their students were fully achieving these goals, with three expressing the belief that many of their students were falling far short.

CP790, *2004 Physics Education Research Conference*, edited by J. Marx, P. Heron, and S. Franklin

PER-Compatible Beliefs and Traditional Practices

Based on the research literature [3] and our own experiences, we developed a list of 8 instructional practices (see Table 1) and 14 instructional beliefs (e.g., What should students learn?, What should be explicitly taught?) that differ between traditional instruction and PER-compatible instruction. Each author independently rated the beliefs of each instructor and then compared. Differences in ratings were minimal and were resolved through discussion.

TABLE 1. Comparison of Practices

Traditional Practices	PER Compatible Practices
Teacher ideas are center of classroom activities, students are allowed to be passive (e.g., lecturing, teacher-centered discussion).	Student ideas are center of classroom activities, students are often required to be mentally active (e.g., individual reflection, small group discussion).
Encourage or support competitive/individualist learning modes (e.g., grading on a curve, only individual assignments)	Encourage or support cooperative learning modes (e.g., grading based on preset standard, expectation of working with others)
External Motivators (grades, testing)	Internal Motivators (connections to student interests or needs)
Assess for quick and accurate performance in solving a set of familiar problems or in recalling a set of facts and principles.	Assess for thinking/problem solving skills in conjunction with meaningful understanding of facts and principles (i.e., open-ended or novel questions)
External definition of success for all students, same instruction for all, diversity seen as a problem.	Treat students as unique individuals with different needs, value all students
Grades/testing used to sort, rank, certify	Assessment for feedback
Major decisions made by teacher (content, focus, how class time spent)	Students contribute to decisions about what/how they learn (e.g., projects).
Explicitly teach only physics content.	Explicitly teach learning, thinking, and PS skills in addition to physics content.

We found that instructor beliefs were generally PER-compatible or mixed (instructor expressed both types of beliefs). At least one other study has also found that instructors have beliefs about how students learn that are more compatible with PER than with traditional instruction [4]. In contrast, three of the instructors had a majority of self-described teaching activities consistent with traditional practices. The fourth instructor described a mix between traditional and PER-compatible teaching activities. This closer alignment of practices to beliefs, however, occurred recently and is described in a separate paper [5].

For example all of the instructors expressed the belief that one of their main goals was developing students' problem solving and thinking skills. They also all agreed that the best evidence of problem solving skills (as well as an understanding of physics principles) is a student's ability to solve novel problems. In practice, however, most of the instructors explicitly taught only physics content and wrote exams that contained problems very similar (or identical) to ones students had already seen.

This can be seen in "Gary's" interview. When he was asked to describe his main goal for the course, he said: *"I think I'm teaching problem solving. And I'm probably teaching in the broader sense, I'm teaching problem solving in life as much as physics, physics is kind of incidental, almost."* Later, when he was asked how he knew when his students had developed their problem solving skills he responded: *"If they are encountering a new application they're showing problem solving skills in physics, other than just repeating a solution they've done before, they've seen me do before."* Gary had already mentioned that most of his exam questions come directly from a study guide he provides. When probed on this issue, Gary recognized the inconsistency in his belief and practice and commented: *"I know that most of my students are not learning problem solving. If I change the situation they think it's a whole new problem."*

This type of inconsistency was common. Although the instructors held predominantly PER-consistent beliefs and goals, their practices were predominantly traditional.

Instructional Change and PER

If the goals and beliefs of these instructors are generally compatible with the results of educational research, what is preventing them from using more research results in their practice? Part of this inconsistency between beliefs and practices arises, no doubt, from the difficulties involved in translating abstract beliefs and goals into concrete instructional actions. PER, however, provides many examples of how this can be accomplished.

This lack of more fully PER-compatible practices cannot be explained by unfamiliarity with PER. Three of the instructors were reasonably familiar with PER, They were aware of the names and basic practices involved with innovative curricula such as Peer Instruction, Workshop Physics, Physlets, University of Washington Tutorials, and several other research-based strategies. The fourth instructor, while not explicitly familiar with PER had been exposed to general research-based teaching techniques through a residential grant-sponsored program. All instructors were similar, however, in that none use educational research as recommended by curriculum developers. It is always modified. Sometimes these modifications are consistent with the underlying workings of the innovation, but often they are not. For example, Terry believed that *"every student has a different way of learning and my challenge, every teacher's challenge is in fact to find a variety of ways that can engage those students."* He knew about Peer Instruction [6] and described using ConcepTests in his teaching. However, he did not implement the peer interaction aspect of Peer Instruction. Thus, although he was familiar with Mazur's book [6] in which Mazur carefully describes how to manage the class and argues that peer-to-peer interaction is a good way to handle diverse student learning needs, he was not convinced enough of the usefulness or practicality of peer interaction to incorporate it into his instruction.

PER is not always well received: Although all of these instructors reported having changed their instruction, often in ways compatible with PER, it became clear in the interviews that instructors' perceptions of PER and the PER message may be preventing more rapid dissemination of PER. This section will present results from the three instructors who indicated explicit familiarity with PER.

All three instructors expressed the opinion that PER sends the message that there is only one way to teach and fails to account for different instructional situations and instructors. For example, Terry said that *"All of those people [PER researchers] seem to think that their way is the only way. . .That the only way that a student's going to learn is if I stop doing this and start doing that. And I argue that in fact that's unfair to both teachers and students. I think that . . . teachers teach well in many different ways . . . and I think that the one size fits all is not very good for the whole physics community."*

All three instructors expressed a lack of trust in certain PER results. There were three types of reasons that they gave for this lack of trust.

1) PER uses non-scientific research methods: *"Some of the research I find very useful and well done and some of it I find, uh, almost counterproductive because it seems it was done in such a fashion as to throw doubt on all the research." "All of these studies tend to be done with people who are very concerned about teaching, very interested in it, and are putting a lot of effort into their teaching at the time because they are making changes. And, all of those things alone could make a really big difference in how effective a class is."* - Mary

2) PER places too much emphasis on conceptual inventories like the FCI.

3) PER curriculum developers only present good results and do not present difficulties: *"I went to [a particular university] and learned what [the PER] group does. None of that really seemed applicable frankly. . . You know when I'd asked the graduate students who'd been teaching them, they didn't think they were all that great for that audience."* -Harry

PER may also make some instructors feel disrespected by insinuating that they are bad teachers who are harming their students. Threats to self-esteem have been identified in the literature to cause resistance to new ideas [7]. For example, Terry said that *"The first word out of their [a typical PER presenter] mouth is you're not doing things right." "If you tell me that you think my teaching is bad that automatically sets up a barrier. If I tell you that the only really good way to teach introductory physics is X, I've again set up some kind of barrier. . . I think there's just too much of that going on right now."*

Beware of language use: Another related issue that may hinder communication between PER and instructors is vocabulary. In analyzing the interviews we found several cases where instructors were using educational words with clearly different meanings than educational researchers. Words such as "coaching", "asking questions", "discussing", and "interactive" were all used to describe instructional practices that were more traditional than the language would suggest to an educational researcher. For example, Terry describes his interactive teaching style: *"I stay out in front of the bench and not behind the bench and I wave my hands at them all the time, I mean you know it's a very interactive class."* Other instructors used the word "interactive" in a similar manner that appeared to mean that the instructor was interacting with their students. It did not necessarily mean that the students were interacting back or with each other. A possible explanation for this phenomena is that these instructors are unconsciously misinterpreting their instructional practices to be more in line with their PER-compatible image of desired practices. Regardless of the cause, it suggests that educational researchers need to be aware of this issue when communicating with instructors.

DISCUSSION

Although these instructors should be ideal consumers of educational research (compatible goals, many compatible beliefs, dissatisfaction with current outcomes, and familiarity with alternatives) they indicate only modest influence of this research on their teaching. Often it is hypothesized that instructors' strong traditional beliefs about teaching and learning are the dominant factor in their resistance to implementing PER-based curricula. These instructors, however, had beliefs that were more consistent with PER than their practices. At least one other study [8] has suggested the possibility of a similar "disjunction between the stated aims (promotion of critical thinking) and educational practice (unimaginative coverage of content and testing of factual recall)" of college faculty (p. 110). We hypothesize that, while traditional beliefs do appear to play a role, they are not the dominant resistive factor. As described in this paper, more complete incorporation of PER appears to be hindered by instructors' either misinterpreting or having a low opinion of the trustworthiness of educational research results. In addition, instructors must also contend with significant external constraints on their instructional practices. These are discussed in a separate paper [5].

These instructors have beliefs and goals that are largely compatible with PER. Thus, there is no need for PER researchers to convince these instructors of the weaknesses of traditional lecturing. They are already convinced. It might be better to affirm their PER-compatible beliefs and goals and help them find ways to put these beliefs into practice. As Terry put it: *"I think . . . that we've set up in some ways a straw man. That what everybody talks about is the poor bedraggled faculty member and all that that person does, he or she, is stand at the blackboard and writes on the blackboard for 50 min of every class period. And I'll be honest, I've never taught in a place where that's happened. So I don't know if that person exists."*

These instructors were critical of the way PER results were presented and of some of the results themselves. Although it is easy to dismiss critical comments as being uninformed we would be wise to do our best to understand and correct this image problem if we expect instructors to expend the time and energy necessary to understand and use our research results.

These faculty often misinterpreted the PER message. We should be careful about trying to oversimplify the PER message in an attempt to communicate with instructors. We should also not assume that just because they use the same words as we do that they mean the same things.

Finally, these instructors did not take PER curricula and implement them "as is". They took pieces from a variety of sources that they felt were valuable and compatible with their unique instructional situation and personal conditions. Thus, we should not insist that they use the curricula that we developed, but give them information about a wide range of PER-based ideas to help them design their own instruction. Redish provides a good example [9].

Of course, the findings of a small exploratory study such as this one are appropriately used primarily to develop hypotheses for testing in larger, more focused studies. We intend to use the results in this manner.

ACKNOWLEDGEMENTS

We wish to thank the four instructors who gave their valuable time to participate in this study.

REFERENCES

1. See, for example (a) Fullan, M. *The New Meaning of Educational Change*, Teachers College Press, New York, 2001; or (b) Rogers, E. M. *Diffusion of Innovations*, Simon & Schuster, New York, 1995.
2. C. Henderson, K. Heller, P. Heller, V. Kuo, and E. Yerushalmi, "Instructors' Ideas About Problem Solving: Setting Goals" *PERC Proceedings*, 2001, pp. 79-82.
3. See for example (a) R. Barr and J. Tagg, From Teaching to Learning - A New Paradigm for Undergraduate Education, *Change*, November/December, 13-25 (1995); (b) P. Black and J. M. Atkin, *Changing the Subject: Innovations in Science, Mathematics and Technology Education,* Routledge, London, 1996; and (c) M. Prosser; and K. Trigwell, *Understanding Learning and Teaching: The Experience in Higher Education*, St. Edmundsbury Press, Great Britain, 1999.
4. P. Heller, K. Heller, C. Henderson, V. Kuo, and E. Yerushalmi, "Instructors' Beliefs and Values About Learning Problem Solving," *PERC Proceedings*, 2001, pp. 71-74.
5. M. Dancy and C. Henderson, "Beyond the Individual Instructor: Systemic Constraints in the Implementation of Research-Informed Practices," This Volume.
6. E. Mazur, *Peer instruction: A User's Manual*, Prentice Hall, Upper Saddle River, New Jersey, 1997.
7. G. Zaltman and R. Duncan, *Strategies for Planned Change*, John Wiley and Sons, New York, 1977.
8. K. Samuelowicz and J. D. Bain, Higher Education **24,** 93-111 (1992).
9. E. F. Redish, Teaching Physics with the Physics Suite, John Wiley & Sons, Hoboken, NJ, 2003.

Beyond the Individual Instructor: Systemic Constraints in the Implementation of Research-Informed Practices

Melissa H. Dancy* and Charles Henderson**

** Department of Physics, University of North Carolina at Charlotte, Charlotte, NC, 28223, USA*
***Department of Physics, Western Michigan University, Kalamazoo, MI, 49008, USA*

Abstract. Anecdotal evidence suggests that findings of educational research and resulting curricula are, at best, only marginally incorporated into introductory physics courses. Based on interviews with four non-PER physics faculty we investigated why incorporation of research-based curricula is uncommon. Elsewhere [1], we report that these instructors have PER-compatible beliefs about teaching and learning, but largely traditional instructional practices. In this paper we explore the significant role that systemic influences play in this apparent discrepancy and present a theoretical model to describe the interplay between individual beliefs and systemic influences.

INTRODUCTION

Physics education research (PER) is a thriving field. Although we still have much work to do, our collective knowledge about how students learn (and fail to learn) physics is well formed. We now have numerous research-based and class-tested curricular packages for introductory level physics. Yet, despite evidence that many physics faculty are familiar with the products of PER [1,2], we still have not witnessed wide-spread adoption of PER-informed practices [3]. Dissemination efforts in PER have concentrated on holding workshops, distributing curricular material, publishing papers and books, and giving talks. All of these efforts focus on communicating ideas to individual instructors. The implicit assumption is that research data can convince instructors of a better way to teach, and that by providing them with ready to use curricula they will be able to incorporate the results of this research into their own classrooms. In this paper, we propose that this assumption is incomplete because it fails to acknowledge that instructors do not exist in isolation, they are part of a larger system. That system exerts forces on instructors that can support, but usually resists, a shift from traditional practice to practice consistent with PER.

DATA AND ANALYSIS

Semi-structured interviews were conducted with four senior, well-respected, physics faculty who are thoughtful teachers and put significant effort into their teaching. These faculty represent the kind of faculty that would be expected to easily adopt PER methods. Elsewhere [1] we describe our finding that, although these instructors have beliefs about teaching and learning and instructional goals that are largely consistent with PER, their self-described instructional practices are largely traditional. We hypothesize that one factor impeding complete incorporation of PER is instructors' either misinterpreting or having a low opinion of the trustworthiness of educational research results. In this paper we examine this mismatch between beliefs and practices from the perspective of the situation in which these instructors teach.

Belief vs. Practice: A Jumble of Inconsistencies

Working from interview transcripts, we identified factors that either promoted or retarded the instructors' teaching in a PER-compatible manner. We found that, although there were individual factors resisting an alignment between these instructors' beliefs and practices (e.g., inconsistencies in their belief system, lack of knowledge about PER), many of these factors were situational.

For example, "Harry" described his belief in the value of having students work in groups. *"I like the idea of dividing the class into smaller sub-groups and*

CP790, *2004 Physics Education Research Conference*, edited by J. Marx, P. Heron, and S. Franklin

work independently on projects…I want to try to turn the lecture into sort of a mini tutorial at various points. I think that has promise. In all the times in the past when I've done that, when I've gotten students to organize into small groups and talk to each other, at least they're talking physics to each other. You can see that there is some understanding going on, some transfer of knowledge taking place."

Although Harry believed it was beneficial for students to work in groups and had positive experiences with this method, he did not use the method in his practice as often as he would like: *"Because I was racing to get through the curriculum I had to pretty much drop [the group work]."*

Harry repeatedly talks in his interview about feeling pressure to present material, rather than use interactive methods, due to a need to cover content. He also talked about students' resistance to working in groups and a lecture room with fixed seating that is *"not ideally configured for group work"*.

The finding of inconsistencies between beliefs and practice in and of itself is not particularly surprising. Sociologists [4] and educational researchers [5] have long been aware that beliefs are generally a poor predictor of practice. Based on a theoretical model developed by Warner and DeFleur to understand the discrepancies between stated attitudes and behaviors related to racial discrimination [4], we suggest a similar theoretical framework for understanding these instructors' inconsistencies (Figure 1). In this model, practice is consistent with belief when situational variables support the practice but may be inconsistent when situational variables are in opposition to a particular practice. For example, Harry has beliefs that might be characterized as being moderately PER-compatible. Given that his practice is often traditional, and inconsistent, the model predicts the existence of the incompatible traditional situational variables described above.

It was striking that in every case, when there was an inconsistency between belief and practice, the instructor had PER-consistent beliefs but traditional practice. We found no instance of traditional beliefs and PER-consistent practice. This suggests that these instructors work in environments that are strongly supportive of traditional instruction and resist movement toward PER-consistent practice.

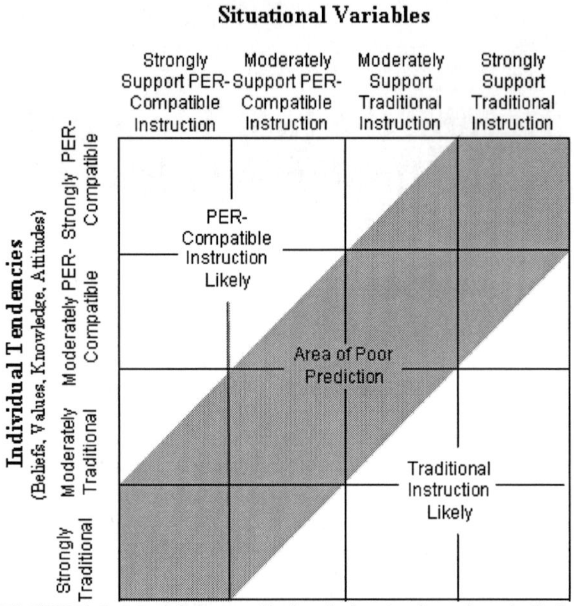

FIGURE 1. Model for predicting behavior based on beliefs and context. Adapted from Warner et. al. [4] p. 168.

Main Systemic Forces

To better understand an instructor's ability to implement PER-compatible instruction, we asked the interviewees to describe actual or attempted instructional changes throughout their careers. This allowed us to create a "free-body diagram" for each instructor identifying the main forces that either supported or resisted their attempts to change their instruction. Many of these forces were specific to the instructor, for example whether they held transmission ideas about learning or ideas based on constructivism. But many of the forces were systemic.

It is important to note that our data only illuminates self-reported resistive forces. It is likely that there are large resistive forces that are not noticeable because they are so pervasive. For example, the process of grading commands considerable time and attention in most classrooms and the requirement that an instructor give a final grade to each student must therefore significantly affect instruction. However, because the practice of giving grades is so pervasive and generally unquestioned, most instructors probably have not considered how this systemic requirement affects their practice. Since the system was built around and supports traditional practices, systemic constraints are likely only noticed by instructors when they attempt to move out of the traditional mode of instruction.

A discussion of the most common systemic forces found in our interviews follows.

Student Resistance: Students often do not support research-based methods. In particular, they do not like to interact with each other and are not prepared to think independently.

"What I want to do is to turn the class into a real working session. Where it's just not possible for them to come there and sleep. That may turn off students and decrease enrollment, they may switch courses. I'm a little worried about attrition. That's another aspect" .- Harry

Time Structure: Semesters are of a fixed length of time and do not allow for individual differences in learning needs. Also, since students are taking other courses the time they have available for one course is limited.

"I think time students can spend on a particular course is one thing [that prevents me from reaching my goals]. Time for every student in the course to reach the same level because they all start at different levels and they would all take different amounts of time to get to whatever level you want to call understanding." - Terry

Departmental Norms: If other members of the department are integrating research-based methods it is easier for instructors to do so as well. It is much more difficult if traditional methods are the norm and there are no local role models to follow or be supportive.

"I am more comfortable with being more interactive and, of course, since we've started [a grant supported departmental reform]. I'm much more comfortable having them do group work in class, and feeling that that's a valid way of spending time in class. And I'm more comfortable asking conceptual type questions instead of just problem solving type questions because you know there's that extra validation of having a group of people doing this and that it is a grant and it's a research project."- Mary

Expectations of Content Coverage: Instructors may forgo research-based methods that are geared toward deep understanding if they feel they must cover a lot of material. Likewise, they may change their instruction if this expectation is diminished.

"The fact that we cut out a lot of the material that we need to cover. Because before, I'd think gee if I don't cover fluids and the next instructor is expecting it I'm really crippling these students, handicapping them. But as a whole department we said OK, it's alright for us to cut this material out and spend the time on what you feel is necessary to go more in depth on.... And so the pace was so much quicker that to take a whole class period and potentially have them be a little floundering with group work was just so big of a risk. You know I would have them do some, but it was much more focused and shorter periods of time and I was still much more tentative about how many of them I ended up doing." - Mary

Lack of Instructor Time: Instructors are sometimes too busy with large teaching loads and/or research responsibilities to have the time to learn about and integrate new techniques.

"It kinda depends on how lazy I am, I will try to write those [test questions that students have not seen before] as much as possible. If I'm in a hurry then I will tend to pick more from the old questions." - Gary

Systemic Change Affecting Practice

If our theoretical model (Figure 1) is correct, when systemic resistive forces are removed, then instructors with PER-compatible beliefs should change their practice. We found evidence of this in our interview with Mary, who indicated that she had always held predominantly PER-compatible beliefs but was better able to follow through with them when the structural forces changed. Her two comments in the previous section illustrate her increased confidence when "Departmental Norms" changed due to a new program and her increased ability to implement changes when "Expectations of Content Coverage" changed.

This instructor described her changes in instruction as being precipitated by situational changes, rather than changes in personal beliefs.

"I would say that it's not just one thing. There've got to be at least three things. It was the release of time so that I had more flexibility in how to cover a lesser amount of material more in depth. Two that there is a group here doing it. And three that I was exposed to more research on how [cooperative learning] works." - Mary

IMPLICATIONS

If a goal of PER is to impact mainstream physics teaching, it is not enough to simply research how students learn physics, develop curriculum based on that research and then disseminate the results. We found that even if instructors believe in PER-informed practices, there are systemic forces that resist the adoption of those practices. This result is consistent with others reports [6].

Our interviewees held moderately to strongly PER-compatible beliefs. Thus, the theoretical model (Figure 1) predicts PER-compatible instruction if they are in a setting where the situational forces are at least neutral. As discussed earlier, we saw evidence of such a change in Mary's instructional practices when her situational variables became more PER-compatible.

The PER community has focused a majority of its efforts on moving instructors' individual tendencies to become more PER-compatible. For instructors like the ones that we interviewed, we argue that changing their beliefs from moderately to strongly PER-compatible is not only very difficult but also no guarantee of change in practices. We suggest that some of the emphasis be placed on attempting to understand, classify, and change the systemic forces that inhibit changes in instructor practices.

History [7] and the current situation tell us that the structure of educational systems is often not strongly based on educational research, but rather on "such factors as real or alleged economic considerations, national ideals, social change, the way in which schools are structured, gender, racial, and class distinctions, as well as symbol and ritual" (p. 157). Since successful systemic changes are more likely if all of these factors are considered, we will conclude this paper by offering some brief thoughts based on a historical [7,8] and structural analysis.

Historically, educational reform movements in the United States have followed major shifts in the economic and social structure of society. Public schools were formed, in large part, to address concerns about the "proper" socialization of the large immigrant population and also to meet the need for factory workers. Therefore, early pedagogical methods were often authoritarian and teacher-centered in nature.

Around the turn of the century, there were two competing educational reform movements that parallel our current situation. The progressive movement (associated with John Dewey) which called for more humanized, child-centered education and the social efficiency movement which pushed for structures such as testing and tracking. The social efficiency movement was backed by business leaders and directed at changing policy. In contrast, the progressive movement (like PER) did not focus directly on policy, but worked to change the practice of individual teachers. In the end, the progressive movement had a considerably smaller impact on actual practice.

It may be argued that the system is changing ... just slowly. However, the historical account indicates strong connections between educational policy (and therefore practice) and economic reality. Others have explored this connection more in depth [8a, 9] and argue that our current system is the result of neither incompetence, accident, nor slow changes.

Our finding, that systemic forces play a large resistive role in research-based change, indicates that the nature of "traditional" structures must be critically questioned. It is not enough to develop and disseminate curriculum, we must work toward understanding and changing the structures that oppose research-based reforms.

REFERENCES AND NOTES

1. C. Henderson, and M. Dancy, *Teaching, Learning and Physics Education Research: Views from Mainstream Faculty*, This Volume.
2. H. Demarest, Examining Teacher Expectations about Physics Homework, 2004. Available at www.colorado.edu/physics/EducationIssues/papers/Heather_Thesis_Final.pdf
3. It is not known exactly how integrated PER practices are into the mainstream. We intend to survey physics faculty to find out, among other things, what practices they are aware of, what they have tried, and what they still use.
4. L. G. Warner, and M. L. de Fleur Attitude as an interactional concept. *American Sociological Review* **34** p. 168, 1969.
5. J. Calderhead, Teachers: Beliefs and knowledge. In D. C. Berliner and C. R. Calfee (Eds.), *Handbook of Educational Psychology*. New York: Prentice Hall, pp. 709-725, 1987.
6. A. Lumpe, C. Czerniak, and J. Haney, Science teaches beliefs and intentions regarding the use of cooperative learning. *School Science and Mathematics* **98** (3), 123-135, 1998.
7. H. M. Kliebard, Constructing A History of The American Curriculum, in P. W. Jackson (Ed.), *Handbook of Research on Curriculum*. New York: Macmillan, pp. 157-184, 1992.
8. For more complete historical accounts see: (a) M. Carnoy, and H. M. Levin, *The Limits of Educational Reform*. David McKay Company: New York, 1976; and (b) J. L. Rury, Education and Social Change: Themes in the History of American Schooling. Mahwah, NJ: Erlbaum, 2002.
9. (a) S. Bowles, and H. Gintis, *Schooling in Capitalist America*. New York: Basic Books, 1976; or (b) S. Rosenberg, *Questioning Assumptions about the Role of Education in American Society: A Review of Schooling in Capitalist America*, PERC Proceedings, 2003.

Using the System Schema Representational Tool to Promote Student Understanding of Newton's Third Law

Brant E. Hinrichs

Department of Physics, Drury University, 900 N. Benton Ave., Springfield, MO 65802

Abstract. The Modeling Instruction program at Arizona State University has developed a representational tool, called a system schema, to help students make a first level of abstraction of an actual physical situation [1]. A system schema consists of identifying and labeling all objects of interest from a given physical situation, as well as all the different types of interactions between the objects. Given all the relevant objects and their interactions, students can explicitly identify which are part of their system and which are not, and then go on to model the interactions affecting their choice of system as either (i) mechanisms for energy transfer, or (ii) forces being exerted. In this paper, I describe the system schema tool, give examples of its use in the context of forces, and present some evidence on its effectiveness in helping students understand Newton's Third Law.

INTRODUCTION

A strength of the Modeling Instruction (MI) curriculum developed by Arizona State University (ASU) is that it provides students with some basic representational tools for modeling physical objects and processes [1]. Quality tools are seen as vital for helping students build quality scientific models, a central goal of MI [2]. In addition, students who are comfortable using a range of tools in problem solving better approximate physics experts, who routinely create many different representations (e.g. mathematical, graphical, diagrammatic) in the analysis of a single problem [3,4].

Examples of representational tools include: graphs, motion maps, vector diagrams, system schema, pie charts, bar charts, free-body diagrams, etc. [3]. This paper focuses on a description, implementation, and evaluation of just one, the system schema. The second section of the paper describes system schema and gives several illustrative examples. The third section discusses how system schema were implemented in a calculus-based introductory physics course taken by all science majors at Drury University. The last section presents and discusses data from a portion of the Force Concept Inventory (FCI) [5,6], as evidence for the effectiveness of the system schema in helping students understand Newton's Third Law. See also [7] (and especially Jiménez 1999 and 2001 in that paper) for other recent related work in this area.

WHAT IS A SYSTEM SCHEMA?

A system schema is used to represent an actual physical situation and is the first level of abstraction after a pictorial representation. It serves as a conceptual bridge for students to more abstract representations like free-body diagrams and Newton's Laws. The Physics Teacher has a recent article that details typical implementation of system schema in high school modeling instruction [8]. This paper will instead focus on how system schema have been used in university level introductory courses [9]. There are differences between the two approaches (see below) but it is not the purpose of this paper to highlight them.

Figs. 1 and 2 show typical introductory university level physical situations, corresponding system schema, and free-body diagrams for specified systems within each schema. Figs. 1 and 2 show a static and dynamic situation, respectively.

Although it is not an exhaustive list, here are many of the basic rules that guide students in constructing system schema from a verbal or pictorial representation. (1) All objects are represented without any details of their shape or structure. (2) Two objects interact if they influence one another [10]. For example, in Fig. 1, the book influences the brick (it holds it up) and the brick influences the book (it squashes it a little). (3) Therefore, interactions between objects are represented by two-headed arrows, and labeled with the type of interaction. If object X influences object Y, object Y also influences object X.

CP790, *2004 Physics Education Research Conference*, edited by J. Marx, P. Heron, and S. Franklin

This is a built in mechanism that later helps students "explain" Newton's 3rd Law. (4) The introductory level course only deals with four interactions, namely three types of non-contact (gravitational "g", electric "e", and magnetic "m") and one type of contact ("c"). All contact interactions (i.e. normal, frictional, etc.) are lumped into one type, as compared to reference [8] which separates them out. A single type of contact is preferred because it simplifies the process of constructing system schemas, simplifies using them to help construct free-body diagrams (see (10) below), and simplifies the physics model of what an interaction is. (5) Double-headed interaction arrows are drawn as solid lines if the interaction persists for the entire time interval the system schema is of interest; otherwise they are drawn as dashed lines (See Fig. 2b). (6) The gravitational interaction only occurs between a huge object, like the earth or moon, and any other object. (7) A system boundary is represented by drawing a dashed line around one or more objects.

If forces are to be studied, here are rules students use to construct free-body diagrams from a completed system schema. (8) Each object or set of objects for which Newton's Laws are to be written must be identified as a separate system [10] (see Fig. 1). (9) Define Force as a description of an interaction between two objects. (10) For each interaction that crosses a system boundary, there should be one and only one force on the corresponding free-body diagram for that system. Students find this rule enormously helpful especially for complicated systems involving multiple objects with multiple interactions. (11) The label for a force vector in a free-body diagram (e.g. $\vec{F}^{c}_{B\rightarrow R}$) is constructed using the relevant labels from the system schema. See the caption of Fig. 1c for an illustrative example.

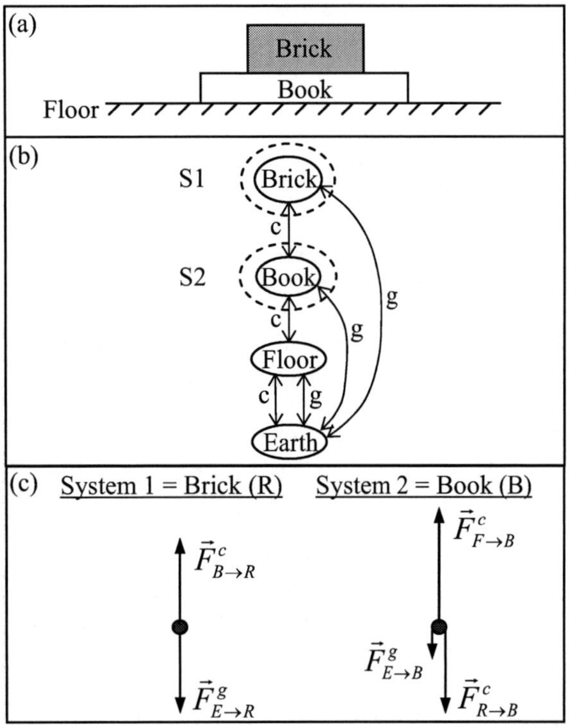

FIGURE 1. (a) Pictorial representation of a physical situation. All objects are at rest. (b) System schema of this physical situation, with two of many possible systems identified [10]. The dashed ellipses represent system 1 (S1) and system 2 (S2) respectively. "c" labels a contact interaction, and "g" labels a gravitational interaction. (c) Free-body diagrams for the two systems identified in (b). In a force label "c" means contact, "g" means gravitational. Also, for this particular scenario "B" means book, "R" means brick, "E" means earth, and "F" means floor. For example, the symbol $\vec{F}^{c}_{B\rightarrow R}$ means a contact force by the book on the brick. Note the mass of the brick has arbitrarily been chosen to be three times the mass of the book, thus $\vec{F}^{g}_{E\rightarrow R}$ is three times the length of $\vec{F}^{g}_{E\rightarrow B}$.

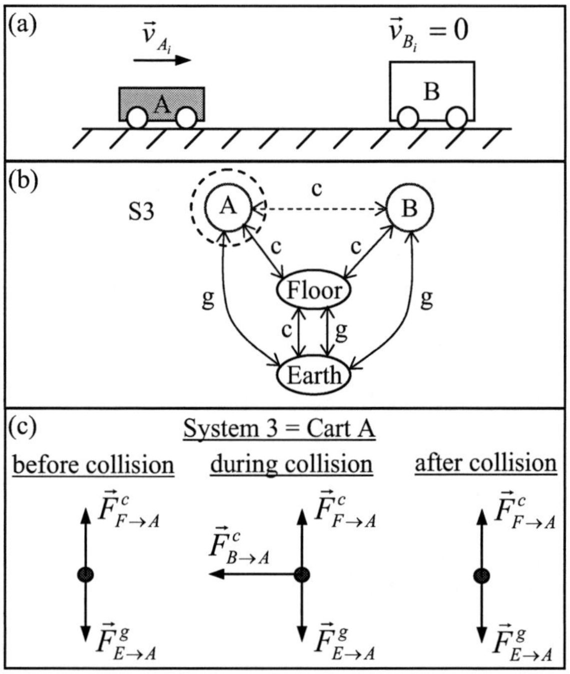

FIGURE 2. (a) Pictorial representation of an experiment with two low friction carts of differing masses. Cart A is launched toward cart B, which is initially at rest. After the collision (not shown) cart A has reversed direction (not shown). (b) A single system schema for this complete experiment, with one of many possible systems identified. The dashed circle represents system 3 (S3). Note that in contrast to Fig. 1b, this system schema is for a physical situation that changes over time. Before, during, and after the collision is represented here by a single system schema. Therefore, the contact interaction between cart A and cart B is drawn as a dotted line rather than a solid line to indicate that it is not present for the entire time interval of interest of the schema. (c) Free-body diagrams for System 3 at different instants in time in the experiment. Force labels follow the same conventions described in Fig. 1c.

TABLE 1. Individual class performance on the Force Concept Inventory (FCI) with and without use of system schema.

System Schema	Fall Semester	Number of Students (N)[1]	Matched Pre-test Avgs FCI (29)[2]	Matched Pre-test Avgs FCI 3rd Law (4)[3]	Matched Post-test Avgs FCI 3rd Law (4)[3]
no	1999	17	11.1 ± 3.7	1.2 ± 1.1	2.8 ± 1.2
no	2000	14	12.1 ± 5.6	1.1 ± 1.0	2.7 ± 1.2
transition	2001	9	7.3 ± 3.1	0.9 ± 0.8	2.3 ± 1.4
yes	2002	15	10.2 ± 2.7	1.1 ± 1.0	3.6 ± 1.1
yes	2003	13	11.3 ± 3.3	1.1 ± 1.0	3.8 ± 1.0

CLASSROOM IMPLEMENTATION

In 1999, I began teaching at Drury University and used the Workshop Physics (WP) curriculum [11,12] for the first time. Because I was disappointed in my student's understanding of the 3rd Law, I worked especially hard in 2000 to help them improve, but their scores remained flat (see Table 1, years 1999, 2000).

At a summer 2001 workshop at ASU I was introduced to modeling instruction for the university-level introductory course [9], discourse management [13], and system schema. Fall 2001 I switched from WP to using those methods. As it was a transition period in my teaching, FCI data from that year are not included in any subsequent composite analysis.

The last two years (2002-03) I used the following procedure in teaching Newton's Third Law. Shown the physical situation of Fig. 1a, students groups were asked to predict how the contact force by a 10kg (lead) brick on a 3kg book compared to the force by the book on the brick. They defended their answer with a system schema. Occasionally one group but frequently none came up with the "physics" answer. Most predicted the brick would exert a greater force because it was so much heavier. An instructor demo using force plates to measure the two forces revealed that they are in fact equal in magnitude but opposite in direction. At first, students were puzzled, but then at least one group always noticed that "this makes sense since $\vec{F}^{c}_{R\to B}$ and $\vec{F}^{c}_{B\to R}$ are the same interaction." The rest of the class quickly agreed and most seemed satisfied. Here then is the power of the system schema. It gives a visual and conceptual representation, in terms of objects and interactions, for the experimental results showing the 3rd law.

In a later unit on collisions, student groups were shown physical situations similar to Fig. 2a and asked to predict how the force by cart A on B compared to the force by cart B on A. They were asked to make predictions for a range of scenarios, varying both the masses and initial velocities of the two carts. Usually a quarter to a third of groups predicted the "physics" answer. Often these were groups that used a system schema to guide their thinking. Groups then performed experiments using a track, low friction carts, force probes, and a computer interface. Students who predicted incorrectly were initially puzzled by data indicating equal but opposite forces. But when asked whether or not their data was consistent with a system schema, they would draw one, and a light bulb would go off: "System schema even work for collisions! They've got to be the same size because they're the same interaction." Here is, I claim, a conceptual understanding reinforced by a representational tool. Lastly, though most students seemed to understand their data, to reinforce their "aha" moments, the instructor led a short class discussion on Elby's idea of refining our intuition about "reacts more" in a collision to correspond with change in velocity rather than with force [14].

RESULTS AND DISCUSSION

Student understanding of Newton's Third Law was measured using the Force Concept Inventory (FCI) [5,6]. The FCI was given as a pre-test the first day of class, and then as a post-test at the end of mechanics on the final exam, or on an exam sometime in the second semester during electricity & magnetism. Only data from students who took both the pre-test and post-test are reported here (i.e. the data is matched). All data is from the same instructor, and from the calculus-based introductory physics course taken every year by all science majors at Drury University.

Table 2 shows two composite sets of data. The first row is for classes from 1999 and 2000 that used Workshop Physics (WP) (N=31). The second row is for classes from 2002 and 2003 that used modeling, system schema, and discourse management (N=28).

The pre-test data for both the entire FCI, as well as the four 3rd Law questions, show that students in both populations were statistically similar at the start of the semester. Post-test data show that students in the second population performed much better than the first

[1] Class size each year was 19, 14, 18, 18, and 19 respectively.
[2] For historical reasons, Drury uses the original 29 item FCI found in the appendix of [5].
[3] Four FCI Questions - 2, 11, 13, and 14 - test Newton's Third Law.

TABLE 2. Composite student performance on the FCI, with and without use of system schema in instruction

System Schema	Fall Semester	Number of Students (N)	Matched Pre-test Avgs FCI (29)[1]	Matched Pre-test Avgs FCI 3rd Law (4)[2]	Matched Post-test Avgs FCI 3rd Law (4)[2]
no	1999, 2000	31	11.6 ± 4.6	1.2 ± 1.0	2.8 ± 1.2
yes	2002, 2003	28	10.7 ± 3.0	1.1 ± 1.0	3.7 ± 0.8

on the four 3rd Law questions. Levene's Test for Equality of Variances indicated that the variances of the post-test data were different [15], so a 1-tailed, unequal-variance student's t-test was run in Excel. It indicated that we may safely reject the null-hypothesis that the difference in the means is due to random chance (p=.0003 at the 95% confidence level). That is, the difference *is* statistically significant.

Table 2 shows that modeling, discourse management, and system schema (MDMSS) can lead to excellent scores on the FCI 3rd Law questions. It's possible that students did better in 2002-03 than in 1999-2000 because I was less experienced and confident in my teaching my first years at Drury. However, I was unable to make any improvements in two years of using WP, but I was able to with MDMSS (see Table 1, years 2001 and 2002). Perhaps I just "get" MDMSS better than I "get" WP. Indeed, some recent unpublished data suggest that more experienced WP instructors have data whose means are midway between the two reported here [16].

Although Table 2 does not show if system schema were the primary reason that students did better, a case can be made that it in particular helps students learn Newton's 3rd Law. This is because WP and MDMSS are very similar except for the emphasis in MDMSS on use of representational tools, and especially system schema. Both are similar in their approach to the classroom, emphasizing students working in groups of 3-4, real-time data collection to build models, and a 3-cycle pedagogy of predict, experiment, and resolve-differences. And both are similar in their approach to teaching the 3rd Law as well. Activities based on Fig. 2 in MDMSS are almost identical in style and substance to those in WP. A small difference is that MDMSS also includes the activity based on Fig. 1.

A main difference between MDMSS and WP is an emphasis on system schema. Students start in the third week of the class using system schema to qualitatively model energy concepts [3]. Long before the concept of force has been introduced, students get comfortable talking about objects and interactions. Further, I suggest that spontaneous comments by students about how 3rd Law force pairs are the same size because they are "from the same interaction" indicate something more is going on than just appeal to experimental results. They suddenly seem more satisfied with their experimental results, perhaps because there is a "reasonable" mechanism to "explain" them. Perhaps by the time forces are studied, the system schema has become a valuable epistemological resource to them [14]. It would be an interesting experiment, now that I am more experienced, to return to teaching with WP and see how well my students did with the 3rd Law. And then try teaching with WP but including extensive use of system schema as well.

ACKNOWLEDGMENTS

Kathy Coy, Kerry Leibowitz, Dave Yount helped mightily with statistics. Miki patiently encouraged. Mani supported and provided extensive critical review.

REFERENCES

1. M. Wells, D. Hestenes, and G. Swackhamer, "A Modeling Method for High School Physics Instruction," *Am. J. Phys.* **63**, 606-619 (1995).
2. D. Hestenes, "Toward a Modeling Theory of Physics Instruction," *Am. J. Phys.* **55**, 440-454 (1987).
3. E. T. Brewe, "Inclusion of the Energy Thread in the Introductory Physics Curriculum: An Example of Long-Term Conceptual and Thematic Coherence," Ph.D. Thesis, Arizona State University, 2003.
4. F. Reif and J. I. Heller, "Knowledge structures and problem solving in physicists", *Educ. Psychol.* **17,** 102-127 (1982).
5. D. Hestenes, M. Wells, and G. Swackhamer, "Force Concept Inventory," *Phys. Teach.,* **30**, 141-158 (1992).
6. http://modeling.asu.edu/R&E/Research.html
7. A. Savinainen, P. Scott, and J. Viiri, "Using a bridging representation and social interactions to foster conceptual change," *Sci. Educ.* **89**, 175-195 (2005).
8. L. Turner, "System Schemas," *Phys.Teach.* **41,** 404-408 (2003).
9. http://modeling.asu.edu/rup_workshop/
10. A. B. Arons, *Teaching Introductory Physics,* (John Wiley & Sons Inc., New York, 1997), Part III, pp. 45-47.
11. P. Laws, "Millikan Lecture: Promoting Active Learning Based on Physics Education Research in Introductory Physics Courses," *Am. J. Phys.* **65**, 14-21 (1997).
12. P. Laws, *Workshop Physics Activity Guide,* (John Wiley & Sons, New York, 1997). 855 pp.
13. D. M. Desbien, "Modeling Discourse Management Compared to Other Classroom Management Styles in University Physics," Ph.D. Thesis, ASU, 2002.
14. D. Hammer, "Student Resources for Learning Introductory Physics." *Am. J. Phys.,* **68**, S52-S59 (2000).
15. K. Coy, private communication, July 2004.
16. B. Callen, private communication, July 2004.

Representational Format, Student Choice, and Problem Solving in Physics

P. B. Kohl and N. D. Finkelstein

Department of Physics, University of Colorado, Campus Box 390
Boulder, CO 80309

Abstract. Student problem-solving ability appears to be tied to the representational format of the problem (math, pictorial, graphical, verbal) [1]. In a study of a 367-student algebra-based physics class, we examine student problem solving ability on homework problems given in four different representational formats, with problems as close to isomorphic as possible. In addition, we examine students' capacity for assessing their own representational competence by giving follow-up quizzes in which the students can choose between various problem formats. We report student performance and consider factors that may influence their ability or choices. As a control, part of the class was assigned a random-format follow-up quiz where students received quiz formats at random. We find that there are statistically significant performance differences between isomorphic problems. We also find that allowing students to choose which representational format they use improves student performance under some circumstances and degrades it in others.

INTRODUCTION

Student competence with different representational formats is a popular topic in modern science and mathematics education. By 'representational formats' we refer to the many ways in which a particular concept or problem can be expressed. Scientists have to be able to interpret all of these formats effectively and integrate and translate among them. As a result, a possible instructional goal is to instill this representational facility in science students. In physics education research, there have been several studies in which students are explicitly taught to handle multiple representations of the same topic [2,3,4]. Other studies have addressed students meta-representational competence [5], in particular student generation and critique of science and math representations [6,7].

In this study, we directly compare student performance on different representational formats in the style of Meltzer [1]. We also attempt to broaden the examination by investigating whether students can assess their own representational competence, what motives they have for handling a problem in a particular representational format given a choice of formats, and whether providing this choice affects their performance.

One can categorize representations according to whether they are formal or informal, abstract or concrete, or text-based versus graphics-based, to name just a few. Studies involving representational formats have taken many approaches to this division, including comparisons of mathematical problems couched in words to those stated primarily in equations [8], comparisons of learning environments that are virtual to those that are concrete [9], and comparisons between verbal, mathematical, graphical, and diagrammatic formats [1]. Here we take the last of these approaches and divide our study problems into verbal, mathematical, graphical, and pictorial formats.

METHODS

We administered our study in recitation sections to a large (367 student) traditionally taught second-semester algebra-based physics class at the University of Colorado at Boulder. We performed the study in two different subject areas: wave optics and atomic physics. In each subject area, the students were assigned four multiple-choice homework questions that covered the same concept in four different representational formats. They were then assigned a one-question multiple-choice quiz in recitation. Approximately 170 students in nine

CP790, *2004 Physics Education Research Conference*, edited by J. Marx, P. Heron, and S. Franklin

Question 3 – Graphical

An electron in a Bohr hydrogen atom jumps from the n=3 orbit to the n=2 orbit. The following graphs show the orbit radius r as a function of the orbit number n. Choose thegraph that best representsthe relative locations of theelectron orbits

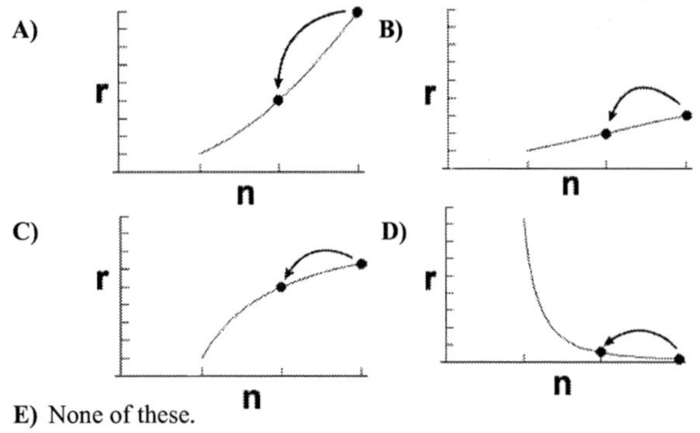

E) None of these.

Question 4 – Pictorial

An electron in a Bohr hydrogen atom jumps from the n=3 orbit to the n=1 orbit. Choose the picture that best representthe relative locations of the electron orbits.

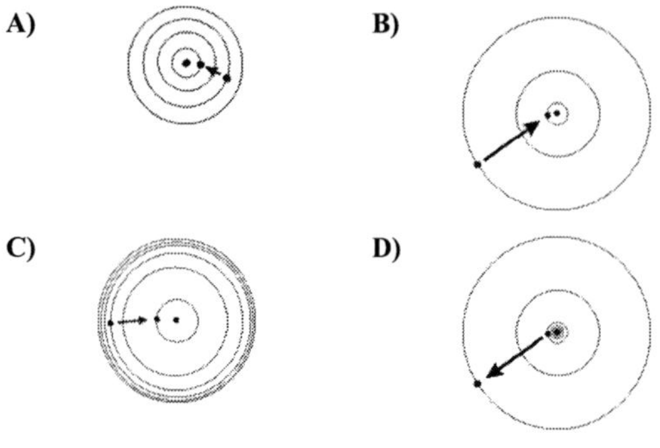

E) None of these.

FIGURE 1. Isomorphic homework problems (in graphical and pictorial/diagrammatic formats) regarding Bohr-model electron orbit radii.

sections were allowed to choose a representational format for their quiz without examining the quizzes beforehand. As a control, seventy students in four sections received quizzes in random formats.

We attempted to make the homework and quiz problems as isomorphic as possible across representational formats; though the mapping between formats cannot be perfect or the problems would be identical. It is the case that for some representation pairs the mapping is more complete than for others. It is also worth noting that we use the word 'isomorphic' to mean isomorphic from the point of view of a physicist. A student may have a different view of the similarity (or difference) between these problems [10].

DATA

The above trials provided a wealth of data, primarily in the form of student success rates on the different problems. Various examinations are possible, including but not limited to comparisons of student performances on isomorphic problems in different formats and comparisons of student performance in choice and random-assignment (control) recitation sections. We consider some of these below, with full analysis reserved for a longer paper (in preparation).

Performance Across Representational Format

We shall first look at examples of varying student performance on problems of different representational format that are otherwise very similar. Consider the pair of questions shown in Figure 1. Both require knowledge of how the electron orbit radius varies with the principal quantum number in the Bohr model. The questions differ only in which specific transition is being presented and in whether the problem and solutions are expressed in graphs or pictures/diagrams. Of the 218 students who answered both problems, 77% answered the graphical problem correctly (A) and 62% answered the pictorial problem correctly (B). This difference is statistically significant ($p = 0.006$, 2-tailed binomial test) and is particularly interesting in that the graphical representation is a rather non-standard one. Students had not yet seen any graphs of orbital radius versus quantum number, but the pictorial representation of electron orbits should be somewhat familiar since it is featured in both the textbook and the lectures. Further examination of the individual student answers on these two questions indicates that this performance difference can be attributed almost entirely to the 36 students who answered the graphical problem correctly and missed the pictorial problem by choosing the distractor C. This distractor bears a strong resemblance to the energy-level diagrams seen in the Bohr model section of the text and lectures. It appears that in this case representation-based performance variations may be traceable to a very topic-dependent cueing on visual features of one of the problems.

Another example of performance variation across isomorphic problems exists in the second quiz given, which deals with the emission spectrum of a Bohr-model hydrogen atom. The students were prompted to recall the spectrum of hydrogen, and were asked how

Spectroscopy Problem -- Pictorial Format

The Balmer series of spectral lines is shown below, as seen through a spectrometer:

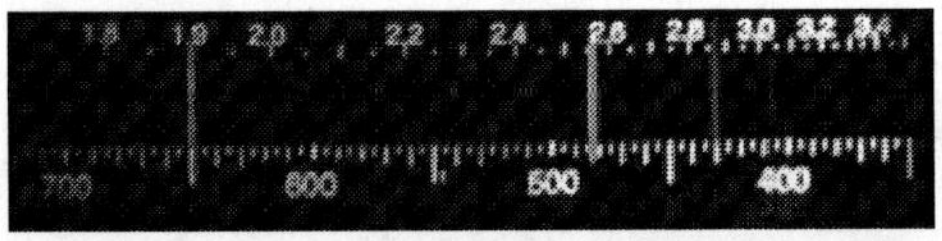

Now suppose we are in a world where electric charges are weaker, so the electron is not held as tightly by the nucleus and the ionization energy is 13 eV instead of 13.6 eV. Choose the picture that best represents what the new spectrum would look like.

B)

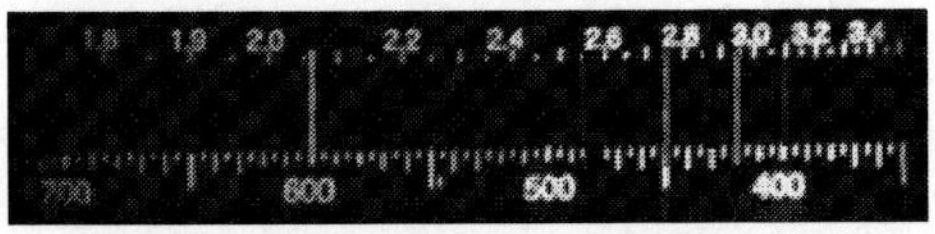

Spectroscopy Problem -- Verbal Format

Consider the Balmer series of spectral lines from hydrogen gas. Now suppose we are in a world where electric charges are weaker, so the electron is not held as tightly by the nucleus. This means that the ionization energy for the electron will be smaller. What will happen to the Balmer lines that we see?

B) The spectral lines will all shift to shorter wavelengths (toward the bluer colors).

FIGURE 2. Setup and second answer choice for the verbal and pictorial format quizzes given in the second trial. The other distractors match up between the formats as well.

that spectrum would change if the binding of the electron to the nucleus were weaker.

Figure 2 shows the problem statement and one distractor for the verbal and pictorial formats. Note that one week previous to the quiz, students had a lab covering emission spectroscopy, and the quiz images match what students would have seen through their equipment. Nineteen students in the control group were randomly assigned a verbal format quiz, and 18 were assigned a pictorial format quiz. 32% of the verbal group answered the question correctly, while 83% of the pictorial group answered correctly. This difference is significant at the $p = 0.0014$ level. Answer breakdowns indicate that the students in the verbal group that missed the question almost always chose the distractor corresponding to the spectral lines moving in the wrong direction (shown in Figure 2), whereas only one student from the pictorial group made this error.

Effects of Student Choice on Representational Performance

We next examine how providing students a choice of representational format affected their quiz performance, as compared to students who received a randomly assigned quiz format. There were a total of eight choice/control comparisons available (two trials with four formats each). Of these eight, six showed a statistically significant performance difference. These data are summarized in Table 1.

These results are notable in that the effects are in some cases quite strong. For instance, 90% of the 'choice' group answered the math format question correctly for the atomic physics quiz, while 13% of the control group answered the same problem correctly. In addition, the direction of the effect can vary. In four of the six cases, giving students a choice of formats significantly increased performance, while in two of the six cases it resulted in a significant decrease. Furthermore, when comparing across content areas we see reversals in the direction of the effect. On the wave optics quiz, students in the choice group do better than the control group on the pictorial representational format and worse on the graphical representational format, while on the atomic physics quiz the students in the choice group do worse on the pictorial representation and better on the graphical representation.

DISCUSSION AND CONCLUSION

Explaining the performance differences between choice and control groups is non-trivial at this point. In breaking down the answers problem by problem, we see that in some cases the performance differences can be attributed a particular distractor, as in the case of the atomic physics homework questions. In other cases, the group that performed more poorly does not show a preference for a particular distractor as compared to the higher performing group. Examining the student comments, we do see that the students who chose the pictorial format were much more likely to cite a recent lab in making their decision, despite the fact that the labs included all four of the representations from each quiz. However, this also varied by subject, as students taking the pictorial quiz

TABLE 1. Fraction of students answering a quiz problem correctly, broken down by representational format and choice/random assignment section. P-values compare success rate of choice and control groups (binomial test, 2-tailed). Quizzes that had no statistically significant performance differences are not shown.

	Wave Optics			**Atomic Physics**		
	Graphical	Pictorial	Verbal	Math	Graphical	Pictorial
Choice	0.04 (N=26)	0.82 (N=72)	0.81 (N=21)	0.90 (N=42)	0.96 (N=28)	0.39 (N=58)
Control	0.25 (N=16)	0.53 (N=19)	0.32 (N=19)	0.13 (N=15)	0.53 (N=17)	0.83 (N=18)
	$p = 0.04$	$p = 0.03$	$p = 0.002$	$p < 0.0001$	$p = 0.0004$	$p = 0.0012$

over diffraction were more likely to reference the lab than students taking the pictorial quiz over atomic physics.

Our results provide examples of performance differences on isomorphic problems. They also indicate that representational performance can vary strongly with subject area and with other factors such as student choice. Judging from these data, it is possible that people have underestimated the severity of the effect of representational format on student performance. Also, while the performance variations we observe are difficult to explain, their arrangement suggests that the effect is more complicated than a simple alignment of student choices with some individual learning style. When we compare the wave optics and atomic physics quiz performances of the same students choosing the same representational formats, we see substantial performance differences. For example, many of the students who chose a pictorial format quiz for the first subject area chose it again for the second area. Despite this overlap, the choice group substantially outperformed the control group on the wave optics class but was then outperformed by this same control group on the atomic physics quiz. It appears that student representational performance may be heavily influenced by the context of the problems and of the course in which they are presented.

Future work will include student interviews to probe students' problem solving strategies on these quizzes as well as to further elucidate their reasons for choosing the formats that they do. We hope that this and other work will contribute to a more complete characterization of physics students' representational skills and meta-skills, perhaps with implications for building this competence through explicit instruction.

ACKNOWLEDGEMENTS

This work was supported in part by an NSF Graduate Fellowship. Special thanks to the rest of the Physics Education Research group at the University of Colorado at Boulder, and to the PERC Proceedings reviewers.

REFERENCES

1. Meltzer, D. E. (2005). Relation between students' problem-solving performance and representational format. *Am. J. Phys*, **73**(5). 463-478.
2. Van Heuvelen, A. (1991). Learning to think like a physicist: A review of research-based instructional strategies. *Am. J. Phys*, **59**(10). 891-897.
3. Van Heuvelen, A. and Zou, X. (2001). Multiple representations of work-energy processes. *Am. J. Phys*, **69**(2). 184-194.
4. Dufresne, R. J., Gerace, W. J., and Leonard, W. J. (1997). Solving physics problems with multiple representations. *The Physics Teacher*, **35**, 270-275.
5. diSessa, A. A., Sherin, B. L. (2000). Meta-representation: An introduction. *J. Math. Behavior*, **19**, 385-398.
6. diSessa, A. A. (2002). Students' criteria for representational adequacy. In Gravemeijer, K, Lehrer, R., Oers, B. van and Vershaffel, L. (eds.), *Symbolizing, Modeling and Tool Use in Mathematics Education*, 105-129.
7. diSessa, A. A. , Hammer, D., Sherin, B. & Kolpakowski, T. (1991). Inventing graphing: Meta-representational expertise in children. *J. Math. Behavior*, **10** (2), 117-160.
8. Koedinger, K. R. and Nathan, M. J. (2004). The real story behind story problems: Effects of representations on quantitative reasoning. *The Journal of the Learning Sciences*, **13**(2), 129-164.
9. Finkelstein, N. et. al. (2004). Can computer simulations replace real equipment in undergraduate laboratories? (this volume)
10. Chi, M. T. H, Glaser, R. and Rees, E. (1982) Expertise in problem solving. In R. J. Sternberg (Ed.), *Advances in the psychology of human intelligence (Vol. 1)*, 76-74.

Attitudes of Undergraduate General Science Students Toward Learning Science and the Nature of Science

Jeffrey Marx, Shabbir Mian, Vasilis Pagonis

McDaniel College, Department of Physics

Abstract. We investigated general science students' attitudes regarding the acquisition of scientific knowledge and the nature of science by administering a 32-item attitudinal survey. To assemble a representative array of epistemological attitudes at our institution and to determine the impact of instruction, we administered this survey to over 250 students from 19 sections of three general science courses. We characterized the instructional styles for each course using three broad categories: *Traditional*, *Transitional*, and *Learning-centered.* This paper focuses on the impact those different instructional styles had on students' epistemological beliefs.

INTRODUCTION

It has been made clear to the physics education community that confronting students' misconceptions, alone, should not constitute the entire instructional goal of a well-intentioned curriculum. It is also important to positively affect students' feelings about learning science and the nature of scientific inquiry.[1,2] For example, Redish et al. demonstrated that physics courses well-designed to promote conceptual understanding only weakly influence students' epistemological outlooks.[3] Elby showed that achieving attitudinal gains is possible in High School physics, but curricular materials must be designed to address epistemology.[4]

In this paper we discuss the epistemological effects of various instructional styles on students in general sciences courses, the majority of whom are non-science majors. This population is interesting in that it is underreported in the literature; references 1 and 3, for example, report on attitudes in physics classes. Yet, hundreds of thousands of undergraduates enroll in general science courses each year.[5] This motivated us to determine if trends reported for physics students hold for general science students, too.

Laura Lising and Andy Elby, with input from Priscilla Laws and David Jackson, constructed the survey instrument we used for this study by combining and modifying items from the Epistemological Beliefs Assessment for Physical Science (EBAPS) and the Maryland Physics Expectations Survey (MPEX). (These two instruments have a documented history.[6,7]) The survey contains 22 five-point, Likert-scale (strongly disagree – strongly agree) items; five multiple-choice items; and five short dialogues requiring the examinee to side with one person's opinion or the other. The survey has a dichotomous scoring rubric; if the examinee's response matches the favorable (most expert-like) response, the item is graded 1, otherwise 0. (For all but one item there are two favorable responses.) An examinee's overall score is the percentage of his/her favorable responses.

The survey authors largely constructed their tool around five epistemological clusters – the organization of scientific knowledge (*Structure* cluster), accruement of new knowledge (*Nature of Learning* cluster), relationship between classroom science and the real world (*Reality* cluster), evolution of scientific knowledge (*Evolving* cluster), and connections between assiduity and understanding (*Source* cluster). The survey authors omitted six of the 32 items from all clusters, but the accompanying spreadsheet, designed to facilitate grading and analysis, left open the possibility of adding items to and removing items from clusters. To improve the robustness of the clusters, we removed the two-item cluster referred to as *Concepts* and moved those items to the *Structure* and *Nature* clusters. We also added three items to the *Source* cluster and one to *Reality*. In the end we had seven

CP790, *2004 Physics Education Research Conference*, edited by J. Marx, P. Heron, and S. Franklin

items each in *Structure* and *Nature*, six in *Reality*, two in *Evolving*, and seven in *Source*. To determine the cluster score for a group, we calculated the group's average for each item in that cluster and then averaged those items. (Note: Scores for the *Evolving* cluster are difficult to compare because it consists of only two items and one of those is the only item with a single favorable response. As such, the group average on that item tends to be low, which suppresses the *Evolving* cluster's average.)

POPULATION AND COURSE STRUCTURES

McDaniel College is a selective, residential, four-year, liberal arts institute, with an undergraduate population of about 1600 students. The college requires all students to complete two courses under the heading of Natural Science and Mathematics. As such, well over 80% of the students enrolled in our department's General Science offerings are non-science majors fulfilling graduation requirements.

We surveyed students' attitudes from the Spring of 2002 through the Spring of 2004 in 18 sections of three General Science courses – *A World of Light and Color*; *Astronomy*; and *Sound, Music, and Hearing*. (These classes typically have about 15 to 20 students enrolled.) We matched pre-test/post-test data for 255 students. We also have post-test data from an additional section of *Astronomy*.

For the purposes of this paper we broadly characterize these courses into three categories, which depend on the percentage of class time devoted to some sort of research-based curricular materials aimed at addressing conceptual understanding, not developing epistemology.

- *Traditional* courses have little or no research-based pedagogy. Class time is largely filled with lectures and demonstrations; slides and videos; and laboratories that follow prescribed instructions, some of which are computer-based. Traditional classes had some group work. All of the *Astronomy* sections with matched pre-test/post-test data fell into this category.
- *Transitional* courses have approximately half of the classroom time devoted to research-based activities. These activities come in several forms including exploratory laboratories, activities modeled after *Interactive Lecture Demonstrations* [8], and group worksheets designed to reinforce relevant physical concepts and build numeracy. The other half of the time is lecture-based. One section of *A World of Light and Color* and all of the sections of *Sound, Music, and Hearing* were *Transitional*. (We will also present post-test data for a section of *Astronomy* that was *Transitional*.)
- *Learning-centered* courses incorporate research-based curricular materials of the kind discussed above in every class meeting, and lecture time is minimized. *A World of Light and Color* was the only course that fit into this category.[9]

Each course had traditionally-graded assignments, including several homework sets, quizzes, tests, projects, and a final exam. None of the graded assignments were intended to foster or reinforce students' epistemologies.

ANALYSIS

Looking at the overall attitudinal gains achieved by our population, we see that for the entire group of matched pre-tests and post-tests (N = 255) the average score on the pre-test was 57% and the post-test average was 55%. (This is not a statistically significant drop; nevertheless, we hoped for an overall increase.) Only the decrease in the *Source* cluster's score was statistically significant. Table 1 shows the matched data set for the clusters. (In all of the tables, overall and cluster scores are presented as averages of the favorable responses. We do not report the standard deviations for any averages. It suffices to say that the deviations are quite large, typically ranging from 10 to 20%. None of the comparisons we made differed by more than one standard deviation.)

TABLE 1. Cluster Analysis for all students. Statistically significant ($\alpha = 0.95$) changes are in boldface.

Cluster	Pre	Post
Structure	46	48
Nature	54	52
Reality	70	68
Evolving	40	44
Source	**69**	**62**

We realized that combining all of the students together masks the effects of different instructional styles on students' attitudes. To reveal such effects, we divided students by the type of instruction they encountered. (See Table 2.) This showed that the *Learning-centered* and *Transitional* courses resulted in no instructionally significant change in the overall scores. However, the decrease in the *Traditional* courses is instructionally significant. We also see some positive and negative gains in various clusters. The difference between the overall pre-test scores in the

Traditional and *Learning-centered* courses is statistically significant.

TABLE 2. Comparison of various instructional styles. Statistically significant (α = 0.95) changes within a particular style are in boldface.

	Learning-centered		Transitional		Traditional	
	Pre	Post	Pre	Post	Pre	Post
N	(107)		(47)		(101)	
Overall	60	60	57	61	**53**	**48**
Cluster						
Structure	52	55	**41**	**51**	43	39
Nature	56	53	54	58	51	49
Reality	77	75	**71**	**77**	**64**	**58**
Evolving	42	50	36	37	40	40
Source	72	68	**74**	**68**	**64**	**54**

Two courses were taught in more than one instructional setting. *A World of Light and Color* moved from a *Transitional* environment to *Learning-centered,* and one section of *Astronomy* was taught as a *Transitional* course, as opposed to the normal *Traditional* setting. Table 3 holds a synopsis of the post-test data for these courses. Clearly moving *Astronomy* from a *Traditional* to *Transitional* setting had positive impacts on student attitudes. Converting *Light and Color's* learning environment to completely pedagogically-inspired curricular materials seems to produce a mix of positive and negative effects.

TABLE 3. Post-test data for courses taught in more than one style. Statistically significant (α = 0.95) changes within a particular course are in boldface.

	Astronomy		Light and Color	
	Trans.	Trad.	L-c	Trans.
N	(14)	(101)	(107)	(15)
Overall	**58**	**48**	60	62
Cluster				
Structure	**55**	**39**	55	52
Nature	**59**	**49**	53	59
Reality	**81**	**58**	**75**	**83**
Evolving	39	40	50	30
Source	**62**	**54**	**68**	**59**

It is possible that the various instructional styles might affect students with initially weak attitudes differently than those with strong initial attitudes. To examine this, we divided the populations into two groups – *Experts* and *Novices*. For the entire matched set, roughly one-third of the students had pre-test scores at or below 53%. These students became our *Novices*. Likewise, roughly one-third of the students had pre-test scores at or above 64%. These students became our *Experts*. Table 4 shows the impacts the different instructional styles had on these two groups. (Note: The similarity of pre-test scores across each instructional style stems from our categorization.)

TABLE 4. *Experts'* versus *Novices'* overall scores for various instructional styles. Statistically significant (α = 0.95) changes within a particular style are in boldface.

	Learning-centered		Transitional		Traditional	
	Pre	Post	Pre	Post	Pre	Post
Expert						
N	(45)		(15)		(20)	
Overall	**70**	**64**	72	73	**72**	**60**
Novice						
N	(30)		(19)		(57)	
Overall	**47**	**55**	**44**	**51**	**45**	**40**

Finally, we wanted to know if an instructional setting favored one gender over the other. Table 5 displays the results we found when we sorted the various instructional settings by gender. The differences in the overall scores between genders within a particular style are statistically insignificant. Again, the *Traditional* pre-test scores differ statistically from the *Learning-centered* courses.

TABLE 5. Comparison of males' and females' overall scores for various instructional styles. There are no statistically significant changes within any instructional style.

	Learning-centered		Transitional		Traditional	
	Pre	Post	Pre	Post	Pre	Post
Male						
N	(65)		(21)		(54)	
Overall	60	58	56	61	53	48
Female						
N	(42)		(26)		(46)	
Overall	61	62	58	62	53	47

DISCUSSION AND CONCLUSIONS

Overall, there was no appreciable gain in attitudes as determined by our instrument. Looking at the three instructional styles independently, we see no real improvement for the *Transitional* and *Learning-centered* courses, which, in light of published reports, is progress. The scores in *Traditional* classes slid back five points. In the end, students in the *Traditional* sections were about 12 points behind the other students. Even accounting for the initially depressed pre-test scores, the *Traditional* students still lagged considerably behind students in the more research-based curricula.

One positive outcome was that *Novices* fared quite well in the *Transitional* and *Learning-centered* courses. This is encouraging, because it suggests an instructional technique that could help the research-based curricula benefit everyone. By carefully analyzing the clusters (or even specific items) where the *Experts* slipped, one gains insight into the kinds of epistemologically-oriented curricular materials to develop. Hopefully, these targeted materials would prevent the *Experts* from slipping in those clusters, while, perhaps, further bolstering the *Novices'* scores.

Interestingly, we found virtually no difference between males' and females' pre-test attitudes. Differences in gains or losses for males and females within a particular instructional style were not significant, either. Instructionally, though, the question remains open as to whether curricular materials that significantly improve students' epistemological attitudes need to be tailored to males and females, or whether a generic approach will suffice.

We only weakly controlled for the effect of a course's content on students' attitudinal gains. Nevertheless, courses that changed instructional styles followed the patterns we saw across the styles, which helps to establish that our results depend on instructional style and not content.

The *Traditional* students, all of whom were in *Astronomy*, pre-tested at a statistically significant lower level than the *Learning-centered* students. Perhaps this bias made the *Traditional* students less receptive to improving their attitudes about the nature of science and learning science. We did, however, demonstrate that the *Experts* in the *Traditional* sections decreased the most; moreover, *Novices* in *Transitional* and *Learning-centered* environments gained the most. So it seems unlikely that the low pre-test score of the *Traditional* students is indicative of their capacity or willingness to improve.

Unfortunately, we could not probe the influence of the professor on students' attitudes because of insufficient mixing of instructors across styles. In particular, none of the *Traditional* instructors taught in either of the research-based settings. The effect of the instructor on epistemological attitudes remains an open research avenue that needs to be explored.

All of this confirms that, for this group of students (just like the reports of physics students), improving students' attitudes about learning science and the nature of science does not come from curricular materials designed only to address conceptual understanding. Our *Learning-centered* course implicitly addresses issues related to several of the clusters by forcing students to postulate and learn for themselves (*Nature* & *Source*); focusing on fundamental concepts that explain a host of physical phenomena (*Structure*); and having students experience simple, explainable, real-world phenomena in the classroom (*Reality*). Nevertheless, it fails to improve students' attitudes. In order to realize instructionally significant gains in epistemology, it seems we must carefully craft materials demanding students to overtly and critically evaluate how they learn science and the nature of science itself.

ACKNOWLEDGMENTS

This work was supported, in part, through the National Science Foundation (NSF-CCLI grant 0125828), which permitted us to convert *A World of Light and Color* into a *Learning-centered* course. One of us (JM) would also like to thank McDaniel College for the Ira Zepp Teaching Enhancement Grant, which directly supported the investigation of general science students' epistemologies. Finally, we want to thank Gary Toller, Sandy Stemple, and Christopher Kulp for relinquishing some of their classroom time so we could administer the survey to their students.

REFERENCES

1. Elby, A., *Physics Education Research Supplement to the American Journal of Physics* **69**, No. 7, S54 – S64 (2001).
2. Hammer, D., *Physics Education Research Supplement to the American Journal of Physics* **68**, No. 7, S52 – S59 (2000).
3. Redish, E. F., Saul, J. M., and Steinberg, R. N., *American Journal of Physics* **66**, pp 212 – 224 (1998).
4. See Ref. 1.
5. Over 200,000 undergraduates enroll in introductory astronomy each year alone. See Franknoi, A., *Astronomical Society of the Pacific Conference Series*, Vol. 89, edited by J. R. Percy, Astronomical Society of the Pacific, San Francisco, 1996, p. 9 – 25.
6. White, B., Elby, A., Frederiksen, J., and Schwarz, C. Presentation at the American Education Research Association, Montreal (1999). See the EBAPS website: http://www2.physics.umd.edu/~elby/EBAPS/home.htm
7. See Ref 3.
8. Sokoloff, D. R., and Thornton, R. K., *Interactive Lecture Demonstrations: Active Learning in Introductory Physics*, Hoboken, NJ: John Wiley and Sons, Inc. 2004.
9. Marx, J., Mian, S., and Pagonis, V., *Forum on Education* edited by Wolfgang Christian, APS, Spring 2004, pp. 9 – 11.

Preliminary Results of Gender Equity Variations in a Large Active-Learning Introductory Physics Course Due to Laboratory Activity Instructions

Mark L. McKinnon and Wendell H. Potter

Physics Education Research Group, Department of Physics, University of California at Davis, One Shields Avenue, Davis, CA 95616

Abstract. In the UC Davis introductory physics course for life science majors, variations in lab activity instructions were introduced to investigate their influence on gender performance differences. Systematic instructions were compared to unguided, open-ended instructions. On preliminary examination, the performance difference between females and males was unaffected by the different instructions. However, there is some evidence that the open-ended instructions did increase students' conceptual understanding in general.

INTRODUCTION

An important consideration of improving pedagogical styles is to reduce gender bias and performance differences while increasing all students' conceptual understanding of a subject. In an active learning setting, we utilized dissimilar sets of laboratory activity instructions to determine whether instructions that are more open-ended would improve gender equity. If the data shows positive results, the ability to increase retained understanding and decrease gender inequities through the utilization of open-ended instructions would prove a benefit to students and instructors alike.

Much of the research in gender differences has focused on performance on a variety of tests [1]. Differences in cognitive tests that once appeared to demonstrate differences between males and females have diminished considerably, leading many researchers to conclude that cognitive differences do not exist [2,3]. However, males still outperform females in physics and science courses [4,5]. What is not clear is whether the concepts or the methods with which these concepts are taught generated these gender performance differences. Several members of the UC Davis Physics Education Research Group have turned their attention to the differences in gender performance [6]. We have begun to examine the specific wording and structure of activity instructions given to students in the laboratory setting.

Current research indicates that active learning creates a greater understanding among students in science courses than does a passive, lecture-oriented curriculum [7, 8, 9]. In 1996, UC Davis implemented a major restructuring of the introductory physics course for life science majors. We replaced a traditional introductory physics course (a three-quarter sequence labeled 5A, 5B, & 5C) with a cooperative approach that motivates students to become more active in their learning. In this new course (called 7A, 7B, & 7C), students work in small groups to answer questions about an observed phenomenon. Instead of three hours of lecture and 2 1/2 hours of laboratory per week, the students attend a single 80-minute lecture – where they are usually assessed for understanding – and two 2 1/2 hour discussion/laboratories – where they perform experiments and discuss how to analyze and describe what they saw. The lab activities are designed to motivate the student to actively construct new models of a variety of physical phenomena [10, 11].

Two years ago, at the 2002 AAPT National Meeting and PER conference, we presented evidence that demonstrated a tremendous improvement in gender equity when comparing the Physics 7 sequence

CP790, *2004 Physics Education Research Conference*, edited by J. Marx, P. Heron, and S. Franklin

to the Physics 5 sequence [12]. However, while gender equity has been improved throughout the entire sequence of Physics 7, among the three quarters, inequality still exists. Table 1 shows that the distributions of course grades for males and females are nearly aligned in Physics 7A. The distributions for 7B and 7C are less so.

Table 1. Gender Performance Differences in Three Quarters of UCDavis Introductory Physics for Life Science Majors (Physics 7A, 7B, & 7C)

Course	Gender	N	Mean[a]	Std. Dev.
		t-test[c,d]	Diff.[b,c]	Std. Error[c,d]
7A	Female	4187	2.459	0.553
	Male	2449	2.556	0.557
	Compare	-6.861	**-0.097**	0.014
7B	Female	3563	2.424	0.590
	Male	2058	2.617	0.661
	Compare	-10.909	**-0.192**	0.018
7C	Female	2724	2.443	0.549
	Male	1625	2.583	0.590
	Compare	-7.805	**-0.141**	0.018

Notes: a - Mean grade earned based on 0-4.5 scale.
b - Diff. is mean difference for (females – males)
c - Equal variances were not assumed
d - Significance $<10^{-10}$

The gender performance differences may result from dealing with particular concepts or due to variations in pedagogy. Instructors would implement variations in the laboratory activities and assessment tools for each course. One of the noticeable differences between the 7A and 7C lab activities was that the instructions for 7C activities were much more formulaic. While the lab activities still required interaction among the students, the instructions were often systematic and could be performed without much discussion. Our study was to determine if these instructions had an effect on gender differences in understanding the exhibited concept. We generated several activity instructions that required the students to pursue the same conceptual understanding without engaging in a step-by-step format. For example, one activity asks the students to determine what parameters of a pendulum affect its period. The original (*Formulaic*) set of instructions listed the parameters and instructed the students to make a pendulum with a specific length, mass, and amplitude and test specific changes. The *Open-ended* activity merely presents the students with a string and mass and asks them to determine the parameters of a pendulum and find which affect period and what effect they have. The students depend on their group members to help complete the assignment. Our working thesis was that the open-ended instructions would foster greater communication within the lab groups. We hoped that this, in turn, would encourage greater gender equity by decreasing the performance difference between females and males.

METHOD

During the spring quarter of 2004, 471 students completed the Physics 7C course. One group of 185 students received the Formulaic instructions. Another group of 197 students received the Open-ended instructions. The remaining 89 students received a mixture of Open-ended and Formulaic instructions. This arrangement lasted for the first eight of the eighteen class meetings. During the remaining meetings, all students received the same set of instructions, which were a mixture of Open-ended and Formulaic instructions. The physical phenomena discussed in the experimental classes included SHO, 1-D, 2-D, and 3-D waves, and superposition of waves. Phenomena discussed during the remainder of the quarter included nuclear forces and decay, electro-magnetism, and optics. All students were assessed concurrently and with identical assessment tools during the lectures. Assessment for both groups included several short-answer questions that were conceptual, graphical, algebraic or computational in nature.

Sample Set Comments

The student body at UC Davis reflects a socio-economic background that is predominantly middle to upper class. The student body is under-represented with African-Americans and Hispanics. There is an over-representation of Asian and Asian-American students. In this preliminary analysis, ethnic origins were ignored. Nearly all students who take the Physics 7 sequence are Life Science majors, usually pre-med or pre-vet. This population has a far larger number of females than males. In our sample, we had 293 females and 168 males.

Data Analysis

There were four quizzes and three questions on the final that related to the experimental activities. For preliminary analysis, these seven scores were given even weight and averaged. Scores are given on a 4.5 scale (e.g. 4.13 to 4.5 earns an A+). An "acceptable" answer for a particular question earns at least 2.5, with more points given to a more complete, thorough or accurate answer. This score became the dependent variable. The students' course grades for Physics 7A and Physics 7B were covariates and gender and lab group were the factors. We performed the Univariate ANCOVA of this data with SPSS. We eliminated the data for all students (24 students) who were in our section to minimize bias. We also eliminated the data for any student for which we did not have both Physics 7A and 7B grades. Finally, we eliminated students who missed three or more of the assessments. The final numbers of students in each group were 142 in the Formulaic group and 167 in the Open-ended group.

Table 2. Estimated Marginal Means of Quiz Scores by Lab Group & Gender

Lab Group	Gender	Mean[a]	Std. Error	95% Confidence Interval	
Formulaic	Female	2.573	0.047	2.481	2.666
	Male	2.549	0.065	2.421	2.676
Open-Ended	Female	2.595	0.042	2.512	2.678
	Male	2.650	0.063	2.526	2.775

Note: a - Covariates are evaluated at the following:
7A Grade = 2.718
7B Grade = 3.230

Table 3. Univariate Analysis of Covariance: Tests of Between-Subjects Effects on Quiz Scores

Source	Type III Sum of Squares	df	Mean Square	F	Sig.
Corrected Model	95.367[a]	5	19.073	93.559	0.000
Intercept	2.072	1	2.072	10.165	0.002
7A Grade	16.149	1	16.149	79.214	0.000
7B Grade	14.409	1	14.409	70.677	0.000
Lab Group	0.256	1	0.256	1.255	**0.264**
Gender	0.015	1	0.015	0.075	**0.784**
Lab Group & Gender	0.108	1	0.108	0.530	**0.467**
Error	61.771	303	0.204		
Total	2230.676	309			

Note: a - R^2=.607 (Adjusted R^2=.600)

As Table 2 shows, the estimated means due to the covariates, the grades from the first two quarters of the course, are very close. The confidence intervals overlap considerably. As further evidence, the value of the significance of the F-statistic in the ANCOVA shown in Table 3 is very high. This demonstrates that the lab activity instruction had very little effect on the differences in performance due to gender. There seemed to be a quite low level of confidence (sig. = 0.264) that the open-ended instructions improved all students' understanding. However, the null hypothesis that the type of activity instruction did not affect gender performance could not be discarded.

DISCUSSION

The results tentatively indicate that while overall understanding increased, the degree of guidance within the activity instructions had no effect on the performance difference between genders. However, this is only a preliminary look at this data. There are a number of finer points to analyze. A natural next step is to separate the assessment questions into categories (conceptual, graphical interpretation, algebraic and computational). We would like to determine whether the gender performance difference in each category saw the same results. The ethnic breakdown of the student sample might present different results since gender differences do not necessarily cross ethnic lines [13, 14]. Additionally, using a different covariate (e.g. overall GPA) might provide greater accuracy than the 7A and 7B grades, especially if the earlier physics grades disguise a gender inequity. We plan to test the results for repeatability in fall quarter, 2004, in another Physics 7C class. During this second phase of the study, we plan to interview several students from the two groups to capture insight from the participants.

A question that has continued to percolate throughout the running of this experiment is whether the variation in the activities is great enough. Both sets of activities give the student a definite end-point to the activity with the open-ended activity instructions eliminating systematic instructions of how to attain the given end-point. The next step would be to somewhat obscure the end-point itself. This obfuscation might affect the gender performance difference by requiring the students in each lab group to establish greater communication. The concern lies in maintaining overall understanding. There is much to study and there are multiple methods that can be used to analyze these issues.

ACKNOWLEDGMENTS

We would like to express our deep appreciation to several people who are instrumental in this research. Dr. Charles de Leone, CSU San Marcos has provided significant guidance and insight. Dr. Mani Tripathi, David Michaels, and Carrie Prisbrey were instrumental in the actual implementation of the study. Dr. Jon Sandoval, Dr. Cathy Ishikawa, Dr. Ariane McKiernan,

Dr. Jacob Blickenstaff, and Pat Hession also provided tremendous support. Finally, Shirley Brenton, who became our sounding board and kept us focused.

REFERENCES

1. Willingham, W. W., and Cole, N. S., *Gender and Fair Assessment*, Mahwah, New Jersey: Lawrence Erlbaum Associates, 1997, pp. 18-54.
2. Linn, M. C., and Hyde, J. S., "Gender, Mathematics and Science," in *Educational Researcher* **18**(8), 1989, pp.17-27.
3. Linn, M. C., and Petersen, A. C., "Emergence and Characterization of Sex Differences in Spatial Ability: A Meta-Analysis," in *Child Development* **56**(6), 1985, pp. 1479-1498.
4. Baker, D., and Leary, R., "Letting Girls Speak Out About Science," in *Journal of Research in Science Teaching* **32**(1), 1995, pp. 3-27.
5. Cole, N. S., *The ETS Gender Study: How Females and Males Perform in Educational Settings,* Princeton, NJ: Educational Testing Service, 1997.
6. Potter, W. H., De Leone, C. J., Ishikawa, C. M., Blickenstaff, J. A., and Hession, P. L., "Significant Reduction in Gender Grade Disparities in a Reformed Introductory Physics Course," Paper presented at the joint annual meeting of the American Association of Physics Teachers and Physics Education Research Conference, Rochester, NY, 2001.
7. Abraham, M. R., "Research on Instructional Strategies," in *Journal of Research in Science Teaching* **25**, 1988, pp. 185-200.
8. Hake, R. R., "Interactive-Engagement versus Traditional Methods: A Six-Thousand-Student Survey of Mechanics Test Data for Introductory Physics Courses," in *American Journal of Physics* **66**, 1998, pp. 64-74.
9. von Glasersfeld, E., "Cognition, Construction of Knowledge, and Teaching," in *Constructivism in Science Education*, edited by M. R. Matthews, Netherlands: Kluwer Academic Publishers, 1998, pp. 11-30.
10. Potter, W. H., De Leone, C. J., and Coleman, L. B., "Radically Restructured Introductory Physics Course at a Large Research University," Department of Physics, University of California, Davis. 1996, August 13, Retrieved June 15, 2004 from the World Wide Web: http://www.physics.ucdavis.edu/Text/Physics7view/ Physics7intro.html/
11. Allie, S., and Buffler, A., "A Course in Tools and Procedures for Physics I," in *American Journal of Physics* **66**, 1998, pp. 613-624.
12. McKinnon, M. L., Potter, W. H., and Ishikawa, C. M., "Longitudinal Performance Variations in a Large Active-Learning Course," Paper presented at the joint annual meeting of the American Association of Physics Teachers and Physics Education Research Conference, Boise, ID, 2002.
13. Grossman, H., and Grossman, S. H., *Gender Issues in Education*, Boston: Allyn and Bacon, 1994.
14. Yee, A. H., "Asians as Stereotypes and Students: Misperceptions that Persist," in *Educational Psychology Review* **4**, 1992, pp. 95-132.

Development of Scientific Abilities in a Large Class

Sahana Murthy and Eugenia Etkina

Graduate School of Education, Rutgers University, New Brunswick, NJ 08904

Abstract. This paper describes our instructional and research efforts to help students in a large-enrollment (450 students) introductory laboratory course develop abilities used by practicing scientists. We focus on the ability to design an experimental investigation. We provide sample tasks, scoring rubrics and evidence of student improvement.

INTRODUCTION

The practice of science and engineering requires not only content knowledge but also other specific abilities that our graduates have to develop to be successful in the future [1,2]. These scientific abilities include formulating questions, designing and conducting experiments, collecting, representing and analyzing data, modeling, testing hypotheses and solving complex, ill-defined problems [3]. They form as the result of training and practice. We believe that a physics class is an excellent place for students to learn these abilities [4]. However, in terms of helping students develop scientific abilities, most attempts are in middle-school [5,6] with fewer examples at college level such as the SCALE-UP project [7] and the Workshop Physics [8] project. These projects have been implemented in specially designed classrooms or in smaller classes. There have been even fewer attempts to help students develop scientific abilities in large-enrollment courses with a traditional structure. This paper describes our instructional and research efforts to help students develop experimentation-related abilities in a large-enrollment (450 students) introductory laboratory course and investigates the question of whether students develop these abilities, if they design and perform their own experiments.

EXPERIMENTAL DESIGN TASKS

To help our students develop scientific abilities we devised laboratory tasks where students design experiments to test a hypothesis or to solve a problem. These tasks have a number of important features. They are open-ended and some information required to solve them has to be obtained by performing additional experiments or by making informed estimates. Students have to design and describe their own procedure to solve the task. The tasks encourage divergent thinking, as students need to come up with at least two independent methods to solve a problem. A task where students have to solve a practical problem is shown below:

Sample design task: Design experiments to determine the thickness of a strand of your hair using two independent methods. One of the methods must involve ideas about diffraction. Available equipment includes laser pointer, ruler, paper, and holder for strand of hair, Vernier calipers. For each method write in your lab report:

(a) Give an outline of your experimental design.
(b) Draw a labeled diagram of your set-up.
(c) Write the mathematical procedure you will use.
(d) Write how you will measure the physical quantities you need to determine the thickness.
(e) Perform the experiment and record your measurements in a table.
(f) Calculate the thickness, based on your procedure and measurements.
(g) Identify sources of experimental uncertainty. Write the steps you can take to minimize them?
(h) Compare the two values you obtained for the thickness of the hair in two experiments. Describe possible reasons for the difference.

SCIENTIFIC ABILITY RUBRICS

Together with the Rutgers University Physics and Astronomy Education (PAER) group members, we

CP790, *2004 Physics Education Research Conference*, edited by J. Marx, P. Heron, and S. Franklin
 0-7354-0281-7/05/$22.50

developed scoring rubrics to evaluate students' lab reports. The rubrics contain descriptors for individual scientific abilities on a scale of 0 to 3. Table 1 shows a portion of the rubrics that we used to score students' lab reports. The complete rubrics, which contain all the scientific abilities are available on the group's website [9]. The rubrics were extensively tested for inter-rater reliability between the PAER group members. Over a period of two months, we scored many student write-ups and revised our rubrics iteratively until we achieved an agreement of 90-95% in the scores.

We also used the rubrics for another purpose. We divided the design problem into sub-tasks (parts a-h in sample), so that each sub-task reflected one or more scientific abilities that we want students to develop. We structured the tasks to achieve a correspondence between the guidelines to the students and the abilities in the rubrics. For example, ability 1 in Table 1 below corresponds to part (a) in the sample task, and ability 2 corresponds to part (c). Thus our design tasks also underwent an iterative process of revision along with the rubrics. It is important to note that although we provided detailed guidelines we did not provide a recipe for solving the problem. The guidelines provided a template for all experiments. Consequently, the rubrics helped us develop new design tasks.

Although the same rubrics can be used for design tasks with different physics content, to use the rubrics for scoring one needs to have a strong understanding of the content. For example, the ability to design a reliable experiment to solve the problem encompasses physics content knowledge. In the same vein, most sub-tasks for different design problems look similar, yet the process of devising a reasonable procedure to solve the problem involves a thorough understanding of the physics concepts involved.

IMPLEMENTATION OF DESIGN TASKS

These tasks were implemented in the second semester of a laboratory course that accompanied an introductory lecture-recitation algebra-based course for science majors. Almost all students who took the laboratory course were enrolled in the lecture-recitation course. In the first semester of the course, students had performed non-traditional but still rather guided experiments. The guidelines for the experiments were not based on the scientific ability rubrics.

TABLE 1. A portion of the scoring rubrics used for a problem-solving design experiment

Ability/Score	0	1	2	3
1. Is able to design a reliable experiment that solves the problem	The experiment does not solve the problem.	The experiment attempts to solve the problem but due to the nature of the design the data will not lead to a reliable solution.	The experiment attempts to solve the problem but due to the nature of the design there is a moderate chance the data will not lead to a reliable solution.	The experiment solves the problem and has a high likelihood of producing data that will lead to a reliable solution.
2. Is able to choose a productive mathematical procedure for solving a particular experimental problem	Mathematical procedure is either missing, or the equations written down are irrelevant to the experimental design	A mathematical procedure is described, but it is incorrect of incomplete due to which the final answer cannot be calculated.	Correct and complete mathematical procedure is described but an error is made in the numerical calculations.	Mathematical procedure is fully consistent with the design. All quantities are calculated correctly. Final answer is meaningful.
3. Is able to communicate the details of an experimental procedure clearly and completely	Diagrams are missing and/or experimental procedure is missing or extremely vague.	Diagrams are present but unclear and/or experimental procedure is present but important details are missing.	Diagrams and/or experimental procedure are present but with minor omissions or vague details.	Diagrams and/or experimental procedure are clear and complete.
4. Is able to evaluate specifically how experimental uncertainties may affect the data	No attempt is made to evaluate experimental uncertainties.	An attempt is made to evaluate experimental uncertainties, but most are missing, described vaguely, or incorrect.	Most experimental uncertainties are evaluated correctly, though a few contain minor errors, inconsistencies, or omissions.	All experimental uncertainties are correctly evaluated.

During the semester the project was implemented (the second semester), there were 20 lab sections, each with about 25 students. There were 9 Teaching Assistants (TAs): 5 were first year TAs with a first language other than English, and 4 were engineering graduate students. None of the TAs was involved in PER activities. We collaborated closely with the course coordinator who was greatly instrumental in making possible the implementation of our design tasks. One of the authors (SM) led TA training sessions. During an hour-long weekly meeting, the TAs went through a mock process of designing and performing the experiments. We provided TAs model answers based on the criteria in the rubrics, and discussed them in the weekly meetings. TAs used the model answers to grade students' lab write-ups and provide detailed written feedback. This ensured consistency between TAs. In addition, we observed TAs teaching the labs, and provided them feedback.

FINDINGS

We found that in the initial weeks, students had difficulties approaching design tasks. Their lab reports received low scores on scientific abilities, as seen in Figure 1(left). We observed students in the laboratory and noticed negative attitudes and frequent complains that the tasks were difficult. As the semester progressed the students became more enthusiastic about these activities. They spent more time discussing in their groups how to come up with different methods to solve the task. The quality of their write-ups changed. The reports began to resemble experimental reports of practicing scientists. See Figure 1(right) which is a student response to the sample design task. These observations are similar to those of Zou, who implemented design tasks in small classes [10].

We selected four laboratory sections taught by different TAs to sample students' lab write-ups. We included 35 students in the sample. The students were randomly distributed among 17 recitation sections of the lecture-recitation course. The average final exam score for the sample was 78.3, the class average was 75 with a standard deviation of 16. We scored reports of the students chosen above on different scientific abilities. As seen in the bar charts in Figure 2, there was an improvement in the scores for certain scientific abilities. A closer examination using χ^2-analysis revealed that the improvements were statistically significant. Examining each ability we found contributors to the significant changes (see Table 2): For ability 1 there was a significant change in ranking 1; for ability 2 - a significant change ranking 3; for ability 3: - significant changes in rankings 0 and 3; for ability 4 – a significant improvement with no significant contributing cells. We also found no significant χ^2 values when the course grades were compared to a particular ability ranking.

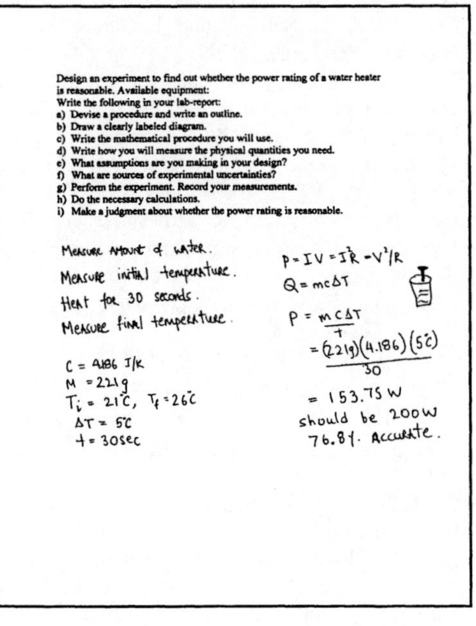

Design an experiment to find out whether the power rating of a water heater is reasonable. Available equipment:
Write the following in your lab-report:
a) Devise a procedure and write an outline.
b) Draw a clearly labeled diagram.
c) Write the mathematical procedure you will use.
d) Write how you will measure the physical quantities you need.
e) What assumptions are you making in your design?
f) What are sources of experimental uncertainties?
g) Perform the experiment. Record your measurements.
h) Do the necessary calculations.
i) Make a judgment about whether the power rating is reasonable.

Measure amount of water.
Measure initial temperature.
Heat for 30 seconds.
Measure final temperature.
C = 4186 J/K
M = 221g
T_i = 21˚C, T_f = 26˚C
ΔT = 5˚C
t = 30sec
P = IV = I²R = V²/R
Q = mcΔT
P = mcΔT / t
= (221g)(4.186)(5˚C) / 30
= 153.75 W
should be 200W
76.8%. Accurate.

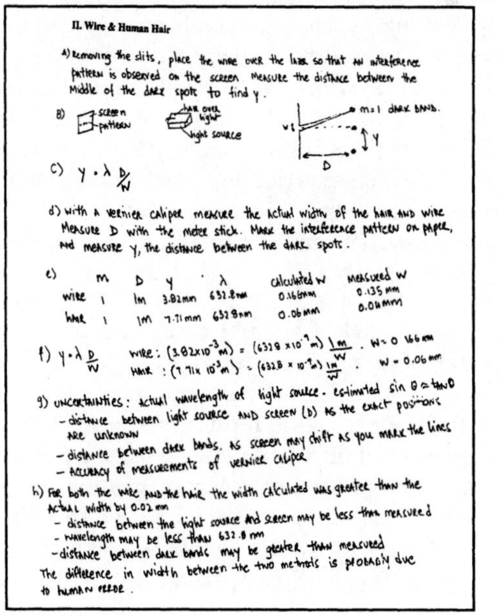

II. Wire & Human Hair
a) removing the slits, place the wire over the laser so that an interference pattern is observed on the screen. Measure the distance between the middle of the dark spots to find y.
c) y = λD/w
d) with a vernier caliper measure the actual width of the hair and wire. Measure D with the meter stick. Mark the interference pattern on paper, and measure y, the distance between the dark spots.
g) uncertainties: actual wavelength of light source. estimated sin θ ≈ tan θ
- distance between light source and screen (D) as the exact positions are unknown
- distance between dark bands, as screen may shift as you mark the lines
- accuracy of measurements of vernier caliper
h) For both the wire and the hair the width calculated was greater than the actual width by 0.02 mm
- distance between the light source and screen may be less than measured
- wavelength may be less than 632.8 nm
- distance between dark bands may be greater than measured
The difference in width between the two methods is probably due to human error.

FIGURE 1. Sample lab reports from students. Left: From week 3. Scores given are as follows: Ability1 -- 2, Ability 2 -- 2, Ability 3 --1, Ability 4 – 0. Right: From week 10, design task is in the text as sample design task. Scores given are as follows: Ability1 -- 3, Ability 2 -- 3, Ability 3 --2, Ability 4 – 2.

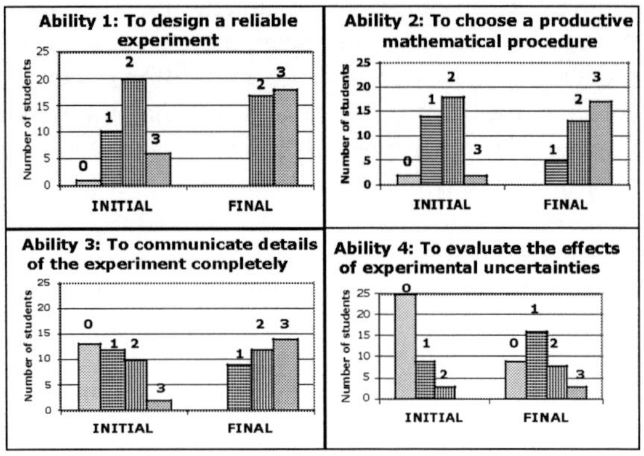

FIGURE 2. Scores of 35 students on the scientific abilities listed in Table 1. INITIAL refers to week 3 and FINAL refers to week 10 in the semester.

Our main conclusion from the data is that students' scientific abilities, measured by our rubrics, improved significantly. There could be multiple explanations for this effect, one of which is that the sub-tasks (parts a-h in sample) provided guidance. We favor this explanation because for ability 4, there were no guidelines and there were no significant standardized residuals. A limitation of our research is that we have not yet performed a controlled experiment to test this hypothesis.

SUMMARY AND FUTURE WORK

We found that is that it is possible to successfully implement and assess open-ended tasks even in large-enrollment classes, without totally revising the course and having highly trained teachers/TAs. We also achieved an important instructional goal: a successful interplay between curriculum and assessment. Descriptors in the rubrics served as goals for writing design tasks Revisions in the rubrics led to changes in the wording of the tasks.

We plan to make the rubrics available to the students next semester. This can serve the purpose of providing both guidance and self-assessment, which is regarded to be a very good form of assessment [11]. We also intend to train TAs on how to use the rubrics, so that they can be used as a part of grading. More tasks from different areas of physics can be found on our website [9]. We hope physics teachers in different settings can use them along with the rubrics.

Acknowledgements: We thank S. Brahmia, D. Brookes, M. Gentile, H. Lopez, M. Milner, D. Rosengrant, Y. Timofeeva, A. Van Heuvelen, and A. Warren for helping in the development of rubrics, and M. Lawrence for help in preparing the paper. We are grateful to G. Alba and R. Fiorillo for support in implementing design tasks. The project was supported by NSF grant DUE 024178.

TABLE 2. χ^2 **analysis**

Ability	χ^2 ($\chi^2_{\text{critical value}}$=7.82)	Standardized residual, R. (R> |2| is a significant contributor to the χ^2 value)
1	17.03	$R_{i,1}$=2.23; R_{f1}=-2.23
2	17.73	$R_{i\,3}$=-2.23; $R_{f\,3}$=2.23
3	25.12	$R_{i\,0}$= 2.55; $R_{f\,0}$=-2.55 $R_{i\,3}$=-2.37; $R_{i\,3}$=2.37
4	15.32	No cells were significant contributors.

REFERENCES

1. See the ABET home page at http://www.abet.org/abethomepage.html.
2. Shaping the Future: New Expectations for Undergraduate Education in Science, Mathematics, Engineering, and Technology," NSF Directorate for EHR Review of Undergraduate Education, May, 1996.
3. Schunn, C. and Anderson, J., "Acquiring expertise in science: Exploration of what, when and how," in *Designing for science: Implications from everyday, classroom and professional settings* edited by K. Crowley, C. D. Schunn, and T. Okada, Mahwah, NJ: Erlbaum. 2001, pp. 351-392.
4. Etkina, E., "Developing scientific abilities in a large-enrollment physics course." Winter AAPT, 2004.
5. Black, P. J. "Can pupils design their own experiments?" *P.E.E. Conference,* Nankai University Tianjin, April 1990.
6. Germann, P. J., Aram, R. and Burke, G., "Identifying Patterns and Relationships among the Responses of Seventh-Grade Students to the Science Process Skill of Designing Experiments", *Journal of Research in Science Teaching,* **33**(1) (1996).
7. See the SCALE-UP project's website at: http://www.ncsu.edu/per/scaleup.html
8. See the Workshop Physics website at: http://physics.dickinson.edu/~wp_web/WP_homepage.html
9. See Rutgers PAER group's website: http://paer.rutgers.edu/ScientificAbilities/
10. Zou, X., Etkina, E. and Van Heuvelen, A., "Experiment designing and hypothesis testing in the introductory laboratory," National AAPT Conference, Rochester, New York, July 2001.
11. Black, P. and William, D., "Assessment in Classroom Learning", *Assessment in education,* **5**(1), pp. 7-75 (1998).

No Single Cause: Learning Gains, Student Attitudes, and the Impacts of Multiple Effective Reforms

Steven J. Pollock

Department of Physics, University of Colorado, Boulder.

Abstract. We examine the effects of, and interplay among, several proven research-based reforms implemented in an introductory large enrollment (500+) calculus-based physics course. These interventions included Peer Instruction with student response system in lecture, Tutorials with trained undergrad learning assistants in recitations, and personalized computer assignments. We collected extensive informal online survey data along with validated pre/post content- and attitude-surveys, and long answer pre/post content questions designed to assess learning gains and near transfer, to investigate complementary effects of these multiple reforms, and to begin to understand which features are necessary and effective for high fidelity replication. Our average [median] normalized gain was .62 [0.67] on the FCI, .66 [0.77] on the FMCE, yet we find we cannot uniquely associate these gains with any individual isolated course components. We also investigate the population of students with low final conceptual scores, with an emphasis on the roles played by demographics, preparation, and self-reported attitudes and beliefs about learning.

INTRODUCTION

Physics education research has documented the need for, and value of, interactive engagement in physics instruction[1,2]. However, the effect of implementing multiple, layered, and coordinated activities is somewhat less well understood. To help facilitate sustainable, structural integration of such activities we investigate necessary features of individual reforms and the dynamics between them. We present here prelimary data from a case study in a large reformed mechanics course. Our initial research questions center on two issues. First, which elements of the course reforms are most closely associated with student learning gains? Second, even with such high average gains, there remains a population of students who fail to develop satisfactory conceptual understanding. Can we identify and characterize such "at risk" students, with a future goal of better targeting and accommodating them?

COURSE STRUCTURE

Physics 1110 is a large introductory calculus-based mechanics course at the University of Colorado, Boulder (CU) with 500+ students in three 50-minute lectures/wk, plus a smaller 50-min. recitation section (with *no* lab). Reforms implemented included the following: we used Peer Instruction[3] in lecture, with individual personal response systems. Roughly 30-50% of class time was spent on ConceptTests. Recitations were converted to engage in Tutorials[4], with grad TA's and undergrad learning assistants. Tutorial homework[4] was assigned, as well as traditional problems graded online with CAPA[5]. There were two texts: traditional and multimedia[6]. A unique feature of the lectures was an increased emphasis on epistemology and metacognition, making explicit what it means to learn physics, the interconnectness of topics, the value of collaboration, and the nature of science. Exams were 75% multiple choice, with a strong conceptual focus, and 25% long answer based closely on the Tutorial instructor's manual[4]. Exams accounted for 60% of the grade, CAPA for 15%, Tutorial homework 15%, Tutorial attendance 5%, and online participation questions 5%. The final grade distribution matched departmental norms: 18% A's, 37% B's, 34% C's, and 11% D/F's.

DATA COLLECTION

We collected extensive student data for two terms, including a pre/post conceptual survey (FCI in Fa'03, FMCE Sp'04), and weekly Tutorial pretests (provided

CP790, *2004 Physics Education Research Conference*, edited by J. Marx, P. Heron, and S. Franklin

online by the U. Washington PEG), with post-tests given as collaborative in-class questions, and later variants on exams. We gave pre/post demographic surveys, along with questions on attitudes and beliefs about learning, and the nature of science (CLASS[7]) We gave frequent informal online surveys probing students' attitudes toward class components. (No data were anonymous except for a SALG[8] survey, the results of which closely matched our informal surveys.) Students in engineering Calculus (approximately half the class) took a math pretest.

RESULTS

We now have an extensive database including individual results on pre/post tests, assignments, exams, and surveys. We first summarize some key initial findings related to the guiding questions of this work. Both semesters, we measured strong learning gains on the FCI/FMCE. (See Fig 1 for results from Sp 04. The pretest average on the FCI was higher but normalized learning gains were similar[9].) These gains lie near the high end of the distribution shown in Hake[2] for interactive engagement courses, and are particularly significant for a large class with no laboratory and limited financial resources. Learning gains for males (who comprise 3/4 of the class) and females are comparable, although females started with a statistically significant lower average pretest score (over 10% lower). Their final course grades are statistically identical.

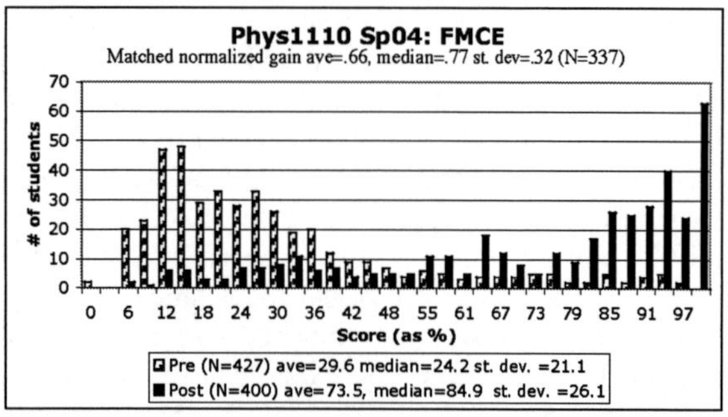

FIGURE 1. Pre/Post results for FMCE exam (Sp 04). Scoring of this exam follows Thornton [9]. All students who took the exam are shown in the histogram, but the quoted normalized gain includes only matched pre/post scores. The most frequent posttest score was 100%.

Table 1 shows learning gains as measured by questions targeting isolated, selected topics beyond those covered on the FCI or FMCE. The table shows a limited sample of our detailed data on topics matching the 12 Tutorials covered. Scores are averaged over both semesters, each with multiple related questions. The "Posttest in-class" column refers to clicker questions in lecture, with peer collaboration allowed. Typical semester to semester variance of data is ±5%.

TABLE 1. Average scores on selected conceptual topics targeted by Tutorials, rounded to nearest 5%

Topic	Pretest online	Posttest in-class	Posttest exams
a at top of ramp	20	70	75
Newton II: tension in string/block series	30	85	75
Galilean relativity	35	35	75

The results in the first row of Table 1 are typical of many topics covered in *both* lecture and Tutorials, showing a strong, rapid learning gain from one lecture to the next with just one Tutorial in between, indicating the significant impact of the Tutorial. The 2nd row shows another common feature: individual scores from exams on difficult topics are often slightly *lower* than in-class results, where peer collaboration is encouraged. This has also been verified on isolated "work alone" in-class questions. The last row shows a less common but interesting result, in which there is *no* apparent immediate learning gain from the Tutorial, but later class/peer discussion and/or homeworks appear to have a measurable impact. This particular topic was covered *only* in Tutorial (and later in Tutorial homework), but not in lecture or reading until the in-class question.

Comparison with U. Washington on common exam questions [10] is shown in Table 2. These questions (and others, not shown) demonstrate remarkable similarity between primary (UW) and secondary (CU) implementation of Tutorials. By this measure, our implementation appears to be of high fidelity.

TABLE 2. Comparison of specific long answer *exam* questions with primary implementation results from U. Washington [10] (Results rounded to nearest 5%)

Topic	UW *no* Tutorial	UW *with* Tutorial	CU *with* Tutorial
Atwood: tension before/after release	25	50	55
Atwood: describe constrained motion	45	70	45 final: 75
Coupled objects: force diagrams	30	90	95
Identify N-III partners	15	70	70

Correlations

It has been well documented elsewhere that Peer Instruction and Tutorials positively affect student conceptual learning. One of our central research questions is: what is the impact of layering these two methods? To this end, we have measured correlations between various course elements and learning gains on standardized (FMCE) exams, shown in Table 3.

TABLE 3. Pearson's correlation coefficients for normalized learning gain (g) on FMCE with specific course and curricular components (Data for Sp '04, results for Fa '03 (FCI exam) are similar)

Correlating gain with:	Correlation coefficient
FMCE pretest	.31
Math pretest (selected students)	.25
CLASS (attitude survey) overall score	.20
Tutorial homework	.22
Conventional homework	.14
Tutorial attendance	.13
In class (clicker) score	.11
Lecture attendance	.02
Average exam score	.63

Although there is a positive correlation of normalized learning gain with each of the specific course components, no single one dominates, nor are the correlation coefficients particularly high. It appears that one cannot isolate any single element (Tutorials, homeworks, attendance, or Peer Instruction) as strongly, individually coupled to conceptual learning gains as measured by the FMCE. The high correlation with average exam score may be a reflection of the conceptual focus of our exam questions, although the exams do cover a much broader range of topics (traditional introductory mechanics) than the FMCE does. But no single measurable course reform element, in isolation, appears to be uniquely associated with the high average learning gains in this course.

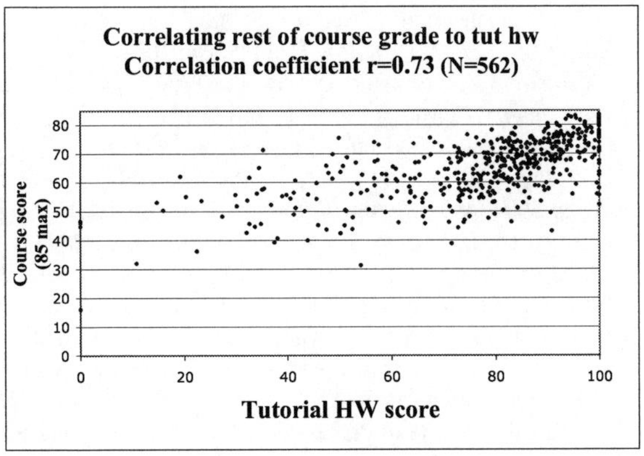

FIGURE 2. Relation of course grade (exams, conventional homework, attendance and participation) to Tutorial homework scores (Fa '03 data, Sp '04 is similar).

We see in Fig. 2 that there *is* a strong correlation (Pearson r=.73) between Tutorial homework score and course grade (excluding Tutorial homework). Since this course grade includes traditional homework, participation, and different types of exam questions (including long answer/explanation) on a broad set of topics extending beyond tutorial materials, Fig. 2 demonstrates the strong coordination of the entire variety of course elements. This is again suggestive that no *single* element is associated with measured learning gains, but rather the gains arise from the coupled aspects of course components, and perhaps the framing of the course itself.

Characterizing Students With Low Final Conceptual Exam scores

Despite strong average FMCE gains for the class, a considerable fraction of students do not achieve final scores above 50%, indicating a lack of Newtonian conceptual understanding[9]. This is seen in the long, flat "tail" of the post-test distribution (Fig 1), which includes roughly 15-20% of the class. These students are *not* generally the ones failing the course, but are more typically C students. (Many D/F students missed one or both of the classes in which the test was given.)

What is most significant about this tail is that their average homework scores, online participation, and attendance (effort based measures) are almost the same as students getting high FMCE posttests. That is, these students are apparently working as hard, participating at nearly the same level, and doing almost as well on homeworks, but are failing to master the conceptual topics measured by the FMCE. Further investigation shows little significant difference in physics backgrounds, or self-reported use of resources (e.g. reading the texts). The tail population has a marginally lower pre-class math diagnostic score, and self-reports a marginally *higher* amount of time spent on physics outside of class (5.7 hrs/week vs. 5.3 hrs/week)

Demographically, women are somewhat more represented in this tail (34% of the tail, vs. 25% for the overall class), and undeclared majors make up a larger fraction of this population (26% of the tail vs. 16% overall) Their average physics pre-test score is lower (13% vs. 32%), but this alone does not characterize the population, because many other students with low pretests do very well - indeed, students in the bottom quartile of pretests overall have a normalized gain of nearly 50%. The normalized gain of the "tail" group is 18% (vs. 78% for those *not* in the tail) We are naturally very interested in this population, because they are the students for whom our efforts at reforming the class are apparently not working well, yet they are not "slackers", and perhaps could be better helped in some other way. Further research is clearly essential.

Student Attitudes and Beliefs

Student attitudes and beliefs about the nature of learning, and of science, are interesting for a variety of reasons. Part of our goal is to help shift students' attitudes from "novice-like" to "expert-like", and attitudes may play a role in learning, performance, and appreciation of the course[11]. We use the CLASS survey as a rough measure of self-reported attitudes and beliefs (AB's). We find no regression in AB's over the course of a semester[7,12], but do find statistically significant differences in AB's between lower and higher performing students (See ref [7] for more detailed discussion). We make no claim regarding cause and effect, but the correlation is itself of interest. E.g., the tail population described above with low final FMCE score began the course with an average overall CLASS score of 60±2% compared to 71±1% for high FMCE scorers. This can also be seen in Fig 3, where students with low FMCE learning *gain* entered the class with lower CLASS pre-scores, and ended even lower. Students with higher learning gains tended to start with more "expert-like" attitudes, and became *more* expert-like after the course.

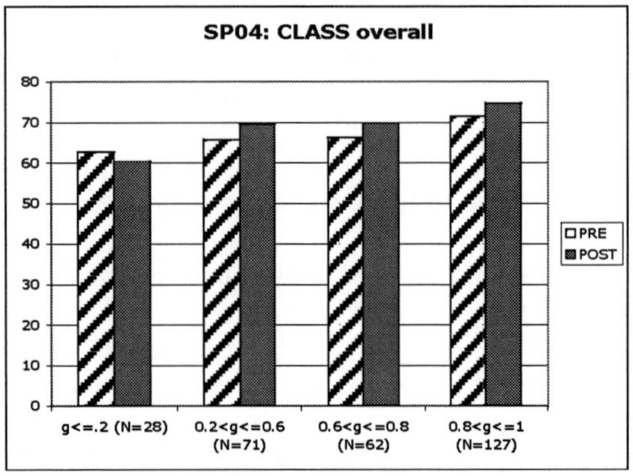

FIGURE 3. Average pre/post overall scores on the CLASS exam[7] which provides some measure of student attitudes and beliefs about learning, and the nature of science. Bins refer to average normalized gain (g) on the FMCE.

CONCLUSIONS

As the physics teaching community becomes increasingly sophisticated at using and integrating multiple proven reforms, questions of how various course components interact and impact students become correspondingly richer and more significant. We have implemented three distinct reforms in a large lecture course: peer instruction, Tutorials, and automated homeworks, and have begun an analysis of how they contribute to student learning and student attitudes. The data shown here focus first on the direct results of the reforms, and on the relationship of course elements to learning gains. We find that no single element alone appears to account for the measured successes, but the combination of course elements has a large impact. We have also identified a population of students for whom this course structure is not effective, and have begun the work of identifying and characterizing them, with the goal of further improving the mix of classroom approaches. No single demographic identifies this population in advance. We do find that student attitudes and beliefs are associated with both grades and learning gains, and believe that further study of e.g. explicit attention to epistemology is warranted.

ACKNOWLEDGMENTS

The author thanks the Physics Education Group at U. Washington for their support as we implemented Tutorials, particularly Peter Shaffer for his exceptional help with data collection, pretests, and exams. Thanks also to the Pew/CASTL program, the NSF STEM-teacher prep grant which supports our learning assistants, CU Physics for its support in implementing Tutorials and Peer Instruction, and all the members of the Physics Education Research at Colorado group.

REFERENCES

1. McDermott, L., Redish, E., *American Journal of Physics*, **67**(9), 755-767, 1999, Redish, E., *Teaching Physics with the Physics Suite*, Wiley 2003
2. Hake, R., *American Journal of Physics* **66**, 64-74, (1998) Hake, R., Conservation Ecology **5**(2):28, (2002)
3. Mazur, E., *Peer Instruction,* Prentice Hall 1997
4. McDermott, L., Shaffer, P., and the PEG, *"Tutorials in Introductory Physics"*, Prentice Hall 2002.
5. CAPA homework system: see www.lon-capa.org
6. Halliday, Resnick, Walker, *Fundamentals of Physics* Wiley 2004, and *Thinkwell Physics*: www.thinkwell.com
7. http://cosmos.colorado.edu/phet/survey/fall2003.html. W. Adams et al., "*Design and Validation of the Colorado Learning About Science Survey*", and K. Perkins et al., "*Correlating student attitudes with student learning using the CLASS*", both in review for Proceedings of the PERC 2004, Sacramento, CA.
8. SALG, the Student Assessment of Learning Gains: www.wcer.wisc.edu/salgains/instructor/
9. Thornton, R., Sokoloff, D., *Am. J. Phys* **B66**(4), 338-352, 1998, also Thornton, R., Kuhl, D., Cummings, K., Marx J., "*Comparing the FMCE and the FCI*", unpublished.
10. McDermott, L., Shaffer, P., Somers, M, *Am. J. Phys.* **62**(1), 46-55 (1994).
11. Elby, A. *Am. J. Phys* **69**(7), (2001)
12. Redish, Saul, Steinberg *Am. J. Phys.* **66** 212-224, (1998)

Teacher and Curriculum Factors that Influence Middle School Students' Sense-Making Discussions of Force/Motion

Cody Sandifer

Department of Physics, Astronomy, and Geosciences
Towson University, Towson, MD 21252
csandifer@towson.edu

Abstract. This study investigated small-group discussions in an inquiry-based middle school science classroom. The purpose of the study was to determine the teacher and curriculum factors that provide support (or not) for students' sense-making discussions. To do this, two student groups were videotaped as they participated in force/motion activities. Analysis revealed that sense-making discussion was influenced by an occasional lack of teacher adherence to the curriculum philosophy, particular types of discussion prompts, the drawing of free-body and energy diagrams, and other teacher and curriculum factors.

RATIONALE AND PURPOSE

The National Research Council [1] recommends that science education should be grounded in collaborative inquiry. Inherently, sense-making discussions are an important aspect of this type of activity.

In my earlier analysis of sense-making discussions in middle school classrooms, it was found that there were a number of group- and student-specific factors that influenced the type and quality of scientific sense-making discussion. [2] The present study builds on this earlier analysis by introducing two additional factor types: teacher factors and curriculum factors. In investigating sense-making discussion (SMD), it is crucial to include these types of factors because of the reality that teacher behavior and curriculum structure are the two factors over which science educators have the greatest control.

Recognizing that particular aspects of teacher behavior and curriculum structure are likely more effective at promoting SMD than others, the goal of the present study is to answer the following question: Which teacher and curriculum factors support (or hinder) middle school students' small-group sense-making discussions?

PARTICIPANTS AND DATA

I videotaped two four-person groups for five weeks as they worked through a portion of the Force and Motion unit of the Constructing Ideas in Physical Science (CIPS) middle school curriculum. [3] As supplemental data, I collected each participant's worksheets, homework, and exams. I also recorded whole-class video and fieldnotes for each class period.

CIPS is an inquiry-based physical science curriculum that engages eighth grade middle school students in constructing meaningful understanding of physical science concepts. CIPS achieves this goal by having the students interact with each other in small groups, participate in whole-class discussions, perform experiments, observe computer simulations, and use notebooks to keep track of the change in their understandings of physical science. CIPS content units are broken into cycles (sub-units). For instance, in the pilot version of CIPS that was investigated here, the Force and Motion unit was broken into Exploring Motion, Pushes and Pulls, Combining Pushes and Pulls, Resistive Interactions, and Gravity. In this study, data were collected on the last 3 cycles of the Force and Motion unit.

The data collected in this study are identical to the data that were collected for my earlier research on

CP790, *2004 Physics Education Research Conference*, edited by J. Marx, P. Heron, and S. Franklin

group and individual factors affecting SMD; the data have been re-analyzed with a new focus on teacher and curriculum factors.

THEORETICAL FRAMEWORK

My six-component framework for SMD is as follows.

1. Predicting a phenomenon or experimental outcome.
2. Clarifying the facts of a phenomenon or experimental result.
3. Describing and explaining a phenomenon or experimental result.
4. Defining, describing, clarifying, and connecting scientific concepts, procedures, processes, and representations.
5. Testing knowledge compatibility.
6. Making a request for any of the above.

Each of these six components of SMD is explained in detail in my earlier report on group and individual factors affecting SMD.

ANALYSIS PROCEDURE

Once the student discussions were transcribed, instances of student sense-making were classified according to the six-component scheme listed above. An overall numerical distribution of these instances was formulated (omitting requests for SMD), which was then statistically compared to the expected distribution (as determined by the distribution of sense-making prompts in the actual curriculum materials). In addition, the percentage of time dedicated to SMD was calculated for each activity.

The final step focused on qualitatively identifying those teacher and curriculum factors that helped to explain the differences in SMD, both in terms of instances of sense-making and sense-making percentages. This identification and analysis [4] of factors was performed by making several analytical passes through the session transcripts and associated videotapes; during these data passes, I carefully checked for supporting evidence, consistency, and alternative explanations.

QUANTITATIVE RESULTS

TABLE 1. Distribution of Sense-making Instances in Group Discussion Compared to the Distribution of Sense-Making Prompts in the Curriculum Materials.

Component of SMD	Instances of Component in Group Discussion	Number of Prompts Addressing Component in Materials
CL	173 (33%)	25 (20%)
P	47 (9%)	18 (14%)
UE	122 (23%)	26 (21%)
DDC	113 (21%)	52 (41%)
TC	71 (13%)	5 (4%)

Abbreviations: CL = Clarifying, P = Predicting, UE = Underlying Explanation, DDC = Defining, Describing, Connecting; TC = Testing Compatibility

χ^2 tests show that all discussion/materials differences in Table 1 are significant at the **p** = 0.05 level, except for differences in underlying explanations.

TABLE 2. Percentage of Time that Groups Engaged in SMD, by Activity.

Activity	%	Activity	%
C3/A1	17	C4/A5	10
C3/A2	17	C4/A6	10
C3/A3	7	C5/A1	9
C3/A4	8	C5/A2	14
C4/A1	17	C5/A3	39
C4/A2	14	C5/A4	22
C4/A3	35	C5/A5	1
C4/A4	44		

Abbreviations: C3/A1 = Cycle 3, Activity 1, etc.

QUALITATIVE RESULTS

There were two teacher factors and five curriculum factors that contributed to the most dramatic sense-making differences found in Tables 1 and 2.

Teacher Factors

Factor 1: Due to the unexpected length of certain activities and the teacher's desire to complete the entire CIPS curriculum, the concluding sense-making sections of some activities were skipped.

In the pilot version of the curriculum, many activities took much longer to complete than the developers intended. As a result, the teacher - who made it known that he felt obligated to go through the entire five-unit curriculum, if possible - would sometimes start a day with a new activity, even though the class hadn't quite finished the end of the previous activity. Of importance here is that the final sections of the activities were key sense-making sections. These concluding ("Making Sense") sections focused on in-depth scientific explanations, and so the skipping of these sections was one of the factors that led to there being fewer DDCs than expected. This factor also contributed to the variation in SMD from activity to activity; had these sections not been skipped, the percentage of time engaged in SMD would likely have been higher in the truncated activities (e.g., C3/A3, C4/A5, C5/A5).

Factor 2: The teacher would, on occasion, fail to adhere to the CIPS curriculum philosophy by revealing in advance the ideas that groups were supposed to construct on their own, and so SMD sometimes became unnecessary.

One of the hallmarks of an effective teacher is the ability of that teacher to outline for the students, without being too specific, the types of science ideas that will be addressed by the day's activities. The notion of being "too specific" is important in CIPS, because of the curriculum's philosophy that idea generation should come from the students, not the teacher. Occasionally, the teacher in this study, in the interest of getting the students properly focused, would violate this philosophy by explicitly telling the students the scientific concept underlying the day's activities.

For example, the teacher introduced the activity about resistive forces (C4/A1) by stating that "friction has now entered the world" - with the unsurprising result that students quickly offered this explanation for the interactions discussed in the activity. Had the teacher not made his introductory statement, it is very likely that student SMD in this case would have been much more in-depth.

Curriculum Factors

Factor 1: Group discussion of concepts not explicitly addressed by the curriculum resulted in a greater number of clarifications of facts than expected.

There were a number of topics that the CIPS developers intentionally omitted from the curriculum in order to ensure that the curriculum could be completed in a single school year. These omitted topics included instantaneous speed, the distinction between friction and drag, and the extent of gravity in the universe. Nevertheless, groups dedicated sense-making time to these topics as students attempted to clarify the facts about speed, drag, and the pull of gravity. Areas of student concern included the following: Can a speed be considered "constant" if an object possesses that speed for a second, or less? What is the exact difference between friction and drag? Is there gravity in the Earth's outer atmosphere?

Factor 2: Groups enthusiastically engaged in SMD in activities that were interesting, personally relevant, and/or asked the students to be creative.

Activity C4/A4 had the highest sense-making percentage of all the activities: 44%. In this activity, students were asked to consider which sports might be playable without friction, and were also asked to give reasons why they would or would not want to live in a frictionless world. These relevant and engaging questions led to creative discussions that were consistently deep and thoughtful. The discussions were carried on at great length, and were unique in the fact that they appeared to be truly enjoyable for the students.

Factor 3: Curriculum prompts that required students to inquire about and record their fellow group members' ideas were effective at promoting SMD.

Over 99% (523 out of 526) of the sense-making instances in this study were direct or indirect responses to questions or other sense-making prompts in the curriculum materials. This demonstrates the crucial role played by the written materials in supporting SMD in middle school classrooms.

A particularly effective type of curriculum prompt consisted of a series of "fill-in-the-blank" statements that prompted students to elicit and record the ideas of their fellow group members. For example, in C5/A3, students were asked to consider whether a hammer on Earth or a hammer on the Moon would hit the ground first if dropped simultaneously, using the following prompt format: "____ thinks that [one of the hammers will drop first] because ____." One final summative prompt was then added to help ensure a rich discussion: "After our discussion, our groups thinks [one of the hammers will drop first] because ____."

Lengthy verbal responses to these prompts played a large role in helping the students achieve the 39% sense-making percentage in C5/A3.

Factor 4: Certain activities dealt with experimental results or scientific explanations that were too obvious, and so SMD was sometimes unnecessary.

In the pilot version of the curriculum, because of the heavily constructivist stance of the curriculum developers, students performed experiments that tested the validity of each and every benchmark idea - even in cases when the ideas of interest were relatively commonsense. From the students' perspective, this translated into a handful of experiments that were deemed unnecessary because the students felt that they already had an intuitive understanding of the ideas explored in the experiment. Such experiments could be found in C3/A1 (combined weight is the sum of individual weights), C3/A2 (a pulley system moves towards the side with more weight), and C5/A2 (all objects are affected by gravity).

The following is a typical student comment regarding these experiments: "Why are we doing this? I mean, we know what the answer is." In these cases, SMD was kept to a bare minimum because group members felt that their understanding of the experimental results (and accompanying scientific explanations) was already adequate.

(Note that the current version of CIPS improves greatly on the pilot version, and so the time and content issues discussed above are no longer problematic.)

Factor 5: Activities that required students to draw free-body or energy diagrams generated a significant amount of SMD.

Conceptualizing a physical system in terms of energy types and transfers is a powerful, yet incredibly abstract process. Due to the high level of abstraction involved, the CIPS students' energy-oriented descriptions and explanations tended to be imprecise, brief sentences in which students confused/ignored the different types of energy transfers, and also did not differentiate between the various parts of the system. However, exceptions to this tendency towards vagueness and a lack of SMD can be found in those activities (e.g., C4/A3) where students were asked to fill out an energy transfer diagram. These energy transfer diagrams, which contained "fill-in-the-blank" spaces for system parts and energy transfers, provided the conceptual scaffolding that supported rich SMD by requiring students to explicitly consider different types of system objects and energies - something that they may or may not have considered doing previously.

Free-body diagrams served a similar function by requiring students to identify and draw individual forces in a quantitatively precise way, and so free-body diagrams also tended to lead to rich SMD in groups.

CONCLUSION

Many teacher and curriculum factors influenced the SMD in this study: the teacher's skipping (for time reasons) of important sense-making activity sections; the teacher's occasional lack of adherence to the CIPS curriculum philosophy; the omission of certain specific science topics; relevant and creative science activities; particular types of discussion prompts; the drawing of free-body and energy diagrams; and the performance of commonsense experiments that students deemed unnecessary.

These factors have implications for classroom practice. Assuming that the teacher is using an inquiry-based curriculum, the teacher should make an effort to consistently support sense-making (both verbally and in the written curriculum) as well as adhere to the curriculum philosophy, while also remaining open to opportunities for improving activities by making them more relevant and grade-appropriate.

REFERENCES

1. National Research Council. (1996). National science education standards. Washington, DC: National Academy Press.
2. Sandifer, C. (2002). Factors influencing middle school students' sense-making discussions during their small-group investigations of force/motion. Annual Proceedings of the Physics Education Research Conference, 117-120.
3. Goldberg, F., Bendall, S., Poel, R., & Heller, P. (2004). Constructing ideas in physical science. Armonk, NY: It's About Time, Inc.
4. Sandifer, C., & Johnson, A. (2002). Using qualitative methods to make and support claims in physics education research. Annual Proceedings of the Physics Education Research Conference, 17-20.

The Role of Evaluative Abilities in Physics Learning

Aaron R. Warren

Dept. of Physics & Astronomy, Rutgers University, Piscataway, NJ

Abstract. This paper outlines ongoing research regarding the role evaluative abilities play in student learning. Evaluative abilities refer to a student's capacity for employing several general strategies for checking, critiquing, and judging information. In the context of physics, such information may include proposed problem solutions, proposed models, and conceptual claims. Strategies for evaluating information include dimensional analysis, checking special and limiting cases, and analyzing the assumptions which underlie a model, solution, or claim. Over the past year I have been developing and testing a number of activities designed to promote and assess students' evaluative abilities. I report preliminary results regarding the correlations between students' evaluative abilities and their understanding of physics, and identify future avenues of research.

MOTIVATION & BACKGROUND

Reports from the AAAS[1] and NSF[2] emphasize that science courses ought to cultivate a wide array of abilities. In particular, these reports agree that students should develop those abilities which allow them to become self-corrective learners. The ability to critically evaluate information is a central component of self-corrective learning. Evaluative abilities allow students to identify mistakes in their thought and work, a necessary first step towards correcting such mistakes.

By the term "evaluative abilities," I mean the ability of a student to use any of several general strategies for checking and judging information[3]. There are several types of information which a student typically produces. Students often propose problem solutions (e.g., while working out homework problems), which entail the use of a model to calculate or predict some quantity. Students may also work to answer conceptual questions in which they must make some claim by employing their qualitative understanding of physics. Finally, students may propose models for specific situations in order to explain or predict the outcome of an experiment.

I focus on a few general strategies for evaluating these types of information. Dimensional analysis is a powerful technique for determining whether a given equation is physically self-consistent. Checking special and limiting cases determines how robust a certain model may be by testing whether it is consistent with other models or conceptual ideas. Similarly, identifying and checking the assumptions which underlie a model allow one to determine how accurately a model can explain or predict a system's behavior in a given situation.

Theoretical Perspective

There are several arguments for why enhancing students' evaluative abilities should enhance student problem-solving performance (on both conceptual and traditional problems), and each argument appeals to a different theoretical perspective of knowledge and learning. One such perspective relies on the notion of schemas. While there are other theoretical perspectives which may be used (e.g., based on neuroscience, or conceptual change theory), I leave such discussions for future publications.

A schema is a pattern of activated knowledge incorporating facts, rules, p-prims, facets, and other knowledge elements. Research suggests that many of the differences between students and physics experts are due to differences in the content and organization of their schemas[4]. Experts have schemas which incorporate knowledge into a hierarchical, globally coherent pattern. Students learn by acquiring new knowledge elements, and by developing or changing

CP790, *2004 Physics Education Research Conference*, edited by J. Marx, P. Heron, and S. Franklin

links between knowledge elements. In this context, the use of an evaluative strategy may be represented as a schema (which I call a *meta-schema*) that is linked to an array of schemas responsible for processes such as problem solving. When a particular schema is activated, an evaluation meta-schema can be triggered. The use of an evaluation meta-schema can lead to the restructuring of the triggering schema, or of links attached to the triggering schema, by causing the student to recognize incoherencies among schema and inconsistencies between the product of the triggering schema (e.g., a proposed problem solution) and a desired result (e.g., a correct problem solution). Evaluation provides feedback which determines whether a schema is inappropriately activated, and whether the schema is sufficient for some task.

Developing & Assessing Evaluative Abilities

Although evaluation abilities are likely to play a significant role in learning[3], they receive relatively little attention in most curricula. Over the past year, I have been developing a set of activities which are designed to develop and assess evaluative abilities among physics students (available for download from http://paer.rutgers.edu/ScientificAbilities/FormativeAssessmentTasks/default.aspx). To maximize the effectiveness of the evaluation activities, I have structured them as formative assessments. Formative assessment activities appear to be among the most powerful instructional intervention methods currently known[5].

Formative assessment tasks generally follow three guiding principles[6]. First, a task must inform students of the learning and performance goals. Secondly, a task should inform each student of their current level of understanding. Finally, a task should help guide each student to achieve the performance goals by developing strategies and skills. The general role of formative assessment, then, is to supply real-time feedback to the students and instructor which identifies a goal state, each student's current state, and a method for each student to progress from her current state to the goal state.

In order to assess and develop students' evaluative abilities, they must first be given something to evaluate. Therefore, each evaluation task will present some proposed problem solution, proposed model, conceptual claim, or experiment report which the student must critically judge. To satisfy the first requirement for formative assessments, I have developed rubrics, guidelines, and examples which inform the students of the nature of the tasks and how their responses will be graded. The rubrics identify four classes of possible responses, with each class indicating a certain level of ability development. Guidelines and examples inform the students of strategies to be used when responding to an evaluation task. By discussing these guidelines and examples with their peers and instructor, students can begin to develop an understanding of each evaluative strategy.

To satisfy the latter two conditions for formative assessment, the students are presented with many opportunities to develop their evaluative abilities. In recitations, students work on evaluative tasks and discuss them with each other and with instructors. In laboratory, students work on evaluative tasks as part of an exit interview process where each student group discusses the tasks among themselves and with the instructor. In lecture, students discuss evaluative tasks in small groups. These activities allow the students to understand how their current level of performance compares with the goal performance, as well as how to improve their performance.

Research Goals

We report on research designed to conduct an initial investigation of the role evaluation plays in student learning and problem-solving performance. This research constitutes a trial study to determine whether more extensive investigation into this idea may be worthwhile. Based on the theoretical discussion above, I expect strong evaluative abilities to correlate with enhanced student learning and better problem-solving performance. As an initial test of this claim, I can measure students' evaluative abilities (using the activities I developed) and determine whether there is a significant positive correlation with student scores on exams.

One may suspect on theoretical grounds that the development of evaluative abilities actually causes students to develop a robust understanding of physics. Although a significant positive correlation would not imply any causal link, it would be consistent with such a hypothesis and therefore indicate that further investigation of this issue is worthwhile.

The validity of any measured correlation relies upon two major assumptions. First, I assume that our activities do indeed measure students' evaluative abilities. Second, I assume that exam scores accurately reflect each student's understanding of physics, at least to some degree. To test both of these assumptions I can conduct student interviews to gather more detailed data on student understanding and compare these data with each student's score on evaluative activities and exams. Such interviews will be conducted during the upcoming academic year.

RESEARCH

Data Collection & Coding

During the 2003-4 academic year evaluation activities were developed and administered to students in an algebra-based introductory physics course for life science majors at Rutgers University. The course had an enrollment of roughly 150 students. Most of the activities were administered during the weekly laboratory sessions. Here, students worked with their lab partners (3-5 students per group) to complete several evaluation activities. The evaluation activities were designed around the subject matter which had been dealt with in that week's lectures, workshops, and labs. When a student group felt it had completed the evaluation activities, they presented and discussed their work with other lab groups and with the lab instructor. These discussions gave students the opportunity to make repeated attempts at understanding and achieving the goals of the activities while incorporating feedback from their peers and instructor, and constituted the formative assessment implementation of my evaluation activities.

As part of this course, students had to take two lab practicals each semester. The lab practicals included short-answer format questions which required a written response. Student responses were photocopied before being graded by the lab instructors, who used the rubrics I developed in order to assign grades. Thus, the lab practicals served as a summative assessment tool.

The photocopied student responses to the first and last practicals of the year (practicals 1 and 4) were used as our raw data. Each evaluation activity on the lab practicals was designed to evoke the use of a particular evaluative strategy. Therefore, student responses to each question were coded in order to measure the level of performance with a specific evaluative strategy. The coding assigned an integer between 0 and 3 to each student response. Similarly, students' answers to the conceptual questions were coded on an ordinal scale from 0-3 using a taxonomy developed by the course teaching assistants.

There were six different topics covered on each practical. Each student was randomly assigned a topic at the practical. Each practical had at least two evaluation questions, and each question required the use of a different evaluative strategy. This design gave an average of 65 students on each practical (this number varied between 64 and 72) for which we could determine the correlation between a specific evaluation ability and performance on corresponding conceptual questions from the practical, as well as the correlation with performance on the multiple-choice exam. This total of ~65 students for each correlation was obtained by combining student scores for a specific evaluation activity regardless of the physical context of the activity. This was done in order to have a sufficient number of data points for calculating correlations.

It is not clear whether such a combination of scores from different contexts should affect the correlations. The possible context-dependence of evaluation abilities is something we will address in future work. While the actual implementation of an evaluation strategy certainly depends on the topic at hand (e.g., needing to know the dimensions of a certain quantity), it may be that the actual correlations between evaluation abilities and conceptual understanding/exam performance are independent of context. Investigation of this matter is necessary to test the robustness of our current results. I am currently repeating this study with more subjects in order to address this issue.

Preliminary Results & Discussion

My coding for each practical was subjected to inter-rater reliability checks with three other members of the Rutgers Physics Department. The overall agreement level was 96%, and the agreement level for each rubric varied between 91% and 99%. To examine the relationship between student evaluation abilities and student performance on exams, I calculated Kendall's Tau-b. Those correlations which were significant (at least at the 0.05 level, 2-tailed) for both the first and fourth practicals are shown in Table 1.

I also note that the correlation between conceptual scores and exam scores was consistently significant (Practical 1, $\tau_b = 0.293$, $p = 0.000$, N=135; Practical 4, $\tau_b = 0.272$, $p = 0.000$, N=169). This supports our assumption that student exam scores reflect their understanding of physics.

TABLE 1. Significant Nonparametric Correlations

Evaluative Strategy Type	τ_b (Practical 1) p (2-tailed)	τ_b (Practical 4) p (2-tailed)	Measurement of Understanding
Limit/Special Case Analysis	0.294 .007	0.323 .000	Conceptual Score
Limit/Special Case Analysis	0.312 .002	0.227 .003	Exam Score
Dimensional Analysis	0.358 .002	0.275 .009	Conceptual Score
Assumption Analysis	0.244 .020	0.357 .003	Conceptual Score
Assumption Analysis	0.292 .003	0.292 .008	Exam Score

The results in Table 1 indicate that evaluative abilities generally correlate with students' understanding of physics. However, dimensional analysis did not significantly correlate with exam score, although it did correlate with conceptual scores as shown. This indicates that dimensional analysis is related to students conceptual knowledge, but is largely independent of their quantitative work. On the other hand, limit/special-case analysis and assumption analysis each correlated with both conceptual and quantitative problem scores. These results seem to suggest that different evaluation abilities correspond to different aspects of students' knowledge and use of their knowledge.

I therefore feel that the structure, function, and evolution of evaluation abilities warrant more thorough study. In particular, I would like to determine whether there is in fact a causal connection between evaluation abilities and learning. Although the current results are consistent with the hypothesis that evaluation abilities are necessary for robust learning, they of course do not warrant such a conclusion because correlation does not imply causation. An alternative explanation may be that there are some confounding variables (e.g., effort, interest, prior knowledge) which are responsible for the measured correlations. I will be conducting research, using both large-N and small-N techniques, to further study the relationship between evaluation abilities and learning.

With the help of Alan Van Heuvelen and Sahana Murthy, I will conduct a large comparison group study between two parallel introductory large-enrollment classes to test whether improving evaluation abilities causes improved exam performance. One course (experimental group) will include my activities as part of recitations, homeworks, and labs, while the other course (control group) will not. Both courses will include evaluation activities on the exams to measure evaluation abilities, and several multiple-choice problems on the same topic. We expect that only students from the experimental group will significantly improve their evaluation scores over the duration of the course (indicating my activities aid the development of evaluation ability). We also expect that as the experimental group's evaluation scores improve, they will significantly outperform the control group on the multiple-choice exam problems (indicating that improved evaluation abilities caused improved exam scores). Videotaped student interviews will be used to triangulate the data.

ACKNOWLEDGEMENTS

I would like to thank Alan Van Heuvelen, Eugenia Etkina, Marina Milner-Bolotin, David Brookes, Michael Gentile, Sahana Murthy, and Jessica Warren for help with the implementation of my tasks, inter-rater checks, and for feedback on this manuscript. I also thank the anonymous referees for valuable feedback. Supported in part by NSF grants DUE 0241078, DUE 0336713.

REFERENCES

1. American Association for the Advancement of Science, Project 2061, *Benchmarks for science literacy.* New York: Oxford University, 1993.
2. NSF Directorate for Education and Human Resources Review of Undergraduate Education, *Shaping the future: New expectations for undergraduate education in science, mathematics, engineering, and technology*, 1996. Recommendations may be seen at http://www.ehr.nsf.gov/egr/due/documents/review/96139/four.htm
3. Anderson, L.W. & Krathwohl, D.R., *A taxonomy for learning, teaching, and assessing: A revision of Bloom's Taxonomy of Educational Objectives.* Longman, 2001.
4. Sabella, M.S., Ph.D. Dissertation. University of Maryland, 1999.
5. Black, P. & Wiliam, D., *Inside the black box: Raising standards through classroom assessment.* London: King's College, 1998.
6. Sadler, R., *Instructional Science,* **18**, 119-144 (1989).

Concerning Scientific Discourse about Heat

D. Brookes, G. Horton, A. Van Heuvelen, E. Etkina

Department of Physics and Astronomy, Rutgers University, 136 Frelinghuysen Rd, Piscataway, NJ, 08854

Abstract. We aim to examine communication in physics from a linguistic perspective and suggest a theoretical viewpoint that may enable us to explain and understand many physics students' alternative conceptions. We present evidence, in the context of the concept of heat, that physicists seem to speak and write about physical systems with a set of one or more systematic metaphors that are well understood in their community. We argue that physics students appear to be prone to misinterpreting and overextending the same metaphors and that these misinterpretations exhibit themselves as students' alternative conceptions. We will analyze physicists' discourse about heat and present evidence of a connection between students' alternative conceptions and the possibility that they are misinterpreting the language that they read and hear.

INTRODUCTION

David Meltzer describes a study in which he gave 32 calculus-based introductory physics students at Iowa State University a closed thermodynamic cycle and were asked whether the net work done by the gas, and the total heat transferred to the gas, were positive, negative, or zero[1]. 69% of the interviewees said (incorrectly) that the total heat transferred to the gas for the entire process was zero. 63% said that the total work done by the gas was zero. Meltzer presents following quotations as a typical student responses.

> "[S1] The net work done by the gas... is equal to zero... The physics definition of work is like force times distance. And basically if you use the same force and you just travel around in a circle and come back to your original spot, technically you did zero work."
>
> "[S21] The heat transferred to the gas... is equal to zero... The gas was heated up, but it still returned to its equilibrium temperature. So whatever energy was added to it was distributed back to the room."

There are many possible reasons why students thought that the total heat and/or work for the cycle must be zero. Meltzer explains that students believe that work and heat are state functions of the system. We wish to investigate this idea further and attempt to explain why students might view heat as a state function. Terms such as "added to" suggest students are viewing the gas/system as a container of heat, and that they are confusing energy and heat as substances and using temperature as an indicator of the amount of heat *in* the system. In contrast, the argument about work seems to suggest a simpler view of work as a state function based on the student's prior experience of work done in a conservative field. Thus we hypothesize that students' confusion with work as a state function comes from inappropriate transfer, while their confusion with heat may have linguistic origins. Our interest is with the role of language in students' reasoning and therefore we will focus on heat in this paper.

We will address two research questions. (1) We want to propose a coding scheme to code language used to describe physics ideas. (2) We will use this coding scheme to analyze textbook language about heat and connect this language to students' reasoning about heat in thermodynamic systems.

THEORETICAL FOUNDATION

There has been considerable theoretical speculation[2, 3, 4, 5, 6, 7] and some practical evidence[7, 8] that language is not just a passive representation of reality. Rather, language also influences what the language user perceives and understands. Physicists and physics education researchers have argued that there is a linguistic component to mastering physics; that part of the difficulty of learning physics may lie in the language used to convey physics concepts. In particular, Arons and Touger have separately suggested that certain common locutions used by physicists may serve to reinforce incorrect ideas[9, 10]. Williams[11] has argued that part of students' difficulty in learning physics may lie in (1) the obscurity of the specialized meanings associated with certain terms, and (2) physicists' sometimes careless and contradictory use of language. Sutton[12, 13] has suggested that part of learning science must involve the reactivation of "dead"

CP790, *2004 Physics Education Research Conference*, edited by J. Marx, P. Heron, and S. Franklin

metaphors. For example, to say that "heat flows..." suggests a metaphor, "heat is a fluid". Such a metaphor started out as an analogical model of heat as caloric fluid. To simply say "heat flows..." hides the applicability and limitations of this model. Itza-Ortiz et. al.[14] have shown a correlation between students' ability to distinguish every-day meanings of the word *force* from its meaning in physics and their performance answering class test questions which included the concept of *force*. Similar correlations were found for the words *impulse* and *momentum*.

When it comes to language about heat, there is consensus that physicists' language can be misleading, but little agreement about why it is misleading or how it can be corrected. Zemansky is content to accept using heat as a noun as long as it is clear that the term refers to the name of a process whereby energy is transferred into or out of a system. He cautions against using the term heat as a verb[15]. Baierlein, on the other hand, says exactly the opposite. To make the process nature of the term heat clear, he argues that one should never use heat as a noun, but rather as a verb[16]. Bauman[17] suggests replacing the term "heat" with "thermal energy" while Zemansky condemns this as a confusing oversimplification[15].

To resolve this confusion and to understand students' reasoning, we will consider language to be a representation of a physicist's model in the same way as a picture, or an equation. Inside any representation is a basic ontology of objects/substances, processes and states[18]. In the basic thermodynamics model the objects are point particles, processes are heating and work (processes of energy transfer) and the system possesses states (energy, pressure and temperature etc...). One of the functions of language is to encode this ontology. Consider (1) "heat flows" versus (2) "energy is added to the system by heating". In the first case heat is functioning as a substance that is moving from one place to another. In the second, "heating" is functioning in the phrase as a process by which energy moves. The key to uncovering the implicit ontology in a linguistic representation lies in grammar of the phrases. This will be explained in the method section below.

METHOD

To study the language physicists use to talk about heat, we analyzed three popular college level introductory physics textbooks[19, 20, 21]. Williams has argued that such textbooks represent a higher standard of linguistic rigor than the regular talk of physicists[11]. Thus a study of textbooks can at least give us an upper bound on the quality of language used to refer to the concept of heat. The goal of the analysis is to present a scheme that can be used to understand the types of meanings that may be construed from the language of physics. To this end, we used an ergative/nonergative model of grammar. The detail of this approach may be found in Ref. [22]. In brief, the model identifies two core constituents of any clause in English, namely a *process* (verb or verb group) and *medium* (noun or noun group) which participates in the process. For example, "heat [*medium*] flows [*process*]". We used this scheme to classify the function of heat in each sentence (that contained the word "heat") in the textbooks, covering the chapters on temperature and thermodynamics. Compound sentences in which heat occurred twice were broken into two clauses and coded separately. This grammatical scheme was then mapped to a basic ontology coding scheme (substances, processes and states) in the following way. If heat functioned as the *medium* of the clause, or as a descriptor of the medium (grammatical *role*) it was classified ontologically as a *substance*. Cases where heat functioned as a grammatical process, descriptor of that process (grammatical *manner*), or as a nominalization (naming) of a process of energy transfer, were coded ontologically as a *process*. In a few cases where the word heat functioned grammatically as a *modifier* (adjective or adverb), it was classified ontologically as a *state*. The full mapping is shown in table 1. Using this mapping from grammar to ontology ensured the reliability and reproducibility of the coding.

TABLE 1. Table of grammar-ontology mapping

Substance	Process	State
medium, agent, beneficiary, range, role	process, nominalizations of process, manner	any modifier

RESULTS

The results of the ontological analysis described above, are presented in figure 1.

We include six definitions of heat from six popular textbooks to serve as examples of the coding scheme. The first four are essentially substance based definitions, the last two are process definitions of heat:

- "Heat is energy that flows from a higher-temperature object to a lower-temperature object because of the difference in temperatures."[23]
- "Heat is the energy that is transferred between a system and its environment because of a temperature difference that exists between them."[24]

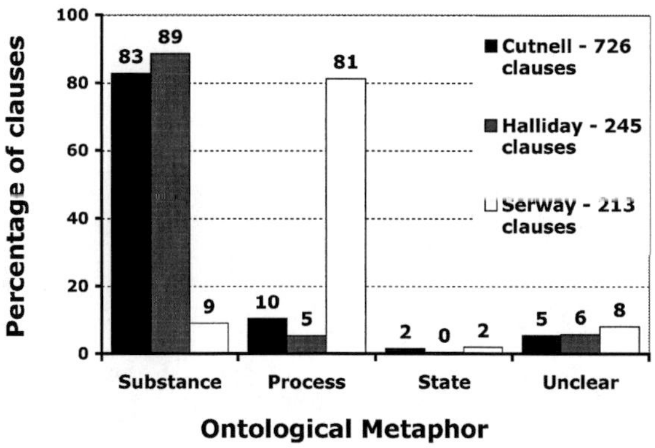

FIGURE 1. Classification of heat clauses into ontological categories.

- "Heat is energy that is transferred from one system to another because of a difference in temperature."[25]
- "...we will define heat as follows: Heat is the energy transferred between objects because of a temperature difference."[26]
- "...scientists came to interpret heat not as a substance, and not even as a form of energy. Rather, heat refers to a *transfer of energy*: when heat flows from a hot object to a cooler one, it is energy that is being transferred from the hot to the cold object."[27]

- "Heat is the transfer of energy from one object to another object as a result of a difference in temperature between the two."[28]

Compare statements, "heat is energy that flows/is transferred..." against "heat is the transfer of energy". In statements that suggest heat is synonymous with energy, the term "heat" is functioning as the grammatical medium and were classified ontologically as a *substance*. In statements that suggest that heat refers to a transfer of energy, "heat" is still functioning as the grammatical medium, but also as a nominalization, making the *process* nature of the term explicit. Giancoli's definition of heat[27] as the name of a process is not upheld by subsequent language in the same textbook. A cursory examination of the remainder of the text reveals a predominantly substance based language when referring to heat. Serway and Beicher[28] are the only authors of the group who use a consistent process-based language and who suggest the possibility of dual meanings associated with heat when they warn the reader: "We shall also use the term *heat* to represent the amount of energy transferred by this method."[29]

In order to further understand student's reasoning about heat, we added a another layer of analysis to this primary coding. Within the category of heat as a substance, heat is described as undergoing processes and the system serving as a receptacle of heat. From a survey of the data (textbook sentences), we identified a scheme of six metaphors (See table 2). Certain words that cue instances of the metaphor were identified. Words like "flow" and "transfer" imply movement of heat. Words like "absorbs", "into", "out" and "reservoir" imply that the system or the gas is a heat container[6]. A typical example of a sentence which would be classified as metaphor 3 is: "Determine whether the heat flows into or out of the gas."[30] The results of this coding are presented in table 2.

DISCUSSION AND CONCLUSION

Although physicists are quite aware that heat should be thought of as a *process* rather than a *substance*[15, 16, 17], our coding shows that their language does not reflect this understanding. Our coding scheme also enables us to decide systematically what language more closely reflects the correct underlying ontology of modern thermodynamics.

Students' arguments in support of zero net heat transfer for a closed cycle seem to include mentioning the volume and temperature of the system, which may reflect the particular aspect of metaphor 3 where the system is viewed as a container of heat. We hypothesize that the fact that students seem to draw on qualities of the linguistic model indicates that their reasoning may, in some cases, be primarily based in the linguistic representation of the model. Or students' misconceptions about heat as a substance come primarily from the manner in which heat is spoken about in every-day language. If so, physicists' language seems to be reinforcing this idea rather than disabusing students of the notion.

Additional evidence for these ideas comes from two students interviewed by D.B., using Meltzer's questions[1]. They confidently argued that temperature was an indicator of how much heat was in the system and that the system was a container of heat.

> [S2] "...[total heat for the cycle is] zero as well since its come back to the same temperature. That means the same amount [of heat] has been subtracted out as was initially added in." He later confirmed: "Temperature's the same, pressure's the same, so I'm saying it started and ended with the same amount of heat."

Interestingly S3 started out with essentially the same reasoning, but later hesitated:

> [S3] "Well, its the term heat...I think I am confusing the term somehow...I think I'm just

TABLE 2. Metaphorical classification of heat clauses

	Metaphor	% Cutnell	% Halliday	% Serway
1	Heat is a substance	13	13	6
2	Heat is a substance that moves from place to place	17	18	1
3	Heat is a substance that moves from system to system and the system/gas is a container of heat	53	58	2
4	Heat is a process	10	5	26
5	Heat is a process which involves the movement of a substance (energy) from place to place	0	0	16
6	Heat is a process in which a substance(energy) moves from one system to another and the system/gas is a container of that substance	0	0	39
7	Unclassified	7	6	10

making it equivalent to temperature... Because its a transfer. Heat is a process, a transfer of energy..."

S3 had previously taken her introductory physics courses from one of the authors. In this course, A.V.H. had modified his language about heat to reflect its status in physics as a *process*.

Using systematic tools of discourse analysis from linguistics, we have shown that physicists talk about heat predominantly as if it were a substance. Thus it is possible that physicists' language may be misleading students, causing them to confuse processes with objects in thermodynamics. This can explain Meltzer's conclusion that students think that heat is state function because students see the system as a container of heat. To test this hypothesis it would be necessary to modify the language used for instruction to reflect the true status of heat as a process rather than a substance and pose the same questions to the students. More investigation is obviously needed.

REFERENCES

1. D. E. Meltzer, *Am. J. Phys.* **72** (11), 1432-1446 (2004).
2. B. L. Whorf, "The Relation of Habitual Thought and Behavior to Language," in *Language, Thought, and Reality. Selected writings of Benjamin Lee Whorf*, edited by John B. Carroll, The M.I.T. Press, Massachusetts Institute of Technology, Cambridge, Massachusetts, 1956, pp. 134-159.
3. B. L. Whorf, "Science and Linguistics," in *Language, Thought, and Reality. Selected writings of Benjamin Lee Whorf*, edited by John B. Carroll, The M.I.T. Press, Massachusetts Institute of Technology, Cambridge, Massachusetts, 1956, pp. 207-219.
4. E. Sapir, "The Status of Linguistics as a Science," in *Culture, Language and Personality*, Selected essays edited by David B. Mandelbaum, University of California Press, Berkeley and Los Angeles, 1957, pp. 65-77.
5. M. J. Reddy, "The conduit metaphor: A case of frame conflict in our language about language," in *Metaphor and Thought*, edited by Andrew Ortony, Cambridge University Press, Cambridge, England, 1979, 2nd ed., pp. 164-201.
6. G. Lakoff and M. Johnson, *Metaphors We Live By*, The University of Chicago Press, Chicago and London, 1980.
7. J. L. Lemke, *Talking Science: Language, Learning, and Values*, Ablex Publishing Corporation, Norwood, NJ, 1990.
8. W-M. Roth and R. Duit, *J. Res. Sci. Teach.* **40** (9), 869-897 (2003).
9. A. B. Arons, *Teaching Introductory Physics*, John Wiley & Sons, Inc., New York, 1997.
10. J. R. Touger, *Phys. Teach.* **29**, 90-95 (1991).
11. H. T. Williams, *Am. J. Phys.* **67** (8), 670-680 (1999).
12. C. R. Sutton, *Leicester University School of Education Occasional Papers* (1978).
13. C. R. Sutton, *J. Res. Sci. Teach.* **30** (10), 1215-1227 (1993).
14. S. F. Itza-Ortiz, N. Sanjay Rebello, D. A. Zollman and M. Rodriguez-Achach, *Phys. Teach.* **41** (6), 330-336 (2003).
15. M. W. Zemansky, *Phys. Teach.* **8** (6), 295-300 (1970).
16. R. Baierlein, *Am. J. Phys.* **62** (1), 15-26 (1994).
17. R. P. Bauman, *Phys. Teach.* **30** (6), 353-356 (1992).
18. M. T. H. Chi, J. D. Slotta and N. de Leeuw, *Learn. & Ins.* **4**, 27-43 (1994).
19. J. D. Cutnell and K. W. Johnson, *Physics*, 5th ed., John Wiley & Sons, New York, 2001.
20. R. A. Serway and R. J. Beichner, *Physics For Scientists and Engineers*, 5th ed. Vol. 1, Saunders College Publishing, Orlando Fl, 2000.
21. D. Halliday, R. Resnick and J. Walker, *Fundamentals of Physics*, 6th ed., John Wiley & Sons, USA, 2003.
22. M.A.K. Halliday, *An Introduction to Functional Grammar*, Edward Arnold, London, 1985.
23. See Ref [19], p.358.
24. See Ref. [21], p.434.
25. P. A. Tipler, *Physics for scientists and engineers*, 4th Ed. Vol. 1, W.H. Freeman and company/Worth Publishers, New York, NY, 1999, p.566.
26. J. S. Walker, *Physics*, Prentice Hall, Upper Saddle River, NJ, 2002, p.499.
27. D. C. Giancoli, *Physics for scientists & engineers*, 3rd ed. Prentice Hall, Upper Saddle River, New Jersey, 2000, p.486.
28. See Ref. [20], p.580.
29. See Ref. [20], p.603.
30. See Ref. [19], p. 452.

Introductory College Physics Students' Explanations Of Friction And Related Phenomena At The Microscopic Level

Edgar G. Corpuz and N. Sanjay Rebello

Department of Physics, Kansas State University, Manhattan KS, 66506-2601

Abstract. We investigated introductory college physics students' explanations of friction and lubrication through semi-structured clinical interviews. Although students were able to construct explanations at the atomic scale, they tended to explain phenomena by using attributes of macroscopic objects. For instance, when they described atoms as balls, they tended to associate attributes of real balls to the attributes of the atoms. In the future our results will guide the design of teaching materials to enable students to construct scientifically correct microscopic models of friction and lubrication.

Keywords: Friction, Microscopic, Students' Conceptions, physics education research, transfer of learning.
PACS: 68.35, 01.40Fk

INTRODUCTION

Numerous studies have documented students' ideas about the particulate or microscopic nature of matter. [1] The importance of such studies can't be overemphasized. We are at the verge of several breakthroughs in nanoscience and technology. Challenging students to think about what is happening at the atomic scale is a great step in advancing the goals of nanoscience education.

We report on students' ideas of friction and lubrication at the atomic level. Friction is a familiar concept to most introductory students and a part of their everyday experience. However, microscopic friction – a hot topic of research, is quite different from macroscopic friction. We believe that it is important that students learn the differences between macroscopic and microscopic friction. To design instruction to help students understand these differences we first investigate their existing ideas about friction.

The research questions of this study are:

- What are the existing models of introductory college students regarding friction and related phenomena at the microscopic level?
- How do students build and use models in explaining common everyday phenomena related to friction?

METHODOLOGY

Two one-hour semi-structured clinical interviews were conducted per student. A total of 11 students enrolled in algebra-based introductory physics classes were interviewed. Almost all students were life science majors. We pilot-tested our protocol with other educational researchers and two students and revised the protocol based on their feedback.

We began by asking students to slide their fingers across a wooden block and sketch the surface starting at the macroscopic scale and zooming in progressively to smaller length scales. Most of the students eventually realized that zooming in would get them to the atomic level. However, they were unsure of the length scale at which that would occur.

Follow-up questions explored ideas about:

- Cause of friction at the atomic level,
- Differences between kinetic and static friction,
- Effect of surface roughness on friction,
- Effect of gravity on friction, and
- Lubricating mechanism of oil.

The interviews were videotaped. We used phenomenographic analysis [2, 3] as per which students' responses are grouped into naturally occurring catego-

CP790, *2004 Physics Education Research Conference*, edited by J. Marx, P. Heron, and S. Franklin

ries based on their quotes. A second layer of analysis combined the categories of responses across different questions to generate themes.

RESULTS & DISCUSSION

Cause of Friction at the Atomic Level

Students were asked to pull a wooden block over a plank. Interviewees talked about friction when explaining why they needed a finite force to start the block moving. Follow-up questions probed students' ideas about the causes of friction at the atomic level. Below are category descriptions.

Intertwining / Interlocking Model

Five of the eleven interviewees explained that friction is the force needed to pull an atom over the bumps due to the interlocking or intertwining of atoms. Figure 1 shows an interviewee's sketch.

"when you set it [the block] on top, it kind of settles in ...like goes into a neutral energy state. When I try to move it I got to pull them out so there will be some friction because there will be some particles getting intertwined (fingers of hand intertwining)."

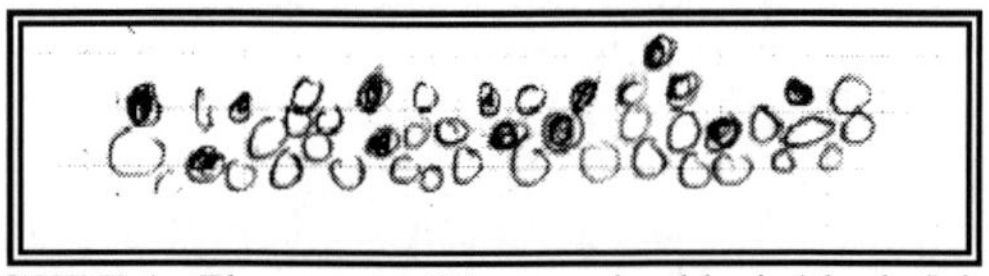

FIGURE 1. The atoms of the wooden block (shaded) interlock with the atoms of the table top (not shaded).

Rubbing / Sliding / Hitting Model

Three interviewees explained that atoms rubbing or hitting against each other is the reason for the finite amount of force needed to start the wooden block moving (see Figure 2)

"They (atoms) don't mesh together at all. They just sit on top of one another...they are touching but they don't interact any more than just the physical contact... one of them is moving and one of them isn't moving so they rub together."

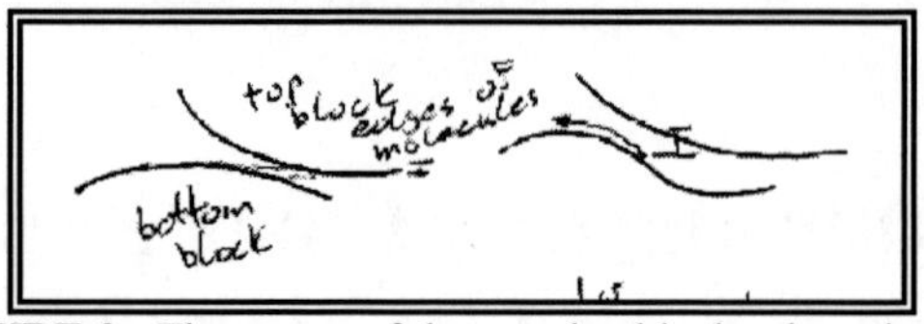

FIGURE 2. The atoms of the wooden block rub against the atoms of the tabletop.

Bonding Model

According to two of the interviewees, friction is due to the force needed to break the bonds between surfaces that come into contact.

"So when it is just sitting, these two would form some sort of bond which make them stick close together...you kinda have to overcome these little bonds enough to break them."

Why Static Friction > Kinetic Friction

In explaining why static friction is greater than kinetic friction the following categories emerged.

Changing Downward Force

One of the interviewees expressed the view that friction is due to the downward force that presses one surface onto the other. When the surfaces begin to move, this downward force decreases causing friction to decrease.

"When it is at rest there's more pressure between the atoms...when it starts moving you have less force pulling down."

Getting Smoother Model

According to most of the interviewees, the reason that friction occurs is that the surfaces are rough even at the atomic level. In explaining why static friction is greater than kinetic friction, one of them described that the surface would somehow get smoother once we started moving one of the surfaces relative to the other.

"I would think that it seems like when it is just sitting there, the surfaces are somehow interacting and making one another almost more rough. The way this works basically is it is more rough when it wasn't moving than when it was."

Broken Bonds Model

This explanation was provided by a student who initially said that static friction between the two surfaces is the force needed to overcome the bonds between the atoms of the surfaces.

"When it has started moving, let's say they might not have enough time to form that (bond)... So there's less number of bonds to be broken."

Skimming Over the Top Model

This explanation was provided by most of the interviewees who previously described friction as due to the intertwining or interlocking of atoms. Once the block has started moving the atoms of the block just skim over the atoms of the other surface (Figure 3).

"It already got some initial velocity... We pulled off the atom. There's still gonna be resistance between the two at the molecular level... but not as much because they are not settled in. When you're moving it they're gonna be not as intertwined."

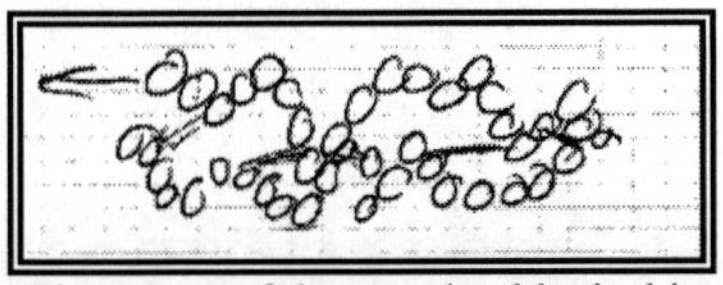

FIGURE 3. The atoms of the wooden block skim over the top atoms of the table surface once it is in motion.

Friction on Perfectly Flat Surfaces

When asked to visualize two perfectly flat surfaces, students typically drew a sketch similar to Figure 4.

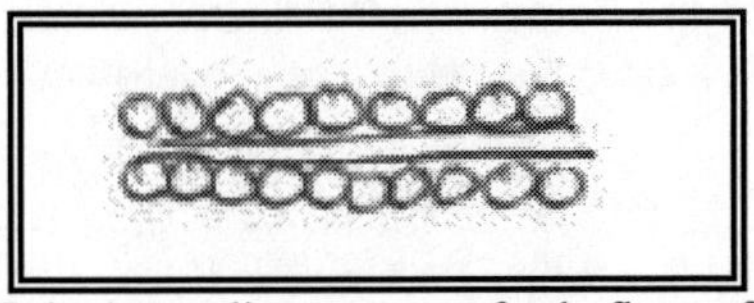

FIGURE 4. Atoms line up on perfectly flat surfaces.

Students were asked whether friction would still exist between these two perfectly flat surfaces. Their responses could be categorized as follows.

Mechanical Interactions Model

According to five (5) of the students interviewed there will still be friction because we still have atoms that interact (mechanically) even when surfaces become perfectly flat.

"There would still be friction because we still have atoms sliding past one another and also hitting each other."

Gravity Model

This is the explanation by the student who earlier had the "changing downward force model" in explaining why static is greater than kinetic friction.

"There will still be friction because there is still force of gravity pushing down. Gravity has to be zero for friction to be zero."

Electrical Interactions Model

Below is a student's description of what happens when two perfectly flat surfaces slide past one another:

"There is still friction because of electrical interactions."

Why Oil Reduces Friction

In explaining the lubricating mechanism of oil, students' explanations could be categorized as follows.

Ball Bearing Model

According to five (5) students, oil reduces friction just like ball bearings.

".. it might be possible that they move past one another easier, but it could be that oil molecules roll."

Weaker Bonds Model

With oil in between the surfaces, one of the students thinks that friction will be less because there is a weaker bond to break.

"maybe they don't exhibit as much intermolecular bonds between each oil molecule than between oil and wood molecules so they can move past one another."

Floating Model

Four (4) students who expressed this view said that friction is reduced because atoms of oil provide a floating barrier for the atoms of the wooden block.

"liquid ... will help separate these bumps and valleys such that they don't have to interact with the full scale interaction."

Emergent Themes

Two themes emerged from the second layer of analysis of the variations in the students' descriptions of friction and lubrication -- mechanical and chemical interactions.

Mechanical Interactions

We observed the persistence of students' responses that friction is simply due to mechanical interactions of the atoms. When they were asked to explain what causes microscopic friction most students alluded to the interlocking/intertwining or rubbing/hitting model. These descriptions are not surprising because these students had learned about friction only in the macroscopic context and had not dealt with friction at the atomic level. Therefore, these students appear to apply their macroscopic ideas to explain friction at the microscopic level. For these students friction is either the force needed to pull an atom over the bumps due to intertwining/interlocking of atoms or the force due to the rubbing/sliding of an atom past another atom. When the five students who expressed these views were asked what happens if the surfaces become atomically smooth they said that friction persists because atoms still physically rub against each other. Similarly, when students were asked to explain why oil reduces friction, most of them likened the oil atoms to ball bearings rolling between the atoms of the two solid surfaces. These explanations of friction and lubrication are clearly based on their experience with macroscopic friction. Thus, for most students, what is true macroscopically must also be true microscopically. This general result is consistent with previous research on student understanding of the microscopic world [4].

"Bonding"

For some of the students, microscopic friction is due to the force needed to break the bonds between the atoms of surfaces that come into contact. When these students were asked to explain why kinetic friction is less than static friction they said that there are fewer bonds to break once the other surface is set in motion than before it begins to move. Similarly when asked to explain what happens when there is oil in between two surfaces, one of these students believed that friction is reduced because it's the weaker bond between oil and wood that needs to be broken instead of the bond between wood and wood

FUTURE DIRECTIONS

Based on the results of this study we are currently designing and testing teaching experiments with the goal of helping students adopt microscopic models that are consistent with those used by scientists. The design of teaching experiments is anchored on the tenets of constructivist conceptual change [5] -- first establish students' existing knowledge and then either modify or build on it to promote conceptual change. The above results have made us aware of the ideas of friction and lubrication that students bring into the classroom. Through the use of discrepant events we will enable students to challenge some of their existing ideas. Alternatively, through reinforcement activities we will help students build on and refine other productive ideas. Thus, we will facilitate students' knowledge construction processes. The teaching experiments can inform us of effective instructional strategies that can contribute to knowledge reconstruction by students.

ACKNOWLEDGMENTS

Supported in part by NSF Grant #0133621.

REFERENCES

1. Hamed, K.M., *Investigating Students' Understanding of Surface Phenomena, Ph.D. Dissertation*, in *Physics*. 1999, Kansas State University: Manhattan, KS.
2. Marton, F., *Phenomenography- a research approach to investigating different understanding of reality.* Journal of Thought, 1986. **21**: p. 29-39.
3. Svennson, L. and T. J., *The Relation between Categories of Description and an Interview Protocol*, in *A Case of Phenomenographic Research.* 1983.
4. Andersson, B.R., *Pupils' conceptions of matter and its transformations (age 12-16)*, in *Relating macroscopic phenomena to microscopic particles: a central problem in secondary science education*, P.L. Lijnse, et al., Editors. 1992, CD-ß Press: Utrecht, Germany. p. 12-35.
5. Duit, R., *Conceptual change approaches in science education*, in *New Perspectives on Conceptual Change*, W. Schnotz, S. Vosniadou, and M. Carretero, Editors. 1999, Pergamon Press: Oxford. p. 263-282.

Transfer Between Paired Problems In An Interview

Kara E. Gray and N. Sanjay Rebello

Department of Physics, Kansas State University, Manhattan KS, 66506-2601

Abstract. This paper will cover a small portion of a larger study designed to address the issue of stability of knowledge in an interview and how it is influenced by transfer of learning. An interview over basic mechanics questions will be used to show how one question can influence a student's answer to another question. Based on this transcript and other data collected during the study, students' ideas appear to be influenced not only by their experiences and the context presented in the question, but also by the context of the question. This analysis was done based on a new model of transfer called the actor-oriented transfer model developed by Lobato that is based on the "personal construction of similarities" by the student between the learning and transfer contexts. This new model will also be discussed in further detail in the paper.

Keywords: physics education research, transfer of learning.
PACS: 01.40Fk

INTRODUCTION

Transfer of learning has become a major focus of education research. Many educators now view transfer as the goal of all teaching and use slogans like "teaching for transfer." [1] These ideas rest on the belief that the purpose of education is to prepare students to take what they have learned in the classroom and apply it to the world around them. This belief seems especially appropriate in physics classes. Most physics professors want their students to walk out of the class with more than simply an appreciation for the brilliance of Newton's laws. Instead the professor hopes that the student is now prepared to apply Newton's laws to everyday situations.

A NEW MODEL IN TRANSFER RESEARCH

While the purpose of transfer is fairly well agreed upon, how to assess, document and teach for transfer are far from settled.[2] Transfer research has typically been done by choosing two tasks, a learning task and a transfer task.[3] The researcher then decides what information, knowledge or skills a student should be able to transfer from the learning task to the transfer task and then documents how many students were able to do this and how many did not transfer the appropriate information. Personal experience suggested that transfer should be a universal process – it should be happening almost everywhere and almost all the time. People spend their lives transferring knowledge learned from previous experiences to the new experiences they encounter. Yet, transfer experiments consistently found transfer to be a rare occurrence. This inconsistency has caused some debate in education research and has led some researchers to abandon the idea of transfer all together. [2]

Other researchers have chosen to redefine transfer by more specifically changing their perspective and research methodology. Joanne Lobato [4] redefined transfer as the "personal construction of similarities between activities." (p 106) Based on this definition Lobato created the actor-oriented model of transfer. [5] This work has immense implications for transfer research. Lobato's definition and subsequent model places transfer squarely back in the mind of the student or interviewee. Transfer theories typically saw transfer only in situations where a student applied a concept in a manner that was consistent with what the expert or researcher expected. [3] Lobato, on the other hand, saw transfer in any situation where a student saw a similarity between two situations and used that similarity in some way. This method requires

CP790, *2004 Physics Education Research Conference*, edited by J. Marx, P. Heron, and S. Franklin

researchers to analyze research data with a more open mind than before. Since transfer is defined by the student, often the concepts transferred by students are not correct physics concepts. This means that transfer and problem solving must be considered independently. For instance, a student may see similarities between two situations and may even use this similarity, but may answer the problem incorrectly because the information transferred was not helpful in correctly answering the problem or other information was needed but not transferred. The following interview was done as part of a larger project unrelated to transfer [6], yet reanalyzing it from the perspective of transfer sheds new light onto the students' understanding and on the usefulness and importance of Lobato's actor-oriented transfer model.

INTERVIEW ANALYSIS

In the following transcript a student works aloud through two questions taken from the Force Concept Inventory (FCI). [7] The FCI is a popular conceptual physics survey that often asks students to choose between a correct physics answer and answers which correspond with their intuition. The first problem, the hockey puck problem (see Figure 1), was question 6 on the original FCI. This problem was intended to focus on Newton's first law and second law in the context of an impulsive force. The second problem, the space ship problem (see Figure 2), was question 24 from the original FCI. This problem was intended to focus on kinematics and Newton's second law in the context of a constant force.

The student interviewed in this transcript was enrolled in an algebra-based physics class at the time of the interview. The interview was intended to be a pre-instruction interview so the topic of forces had not been covered in the class and projectile motion had only just been introduced. This student had taken a previous physics class in high school.

I: Now if you could walk me through how you'd solve this problem.

S: Okay [the student reads the problem aloud]. Okay, it's moving at a constant speed forward which means that it's got inertia and it's already got momentum built up and once it gets hit in, I guess, a perpendicular direction it's going to want to slide at an angle. So, [choice] 4 looks like the best answer. Well. . .

I: Okay. (Pause) Okay, I didn't know if you wanted to look at it some more. If you can do the same thing with this [spaceship] question.

S: Okay [student reads problem aloud]. Ah, it looks like the same deal except this is in space and not under forces of gravity like the hockey puck was, so I think [choice] 4 is also going to be a good answer for this one. Actually, looking at this one I think on the first question [choice] 1 is probably the best answer for that one.

I: Okay.

S: Yeah, I think I mixed up the reasoning. In space the momentum and inertia are going to carry it at an angle to get to its right angle position.

The figure depicts a hockey puck sliding with constant speed v_0 in a straight line from point P to point Q on a frictionless horizontal surface. Forces exerted by air are negligible. You are looking down on the puck. When the puck reaches point Q, it receives a swift horizontal kick in the direction of the heavy print arrow.

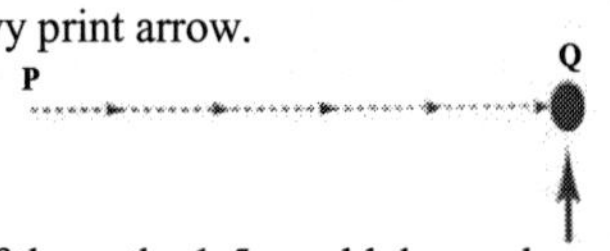

Which of the paths 1-5 would the puck most closely follow after receiving the kick?

FIGURE 1. The hockey puck problem: The first problem presented to the student in think-aloud interview

A spaceship drifts sideways in outer space from point P to point Q as shown below. The spaceship is subject to no outside forces. Starting at position Q, the spaceship's engine is turned on and produces a constant thrust (force on the spaceship) at right angles to the line PQ. The constant thrust is maintained until the spaceship reaches point R in outer space.

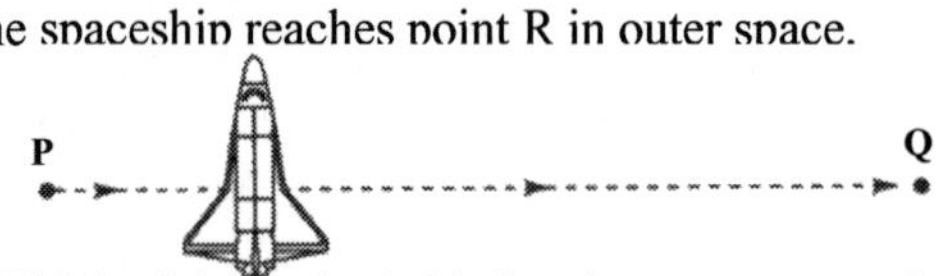

Which of the paths 1-5 below best represents the path of the spaceship between points Q and R?

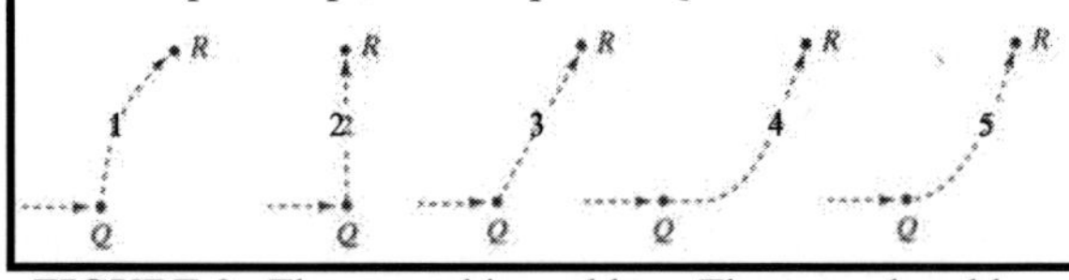

FIGURE 2. The spaceship problem: The second problem presented to the student in the think aloud interview

From a traditional transfer perspective one would look for two transfer goals in the above transcript.

- Does the student use vector addition to solve the two problems?
- Does the student recognize the difference between constant and instantaneous forces and adjust the solution accordingly?

According to the traditional perspective, transfer should take place from prior instruction, specifically in a high school physics class, to the FCI questions. Initially the student does appear to be using vector addition to solve both problems. Although the student does not explicitly indicate this, her responses appear to be consistent with this hypothesis. Her answer to the hockey puck problem is incorrect although she does appear to be correctly adding the vectors in the spaceship problem. While answer 5 is technically the correct answer to the spaceship problem, a careful reading of the transcript suggests that she did not notice the delay in path 4 and that path 4 in the spaceship problem is therefore different than the path 4 shown in the hockey puck problem. This student's decision to change her original answer to the hockey puck problem is unsettling since she switches from an answer that suggests she is at least attempting to add the vectors to one in which no vector addition is used. This student does not appear to recognize a difference between constant and instantaneous forces. Actually she makes no references whatsoever to forces throughout her explanation.

The actor-oriented model of transfer begins very differently from the traditional perspective. It does not begin by defining any transfer goals for the two problems; instead it simply looks at what information the student brings to the problem and where that information comes from. The student appears to be transferring information about inertia and momentum to the hockey puck problem from, most likely, her previous physics class. Again, although the student does not explicitly state this, her responses are consistent with this hypothesis. She then transfers her ideas from the hockey puck problem to the spaceship problem, because she believes that both problems "look like the same deal." After answering the spaceship problem she transfers information from the spaceship problem to the hockey puck problem and uses that information to change her answer, by noticing differences between the two problems. The question that remains is why the student chose to transfer information between these two problems. The above section of the transcript does not really help answer this, but the actor-oriented transfer model defines transfer based on the similarities (and/or as in this case differences) that a student sees between two situations. Therefore, what was transferred and why can be gleaned from the student's response shown below.

[After answering both questions]
I: Okay, now if we can look at both questions. While you were answering the spaceship problem did you find yourself referring back to the hockey puck?

S: Yes.

I: Anything particular?

S: Um, I guess with the spaceship it's the same situation only without gravity and earth forces on it so, yep.

To this particular student the hockey puck and spaceship problems are basically the same question asked in different contexts. This explains her decision to transfer information from one problem to the next. The only difference is the existence of other forces in the hockey puck problem, which cause her to revise her original answer. This student's view of the relationship between these two problems is further clarified when she is specifically asked what the similarities between the two problems are. She responds, "They're both dealing with an object that has a constant velocity in the direction and a quick change in direction and what would happen to its path in the process." According to the problems, only the hockey puck has a quick force applied to it. The spaceship feels a constant thrust. The student appears to have remembered the problems differently. This could be because she misunderstood what was meant by constant thrust, or because she does not see a difference between a constant force and an instantaneous force. Unfortunately this point was not probed during the interview. This discrepancy between the information provided in the problem and what is remembered by the student does help to explain why the student transferred between these two problems. To this student the only difference between the two problems is the existence of "earth forces" not a difference in the type of applied force.

CONCLUSIONS

The usefulness of Lobato's method is that it allows the researcher to see the interplay between the two problems. While previous transfer models would see previous experience or knowledge learned in class as the key source of transferred information, looking at the transcript from the actor-oriented transfer perspective, it becomes clear that the main source of information can be the problems themselves. This student is not simply being uncertain of her answer, she is actually learning from the questions as she works through them. The spaceship question plays an important role in her answer to the hockey puck question. In previous research, not presented here, we found for two mutually related FCI questions, the order in which two questions are asked can affect the information the student reads from the question and how she answers the question. [6]

In the transcript discussed above, the spaceship question caused the student to switch from one wrong answer to another wrong answer. This isn't always the case, sometimes asking another question can lead the student to switch to the correct answer, as was seen in the larger study mentioned earlier. [6] Therefore, asking students to solve a pair of questions can allow them to see and correct errors in their reasoning. The actor-oriented model of transfer provides a framework for recognizing appropriate question sets. The most important aspect of these question sets is that students must recognize the relationship between the two problems. Two problems that appear to be obviously related to an expert may not be related according to a student. This is where Lobato's method becomes so important. This technique is similar to the instructional method used in Elby pairs. [9] Elby pairs ask students to answer two related questions. In one question students are likely to use their raw intuition and incorrectly answer the question. For the other question, students are more likely to use their new physics knowledge or refined intuition and correctly answer the question. The instructor can then lead students through a discussion of why the two answers conflict. The method described above differs from this technique since both questions used in the study can be answered using student's intuition and experience from the world around them. The actor-oriented model of transfer provides a theoretical basis for why these methods successfully allow students to correct their responses or ideas and therefore could serve as a guide to designing productive instructional strategies in the future. Further research is warranted to explore the effectiveness of such strategies.

Lobato's actor-oriented transfer model sheds light on a problem that is usually overlooked. Students usually do not walk into an interview with their answers already planned. Instead they often construct their reasoning as they work through the problems. [8] This means that a student's answer can be influenced by events that occur in the interview, as shown in this transcript. Previously, student knowledge has been considered relatively stable. This is apparently not the case. Students are constantly transferring knowledge from one situation to the next, including knowledge from a previous problem to and from the subsequent problem. This means that by simply asking a question, a researcher has possibly changed the knowledge of a student. This has possible repercussions not only for interviewing situations but also for research surveys and even for exams. [6]

ACKNOWLEDGMENTS

Supported in part by NSF ROLE Grants #0087788 and CAREER Grant #0133621.

REFERENCES

1. McKeough, R.E., J. Lupart, and A. Marini, *Teaching for transfer: Fostering generalization in learning.* 1995, Mahawah, NJ: Erlbaum.

2. Detterman, D.K. and R.J. Sternberg, *Transfer on Trial: Intelligence, Cognition and Instruction.* 1993, Norwood, NJ: Ablex Publishing Company.

3. Reed, S.K., G.W. Ernst, and R. Banerji, *The role of analogy in transfer between similar problem states.* Cognitive Psychology, 1974. **6**: p. 436-450.

4. Lobato, J.E., *Transfer Reconceived: How "sameness" is produced in mathematical activity, Ph.D. Dissertation*, in *Education.* 1996, University of California, Berkeley: Berkeley, CA. p. 254.

5. Lobato, J.E., *How Design Experiments Can Inform a Rethinking of Transfer and Vice Versa.* Educational Researcher, 2003. **32**(1): p. 17-20.

6. Gray, K.E., *The Effect of Question Order on Student Responses to Multiple-Choice Physics Questions*, in *Physics*. 2004, Kansas State University: Manhattan, KS.

7. Hestenes, D., M. Wells, and G. Swackhammer, *Force Concept Inventory.* The Physics Teacher, 1992. **30**: p. 141-151.

8. Itza-Ortiz, S.F., et al. *A framework for the dynamics of student reasoning in an interview.* in *Proceedings of the Annual Meeting of National Association of Research in Science Teaching 2004*. 2004. Vancouver, BC: NARST.

9. Redish, E.F. *A Theoretical Framework for Physics Education Research: Modeling Student Thinking.* in *International School of Physics, "Enrico Fermi", Course CLVI.* 2003. Varenna, Italy: Italian Physical Society.

Sample Exams And Transfer In Introductory Mechanics

Carolann Koleci

Worcester Polytechnic Institute
Department of Physics
Worcester, MA 01609

Abstract. Why do novices use examples so extensively, no matter which subject area? Examples are often helpful because they reveal how to get from problem descriptions to theoretical re-formulations. Moreover, problem solving in semantically rich domains, like physics, requires one to translate a given problem description into theoretical concepts. We address some of the important issues of knowledge transfer, in the context of introductory mechanics, using data from an on-going sample-exam study conducted at Worcester Polytechnic Institute.

INTRODUCTION

The process of transfer is often defined as the degree to which learning contributes to or detracts from subsequent problem solving or the learning of subsequent instruction [1]. The ability to apply a particular skill, or bit of knowledge, to situations differing from those encountered during original learning is a signature of the transfer of learning [1].

It is documented in the cognitive psychology literature that problem solving and transfer are largely domain specific, thereby suggesting that transfer is not likely to occur as the result of general problem-solving instruction [2], but instead requires multiple practice opportunities in a variety of contexts [3].

A prevalent assumption of formal learning is the application of abstract principles and skills across multiple domains, with automatic transfer from the acquisition of these general skills. While some cognitive psychologists disagree that knowledge is entirely connected to the context in which it is learned [4], several researchers have evidenced that knowledge and transfer are strongly linked to context and domain [5-8].

To what extent does context play a role in students' ability to transfer knowledge obtained from an example to a new situation? Sample exams are very popular study aids exploited by students. Do they help or do they hinder the transfer of introductory physics knowledge, or is there no discerning effect?

METHODS

This study was primarily designed to ascertain the impact of examples in the learning of introductory mechanics by science, technology, engineering, and mathematics (STEM) students at Worcester Polytechnic Institute (WPI). A total of 69 STEM students were included in the introductory mechanics portion of this study. Subjects were not randomly selected so as not to deny any students the potential value of supplementary resources and extra credit for participation. This raises the issue of self-selection, and discussion on this topic is provided later in the report.

For both portions of this study, students were offered a quiz incentive for opting to do sample exams. Two days prior to each actual exam, participants were asked to complete the corresponding on-line sample exam and to *describe* the concepts behind each sample exam question (i.e., kinematics combined with application of Newton's 2nd Law). The recitation instructor then corrected all of the exams, offering students formative assessment. The graded sample exams were returned to students the day before the actual exam, and the sample exam was completely reviewed in recitation, thereby ensuring that most, if not all, students received the same amount of content instruction in recitation.

For each of the four exams, as part of the introductory physics recitation grade, students were required to complete a web-survey primarily, but not exclusively, to learn how they reviewed for each exam. As a quantitative form of summative

CP790, *2004 Physics Education Research Conference*, edited by J. Marx, P. Heron, and S. Franklin

assessment, the Force Concept Inventory (FCI) was required (both pre- and post- tests) among all introductory mechanics students [9].

DISCUSSION

Several statistical tests were performed in order to determine the effect of sample exams on exam performance, and in order to address the issue of self-selection. The first table displays course content.

TABLE 1. Content for Introductory Mechanics

Exam One	Exam Two	Exam Three	Exam Four
Vectors and kinematics graphs	Projectile motion and Newton's Laws	Energy and momentum	Equilibrium, rotational motion, and angular momentum

For both portions of this study, a one-way Analysis of Variance (ANOVA) was performed, using normalized gain on the FCI as the dependent variable and choice to do the sample exam as the factor. These tests were performed to learn of any differences in conceptual understanding. For the case of introductory mechanics, statistically significant results were obtained for those students opting to do sample exam numbers one, two, and four. In all three of these cases, those students who complete each respective sample exam fare better on the FCI (normalized gain) than those not doing the sample exam (p ($\alpha = 0.05$) = 0.04, 0.02, 0.04, respectively).

While the above results may suggest that sample exams enhance the transfer of conceptual knowledge, as evidenced by statistically significant FCI scores, there is still the problem of self-selection. If more conscientious and academically astute students are selecting to do sample exams, then it is expected that these students will fare better on conceptual tests like the FCI. To address this issue, data from each of the two class populations was partitioned according to final course grade. In this manner, those students earning an *A* for the course and opting to do the sample exam could be compared with students earning an *A* for the course and *not* opting to do the sample exam. Analogous groupings could be made for *B*, and *C* students, respectively. Since the number of students failing the course is small, comparisons for these populations of the students can not be made.

When partitions are made according to final course grade, no statistically significant results are obtained. This suggests that there is no significant difference in conceptual gain among like populations.

The second table consists of final grade, and average normalized gain, corresponding to the number of students in introductory mechanics.

TABLE 2. Final Grade versus Average Normalized Gain on the FCI

Final Letter Grade	Average Normalized Gain FCI	Number of Students
A	0.43	31
B	0.07	18
C	0.14	17

Do examples enhance the transfer of introductory physics problem solving as evidenced by exam performance? One-way ANOVA tests were conducted, with exam grade as the dependent variable and choice to do sample exam as the factor. In introductory mechanics, statistically significant results were obtained for tests two through four (p ($\alpha = 0.05$) = 0.01, 0.00, 0.00, respectively). These data suggest that in mechanics, sample exams are helpful for improving exam performance, at least 75 percent of the time, and dependent on course content.

To continue to address the issue of self-selection, the data set that was partitioned according to final course grade was used. When those introductory mechanics students earning a *C* for the course and opting to do the sample exam are compared with those students earning a *C* and *not* opting to do the sample exam, a slightly significant result is obtained for *Exam One* only. Those *C* students *not* choosing to do *Sample Exam One* performed slightly better on the corresponding exam.

Additional interesting results transcend from the compilation of web survey feedback. The on-line survey contains a series of questions asking students to compare sample exams with the actual exams, on a question by question basis. In addition to these questions, students are asked to indicate the manner in which they worked on the sample exam. For example, did they work independently, did they consult any of the various course resources for help, or did they work continuously, without interruption?

Survey data continues to support an unsurprising trend. Namely, the student-perceived degree of similarity, or isomorphism, between sample exam questions and that of the actual exam, influences the level of successful problem solving. Students who perceive sample exam one, question one, to be similar to that on actual exam one are more apt to score higher than students who do not perceive the similarity. However, does this necessarily imply that students able to better recognize similar patterns are more apt to comprehend the underlying physical concepts? This question will, in fact, be discussed in a forthcoming publication.

CONCLUSION

There are several interesting research questions to be posed from this study. The first and foremost question is do sample exams enhance the transfer of introductory physics knowledge? Data obtained from this preliminary study suggests that the degree to which sample exams enhance both conceptual understanding and problem solving is context dependent, in support of work by many educational researchers [5-6].

In the context of introductory mechanics, material on sample exam one, two, and four appears significantly helpful for conceptual transfer.

For exam-performance transfer, in the area of introductory mechanics, material on sample exam two through four, inclusive, appears significantly helpful.

When data is further classified according to final course grade and factored according to conceptual gain on the FCI, no statistically significant findings arise in the context of introductory mechanics. When these data are factored according to exam grade, there is some evidence that *C* students not choosing to do sample exams fare better on the actual exams. What exactly do all of these observations translate to in terms of the usefulness and validity of sample exams?

Additional one-way ANOVA tests were performed, with final introductory mechanics grade as the dependent variable and choice to do corresponding sample exam as the factor. Statistically significant results were obtained for introductory mechanics tests one, two, and four (p (α = 0.05) = 0.04, 0.02, 0.04, respectively). Do these results suggest that more academically astute students are opting to do sample exams, or do the sample exams help students to achieve better grades? In both of these courses, exams comprise seventy percent of the final grade.

Data compiled from the on-line survey only begins to address the issue of pattern recognition and similarity. Should the level of difficulty on the sample exam, as perceived by the students, match, falter, or surpass that of the actual exam? Follow-up studies include cross-comparisons between survey responses, FCI scores, GPA (minus final physics grade), and respective exam scores.

Future work will be to untangle the conundrum posed by considering cumulative grade point averages sans physics grades, in addition to interviewing students who do and who do not do sample exams.

ACKNOWLEDGMENT

The author would like to graciously thank the WPI Center for Educational Development, Technology, and Assessment for its generous support of this work. The enthusiastic support of the WPI Physics Department is kindly acknowledged.

REFERENCES

1. Royer, J. M., "Theories of the transfer of learning," *Educational Psychologist* **14**, 53-69 (1979).
2. Larkin, J. H., "What kind of knowledge transfers," in *Knowing, Learning, and Instruction: Essays in honor of Robert Glaser*, edited by L. B. Resnick, Erlbaum, 1989, pp. 283-305.
3. Gagne, R.M., Briggs, L. J., and Wager, W. W., *Principles of Instructional Design (4th ed.)*, Harcourt Brace Jovanovich, 1992.
4. Anderson, J. R., Reder, L. M., and Simon, H. A., Applications and Misapplications of Cognitive Psychology to Mathematics Education, Texas Educational Review, 2000.
5. Bransford, J. D., Franks, J. J., Vye, N. J., and Sherwood, R. D., "New approaches to instruction: Because wisdom can't be told.," in *Similarity and Analogical Reasoning*, edited by S. Vosniadou and A. Ortany, Cambridge University, 1989, pp. 470-497.
6. Bransford, J. D., Sherwood, R. D., Vye, N. J., and Rieser, J., "Teaching thinking and problem solving," *American Psychologist* **41**, 1078-1089 (1986).
7. Brown, J. S., Colins, A., and Duguid, P., Situated cognitions and the culture of learning. *Educational Researcher* **18**, 32-42, (1989).
8. Perkins, D. N., and Salomon, G., Are cognitive skills context-bound? *Educational Researcher* **18**, 16-25 (1989).
9. Hestenes, D., Wells, M., and Swackhamer, G., "Force Concept Inventory," *Physics Teacher* **30**, 141-158 (1992).

Alternative conceptions, memory, & mental models in physics education

Gyoungho Lee, Jongho Shin[1], Jiyeon Park, Sangho Song[2], Yeounsoo Kim[3], Lei Bao[3]

Dept. of Physics Education, Seoul National University, Seoul, Korea 151-742
[1] Dept. of Education, Seoul National University, Seoul, Korea 151-742
[2] Dept. of Education, Andong National University, Andong, Korea 760-749
[3]Dept. of Physics, The Ohio State University, 174 W.18th Avenue, Columbus, OH 43210

Abstract. There are two somewhat independent research traditions, which converge to suggest a form of students' knowledge: alternative conceptions and mental models. However we have little literature that explains what they are different from each other and from memory. This study tried to describe these issues with some thoughts about how cognitive psychology and science education approaches can be best synthesized in order to approach these questions.

INTRODUCTION

Research in cognitive psychology, science education, and developmental psychology during the last decade has shown that students construct intuitive understandings of the world, which are based on their everyday experience. Although different terms have been used to refer to this type of knowledge-such as alternative conception, preconception, misconception, folk theory, naïve theory, intuitive theory, mental model- there is general agreement that this intuitive knowledge provides explanations of natural phenomena which are frequently different from the currently accepted scientific explanations and which tend to be resistant to change. Especially, the term 'alternative conception' has been preferred by many researchers as students' intuitive understandings in science education [1].

In physics education, alternative conceptions (or misconceptions) and mental models are familiar terms for researchers and educators with other terms (e.g., p-prims, facets, coordination class, etc.). However, as a result, different terms are employed, sometimes leading to confusion. As Redish [2] mentioned, it is important to develop a theoretical frame, which is a shared language and shared assumptions and through which different theoretical models of student thinking can be compared so we can accumulate, evaluate, and refine what we learn.

This study is the first step for developing a theoretical frame. Thus, the purpose of this study was to compare three major terms (alternative conceptions, mental models, and memory), which are used for representing students' ideas in physics and to find the relationship among them. For this, we reviewed related literature and tried to synthesize the reviewing results from both cognitive psychology and science education approaches.

ALTERNATIVE CONCEPTION, MEMORY, AND MENTAL MODELS: IN THE PROCESS OF LEARNING

There are two somewhat independent research traditions, which converge to suggest a form of students' knowledge: alternative conceptions and mental models. In the 1970s, in the area of science education a number of researchers published papers describing 'alternative conceptions' in students' understanding of a variety of natural phenomena [1]. There are significant knowledge claims emerging from the research on students' alternative conceptions. Some of these knowledge claims are: "Students come to formal science instruction with a diverse set of alternative conceptions concerning natural objects and events. The alternative conceptions that students bring

CP790, *2004 Physics Education Research Conference*, edited by J. Marx, P. Heron, and S. Franklin

to formal science instruction cut across age, ability, gender, and cultural boundaries. Alternative conceptions are tenacious and resistant to extinction by conventional teaching strategies" [1].

In cognitive psychology, Johnson-Laird (1983) introduced the term 'mental model' to refer to a form of mental representation [3]. The notion of mental model has been used in research in different areas with different meanings [4]. For some researchers a mental model is just a representation of some aspects of the world, whereas for others it is an analogue of objects in the world. Mental models serve as means with which to explain the relation between one's cognitive activity and the world. In this view, mental models are unstable, naturally evolving and incomplete. The views adopted by most researchers may be seen as delimited by these two extreme positions [4].

Recently, one of the main features of science education research is the increasing importance of the concept of 'mental model'. It appears that, with the concept of mental model, science educators are attempting to overcome some limitations of the Alternative Conceptions Movement (ACM), such as: "the frequently context specific character of alternative conceptions; the difficulty of the ACM in developing overall interpretations for domain-specific alternative conceptions; the difficulty in offering theoretically dense approaches to an understanding of such educational phenomena" [5].

Where do mental models (and alternative conceptions) come from? In order to answer this question, at first, we should know what memory is because memory is the place where our knowledge is stored and is activated. Memory can be divided into two primary components: working memory and long-term memory. Working memory is the temporary storage and processing of information that can be used to solve problems, respond to environmental demands, or achieve goals. Working memory is rapidly accessed and severely limited in capacity. Many theorists portray working memory as playing a central executive role, in essence controlling and monitoring an individual's overall memory processes. In a nutshell, working memory is the component in which "thinking" occurs: therefore, we might think of it as the "awareness" or "consciousness" of the memory system [6].

Long-term memory (LTM) is the part of the memory system that retains information for a relatively long period of. It includes both memory for specific events and knowledge that has been gleaned from those events over the years time [6]. Types of memory (in LTM) can be distinguished by the kind of knowledge stored and the way this knowledge is retrieved and expressed [7]. There are two kinds of knowledge, declarative and procedural.

Since conceptions are the principal units of knowledge organization, we can say that students' alternative conceptions are one of the principal units of knowledge as a part of their LTM. There is little literature that explains how alternative conceptions are constructed even though some researchers suggested several sources [8]. Some examples are sensual impression, everyday language, innate structures of the brain, learning in students' social environments, and instructions.

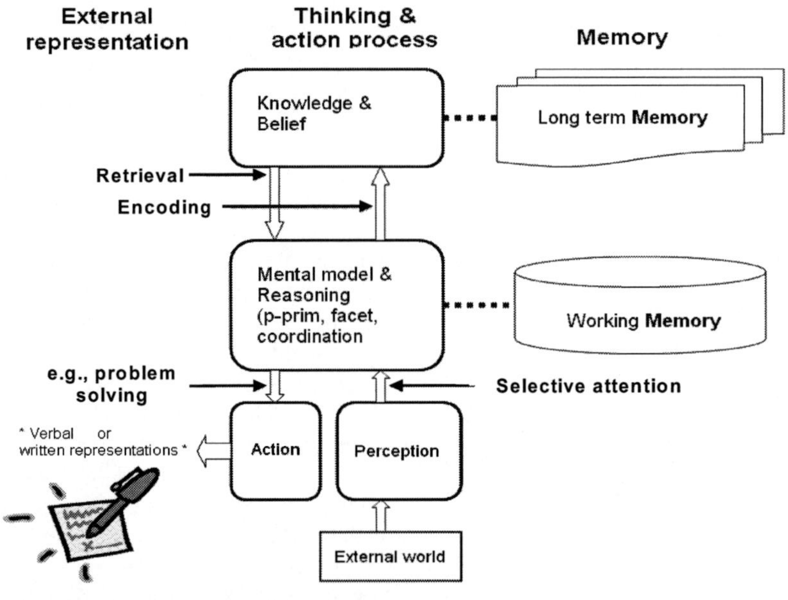

* Researchers usually call external representations as **alternative conception**, misconception, **mental model**, etc.

FIGURE 1. The relationships among alternative conceptions, memory, and mental models in the process of learning

On the other hand, mental model theory has tried to explain students' intuitive understandings of the world from integrated perspectives. For example, Vosniadou [9] explained the construction of mental models using conceptual structure. According to her, mental models are constructed in conceptual structure, which has presuppositions (epistemological, ontological) and beliefs from observations in cultural context. Mental models are constrained by a set of presuppositions which are derived from everyday experiences and which are consistent with their beliefs about physical objects.

Mental models are dynamic representations through integrating external information recognized and individual knowledge. In terms of memory, mental models are constructed in working memory, which involve mental representations of words or images themselves reflect interactions between current sensory data and stored knowledge from LTM. Based on these ideas, we developed a theoretical framework for synthesizing our understandings about mental models (See Fig. 1).

Figure 1 shows the relationships among alternative conception, memory, and mental model in the structure and process of learning. Mental model construct is related to perception processes and to knowledge structures. In fact, perception and knowledge are generally recognized as the principal sources of mental models [10]. The perception of visual, auditive, tactile, or haptic information marks one side of higher cognitive processes that is sometimes called 'bottom up' or 'data driven'. Mental models are contingent on external information insofar as the incoming data as a cue to particular analytical or synthetical subprocesses [11]. Knowledge, on the other hand, marks the other side of higher cognitive processes; the one that is sometimes called 'top down' or 'schema driven'. Mental models depend on individual knowledge insofar as the incoming data are interpreted with respect to context and experience, that is, to results of earlier processing.

Beliefs are also the components of LTM. They serve to constrain the types of knowledge and inferences a person may hold in their mental models. For example, Vosniadou and Brewer [9] stressed the role that epistemological and ontological beliefs play in the formation of mental models. For example, epistemological beliefs include the criteria individuals use to judge what constitutes a phenomenon, the assumptions that phenomena require an explanation, and that causal explanations can be used to explain physical phenomena. Ontological beliefs include the basic beliefs about the nature of objects such as physical objects are solid, stable, and if not supported, will fall down. In this, epistemological and ontological beliefs constrain conceptual knowledge that individuals may acquire from their observations and experience in the cultural context by limiting the inferences individuals make about their observations.

Pintrich [12] extend the logic of Vosniadou and Brewer's theory with suggesting that motivational beliefs about self and learning can play the same role in terms of being a resource to support or constrain the formation of mental model. He argued that "motivational beliefs, as presuppositions or theories about the self and learning, they may influence the types of inferences and belief formation that take place as students acquire knowledge and build their mental models". As Pintrich [12] have noted:

> *For example, certain types of motivational beliefs may inhibit certain types of cognitive processes from occurring, whereas other motivational beliefs may facilitate cognitive engagement. These motivation beliefs and various cognitive and metacognitive strategies are assumed to be "resources" that students can bring to bear on the task at hand to help them to learn (p.34).*

Thus, we believe that mental models depend on individual knowledge insofar as the incoming data are interpreted with respect to various beliefs.

Classroom context can be other factors that affect the formation of mental model since knowledge and beliefs are embedded in various classroom contexts. For example, the task, as a context, that students are confronted with can activate motivational beliefs and related knowledge in problem solving situation. According to the pattern of the activation of knowledge and/or beliefs, there might be different inferences and linked knowledge in the process of mental model formation.

Mental models guide action through reasoning and decision making process. For example, when students read, talk, write, something or take exams, mental models are always involved in that situation preserving information about and interpretations of objects and events in the world.

In summary, we can think about an example of how we form a mental model with figure 1. If we were confronted with an event (like a test), our knowledge and beliefs would be activated. After that, through problem solving, as thinking process, we will form mental model(s) in working memory. Finally, by action (like talking, writing, etc.) we present mental model(s). They become the external representation. In an interview or test situation, we can only see (or listen) the external representation, which students represent. Using the external representation, we can speculate what the students' conceptions, or mental model(s) are in their minds.

We should recognize the differentiations between the external representation, internal/mental representation (that is, mental models), and conceptual knowledge (that is, students' conceptions, alternative conceptions, misconceptions, etc). The places where each term locates are as follows (see Fig. 1, too),

- External representations: in words or graphs on papers, talking, gestures, etc

  (We can see or listen to them. However they are not exactly what students have in their minds)

- Mental models: in working memory

  (We cannot see or listen to them. We can only speculate them from external representation)

- Students' conceptions (alternative conceptions, misconceptions, etc): in long-term memory

  (We cannot see listen to them. We can only speculate them from external representation)

CONCLUSION

"The concept of mental representation is as fundamental to cognitive psychology as force is to physics" [13]. Mental models are dynamic mental representations through integrating external information recognized and individual knowledge. Within physics education, the mental model notion has been invoked in math education, psychology, linguistics, artificial intelligence, etc. Now, mental models have been a crucial term for explaining students' intuitive theory not only because it is located at the intersection of various disciplines but also because it can provide physics education researchers and teachers with valuable information about learners' conceptual framework, that is, their underlying knowledge structures from an integrated perspective.

In this study, we tried to answer the questions, "What mental models are?" and "How mental models are different from other terms, especially, alternative conceptions and memory?" Figure 1 shows our theoretical framework for answering the questions even though this study is the first step to develop the theoretical framework. In the future, we need to further develop our theoretical frame continually by conducting case studies and addressing new theoretical issues.

ACKNOWLEDGMENTS

This work was supported by Korea Research Foundation Grant (KRF-2003-042-B00165)

REFERENCES

1. Wandersee. J., Mintzes. J., & Novak. J., "Research on Alternative Conceptions in Science" in *Handbook of Research on Science Teaching and Learning*, edited by D. L. Gabel, NY: Macmilan, 1994, pp.177-210.
2. Redish, E. F., *A Theoretical Framework for Physics Education Research: Modeling Student Thinking*, The Proceeding of Enrico Fermi Summer School in Physics, 2004.
3. Mishra, P., & Brewer, W.F., *Contemporary Educational Psychology* **28**, 277-303(2003).
4. Borges, A. T., & Gilbert, J. K., *International Journal of Science Education* **21**, 95-117 (1999).
5. Franco, C., Barros, H. L., Colinvaux, D. Krapas, S., Queiroz, G., & Alves, Fatima., *International Journal of Science Education* 21, 277-291 (1999).
6. Ormrod, J. E., *Human Learning*, NJ: Merrill, 1999, pp. 188-195.
7. Westen, D., *Psychology: Mind, Brain, & Culture*, NY: Johnson Wiley & Sons, Inc, pp. 294 (1999)
8. Duit, R., "Students' Conceptual frameworks: Consequences for Learning Science" in *The Psychology of Learning Science*, edited by Glynn et al., NJ: Lawrence Erlbaum Associates, 1991, pp. 65-85.
9. Vosniadou, S., *Learning and Instruction* **4**, 45-69 (1994).
10. Yates, J., *Psychological Review* **92**, 249-284 (1985).
11. Rickheit, G., & Sichelschmidt, L., "Mental Models: Some Answers, Some Questions, Some Suggestions" in *Mental Models in Discourse Processing and Reasoning*, edited by G. Rickheit & C. Habel., North-Holland: Elsevier, 1999, pp. 9-40.
12. Pintrich, P., "Motivational Beliefs as Resources for and Constraints on Conceptual Change" in *New Perspectives on Conceptual Change*, edited by W. Schnotz, S. Vosniadou & M. Carretero., Pergamon Press, 1999, 34.
13. McNamara, T., "Knowledge Representation" in *Thinking and Problem Solving*, edited by R. Sternberg, Academic Press, 1994, 83.

Student Descriptions of Refraction and Optical Fibers

Fran Mateycik*, DJ Wagner*,†, JJ Rivera*, and Sybillyn Jennings§

*Department of Physics, Rensselaer Polytechnic Institute, Troy, NY 12180
†Department of Physics, Grove City College, Grove City, PA 16127
§Department of Psychology, Russell Sage College, Troy, NY 12180

Abstract. This paper reports our research into how students describe and think about optical fibers and the physical phenomena of refraction and total internal reflection (TIR) basic to their operation. The study was conducted as part of the improvement and expansion of web-based materials for an innovative Rensselaer introductory physics course[1] that examines the physics underlying information technology. As we developed the prototype module, we examined students' understanding of the phenomena of refraction, TIR, and optical fibers through the use of clinical interviews. As students discussed refraction and tried to explain how optical fibers work, several patterns emerged. Our analysis of these patterns drives our assessment of the effectiveness of the revised materials in facilitating students' transfer of learning, as well as the development of a multiple-choice diagnostic tool.

[1]Materials found at http://www.rpi.edu/dept/phys/ScIT/

INTRODUCTION

The Science of Information Technology (ScIT) is an innovative introductory course that connects physics with technological contexts of interest to today's students. For example, in ScIT we apply total internal reflection (TIR), a phenomenon commonly covered in traditional physics courses, to optical fibers, a technology eminently relevant to the students. This study is part of the expansion and improvement of ScIT's web-based materials [1] to facilitate their widespread dissemination.

Part of our ongoing development process is to assess student preconceptions associated with different levels of comprehension. The Physics Education Research (PER) literature abounds with studies of preconceptions [2], but few studies have examined how students approach the topics that form the backbone of ScIT: *e.g.*, total internal reflection (TIR) and semiconductor physics. Like many before us [3-6], we used clinical interviews to open a window into student understanding of our target topics – optical fibers, refraction, and TIR. In this paper we discuss our categorization of responses, based on 41 interviews with subjects of diverse physics backgrounds.

METHODOLOGY

In the summer of 2002, we conducted face-to-face (f2f) interviews with Group A, a convenience sample of five in-service teachers, four physics majors, an engineering major, and two physics faculty. These interviews provided a "broad picture" of how individuals think about optical fibers. The following spring, we conducted pre-instruction "e-interviews" with Group B, all 22 ScIT students, followed by post-instruction e-interviews with 8 of those 22. (We chose to e-interview using Chat Rooms on WebCT, since timing and location made f2f interviews impossible. A comparison of the two formats is found in ref. [7].) One pre-interview was incomplete; Table 1 displays the physics backgrounds of the remaining 33 subjects.

All interviews were conducted by the same interviewer (a physicist); many were observed by a cognitive psychologist to gauge their efficacy. Interviews began with participant identification and verification of informed consent. Subjects were then asked, "What does an optical fiber do?" As we probed further, our choice of questions was driven by student responses and included "How might the fiber keep the light from escaping?" "What can you tell me about refraction?" and "How does Total Internal Reflection occur?"

CP790, *2004 Physics Education Research Conference*, edited by J. Marx, P. Heron, and S. Franklin

TABLE 1. Physics Background of Interviewees.

Highest Completed Physics Course	Grp A	Grp B
High School Physics	1	9
College Physics I	0	8
College Physics II	4	3
Upper-Level Coll. Phys.	5	1
Physics Faculty	2	0

We examined the video recordings and transcripts[1] from the 41 interviews for patterns and progressions. We grouped similar student descriptions and drafted network schematics (*e.g.,* Fig. 2). Schematics were roughly outlined at first, then refined to include all interviewee accounts.

RESULTS AND ANALYSIS

Descriptions of refraction, TIR, and optical fibers were as varied as the students providing them. No categorization scheme can completely encompass the diversity, but we have identified some themes. Through this process, we found that optical fibers and refraction needed separate groupings. In hindsight the separation is not surprising. Optical fibers are a technology, and student descriptions fall along a single progression from novice to expert. Whereas refraction and TIR are physical phenomena, and students approach them from a variety of starting points. The following analyses reflect these differences.

Optical Fibers

All interviewees had heard of optical fibers or fiber optics, but few could correctly describe how they worked. We grouped student responses into five "stages," numbered from 5 to 1 according to how close the descriptions were to the "expert" descriptions comprising Stage 5. The stages (and sub-stages) are listed in Table 2 and described below.

Stage 5: Accurate models identify TIR with optical fibers and correctly describe how TIR arises from the bending of light at a boundary between two media.

Stage 4: Incomplete refraction models tie the operation of fibers to refraction but are not complete. Descriptions in Stage 4B indicate that the indices of refraction factor into the operation of an optical fiber, but they do not fully explain how TIR works. Responses in Stage 4C rely upon Snell's Law and TIR but are not entirely correct. *E.g,* they say TIR will

TABLE 2. Number of Descriptions in Each Stage Categorization. Subjects moving between stages during an interview are counted in their initial stage.

Stage	1	2	3				4			5
Sub-Stage			A	B	C	D	A	B	C	
Group A	2		1		3				1	5
Group B Pre	3	2	2	4	5		1	1	2	1
Group B Post						1		1	3	3

occur when light leaves a rarer medium and enters a denser medium. Only one subject suggested refraction as a means to guide light without discussing TIR (Stage 4A). When asked what kept light from escaping the sides of a fiber toy he had previously described, subject B10 replied, "OK, well I would assume that the light was modified some way, either refracted or directed in such a way that keeps the light contained."

Stage 3: Reflective models involve light reflected off the sides of the fibers but make no mention of refraction's role. Stage 3A responses involve the placement of tiny mirrors within the fiber at strategic locations, such as where the fiber must bend. Those in Stage 3B coat the entire inner fiber with a silvered surface. Descriptions in Stage 3C involve reflection from an unspecified coating. Subjects in Stage 3D face a quandary. They realize (generally from having seen a fiber) that the walls are not silvered, but they cannot come up with another way to guide light.

Stage 2: "Containment" models involve the use of an opaque, but not necessarily reflective, coating to keep light from escaping the fiber. This model was the starting point for two subjects; three others entertained this model during their interviews, as seen in Fig. 1.

Stage 1: Unique models lack the information that optical fibers carry data in the form of a beam of light. The six descriptions in this category are very different from each other; here are two examples.

> **A5**: "An optical fiber … filters [incoming] particles out to make their sight or sound or whatever …. Optical meaning light, so I am assuming that it's light particles being broken up and then kind of reassembled. With higher resolution - perhaps these fibers help create better resolution."

> **B17:** "[An optical fiber is] a type of wire connection that uses fibers to make it faster. … Inside the wire there is a type of fiber substance around the wire that helps in the moving of electrons like in TV fiber optics. … Optics deals with sight, so the optical fibers find a way to maximize the speed of electrons with refractors in the fibers to display for sight. … I'm guessing the fibers are made of small mirrors that are attached to insulators in the wires."

[1] Capitalizations and misspellings have been corrected in excerpts of the e-interview transcripts presented here. "…" represents non-essential comments by either interviewee or interviewer.

Subjects moved between stages in 9 of the 41 interviews. This movement is summarized in Fig.1. Movement between stages during an interview was more common in the convenience sample (Group A) interviews because the interviewer did not have to worry that instruction in the interview would corrupt future post-instruction analysis. For example, subject A6 was told that "Optical fibers ... carry transatlantic communication .... How do you think something might be made that ... makes light go the whole distance?" This information helped A6 move from Stage 1 to 3D.

FIGURE 1. Movement between stages. Nine interviewees moved between stages. Subject B20 did not change from one stage to another but put forth two models without choosing between them.

Subject A7 illustrates how interviewees moved between Stages 2 and 3 as they thought about optical fibers. He started in Stage 3C: "From what I understand, it's ... a series of reflections." He next shifted to Stage 3D by rejecting mirrors: "... I think just by a series of ... I don't want to say mirrors, but it's got to be ... a mirror-like substance ... it can't be glass 'cause it's flexible." He visited Stage 2: "maybe it wouldn't need to reflect ... you can't escape the, the insulator, right? ... maybe it can just, shwooo, travel right through. ... it's a plastic substance, I know, 'cause they use it for ... that cable for computers and things." A7 finally ended back in a quandary (3D): "I don't know how you would do it. ... it's gotta be reflecting somehow. I don't know."

The eight post-instruction interviewees were among the weaker students in the class (they agreed to the interview for extra credit at the end of the semester), yet each (to our relief) provided a more expert description than in the pre-instruction interview. The most significant transition was for subject B6, who moved from Stage 1 at the start of the course to Stage 5 in the post-interview. The least change was seen in subject B4. He started the semester in Stage 3C; at the start of the post-interview, he appeared to be in Stage 3D: "I don't believe [the fiber] is coated with a mirror but a material that reflects the light like a mirror...can't recall exactly." When questioned about refraction and TIR, however, this student invoked fiber terminology, apparently moving into Stage 4C: "The critical angle is the two indexes of refraction inside the core ... [When light strikes the surface at less than the critical angle] it is completed reflected within the fiber."

Interviewees who correctly explained optical fibers seemed able to draw upon three key principles: (I) Fibers carry information in the form of a coherent beam of light; (II) Mirrored surfaces are not suitable; [2] and (III) Light can be guided using refractive effects, such as TIR. Subjects in Stages 1 or 2 lack cognizance of I – they have no concept of how data is carried by light. Students in Stages 3A and 3B, and possibly 3C, surmise that fibers rely on mirrors. When they recognize II but not III, they end up in a quandary (Stage 3D) because they can think of no other method to guide light. The ability to correctly apply refraction to the guidance of light distinguishes Stages 4 and 5.

The disconnect between Stages 3 and 4 appears significant. Many subjects deduced principles I and II from having seen optical fibers, but principle III seems difficult to discover without explicit instruction. All interviewees above Stage 3, with the possible exception of B10, had previously "heard" that fibers work through refraction. Explaining that fibers carry a signal in a beam of light could prompt movement from Stage 1 or 2 into Stage 3, but not Stage 4. Even some subjects who knew about TIR appeared to view reflection and refraction as incompatible phenomena. A6 described the TIR ray as "refracted so much that it's actually refracting into ... the fiber," its direction given by Snell's Law rather than the law of reflection.

Refraction

Refraction and TIR were explicitly discussed in 8 Group A interviews and 21 Group B interviews.[3] The term "refraction" was familiar to each of these interviewees, and all but five connected it to the bending of light. Yet subjects evinced diverse approaches to refraction and TIR, drawing upon a variety of resources to explain the phenomena.

Figure 2 illustrates the connections made by interviewees. The goal was a correct explanation of TIR – either by using Snell's Law or through a more conceptual description; experts provided both. Critical to either description is the concept that light bends. Stu-

---

[2] Students in Stage 3D generally reject mirrors because they've seen fibers (*e.g.*, character toys and computer cables) and remember the flexibility and/or absence of a silvered surface. In contrast, faculty pointed out that mirrors absorb light, which makes them unsuitable.
[3] Probing of refraction was constrained by the direction taken during the interview, the subject's background, and time limitations.

dents drew upon specific applications of refraction and TIR, such as prisms, diamonds, and illusions like a pencil "broken" at the surface of water. Other resources, such as the equation for Snell's Law and the association of refraction with a change in speed, were more technical.

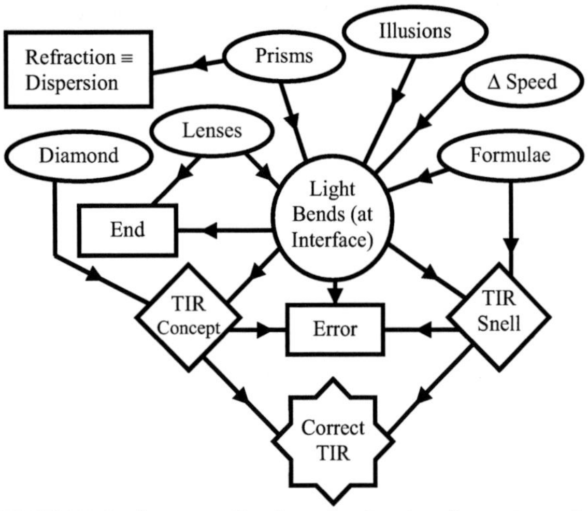

FIGURE 2. Conceptualization of refraction. Items in ovals are resources; rectangles surround dead-ends; and the other shapes represent correct conclusions. Arrows point from tool to conclusion, not always the order in which topics arose.

Knowing that wavefronts bend as they change speed seems essential to a conceptual explanation of refraction and TIR. All 11 subjects who connected a change in speed to refraction also recognized that the path of light bends, but not all who knew light bends mentioned a change in speed. When asked to explain TIR "to someone who doesn't know what a sine function is," subjects like A1 could not get past mathematics and definitions: "Total internal reflection is governed by a, uh, critical angle. …A certain parameter, call it theta c, …. If you have a beam of light, um, at this critical angle, I think that will be the first point which you're going to have total internal reflection." But those interviewees who could correctly explain TIR by both qualitative and quantitative means started from the definition of index of refraction as a ratio of speeds when asked to explain TIR without sines.

Nine students related refraction to prisms; six of them did not differentiate refraction from dispersion. For example, when asked, "How does refraction work?" Subject B1 responded, "Light hits a material and then dissipates into its components based on the various wavelengths." Subject B2 replied, "diffraction is when light passes through a prism and is split into different wavelengths, and refraction is light being recombined." Understanding that bending and dispersion can occur independently evidently requires additional contrasting experiences.

IMPLICATIONS

Analysis of the interviews is influencing the development of ScIT materials in many ways. We have added a photograph of a toy using fibers to explicitly connect to students' prior experiences. Each figure with a refracted ray now shows a reflected ray too, to discourage students from viewing reflection and refraction as mutually exclusive. To discourage students from equating dispersion with refraction, we have moved our dispersion discussion from the refraction module to a signal loss and distortion module. Diagnostic questions involving optical fibers now begin with the explanation that fibers carry information in the form of light pulses and incorporate descriptions from interviews. Subsequent studies with a larger sample, using multiple-choice questions as well as clinical interviews, are necessary to test the generality of these models, to refine our diagnostic tool, and to gauge how different instructional approaches affect students' understanding of refraction, total internal reflection, and optical fibers.

ACKNOWLEDGMENTS

The Rensselaer portion of this work was supported in part by NSF CCLI Program under grant # DUE-0089399. Special thanks to Leo Schowalter, the rest of the ScIT Advisory Committee at Rensselaer, and all the students who participated.

REFERENCES

1. All materials are web-based, and many can be found at the ScIT website: http://www.rpi.edu/dept/phys/ScIT.
2. See, for example, E. F. Redish and L. C. McDermott, "Resource Letter PER-1: Physics Education Research," *Am. J. Phys.* **67** (9), 755-767 (1999).
3. E. Hunt and J. Minstrell, "A Cognitive Approach to the Teaching of Physics," in *Classroom Lessons: Integrating Cognitive Theory and Classroom Practice*, edited by K. McGilly (Cambridge, MA: MIT Press, 1994).
4. L.C. McDermott, "Research on conceptual understanding in mechanics," *Physics Today*, **37** (7), 24-32 (1984).
5. I. A. Halloun and D. Hestenes, "Common sense concepts about motion," *Am. J. Phys.* **53** (11), 1056-1065 (1985).
6. E. F. Redish and R. N. Steinberg, "Teaching Physics: Figuring Out What Works," *Phys. Today* **52** (1), 24-30 (1999).
7. We report on the e-interview experience in D. Wagner, J. Rivera, F. Mateycik, and S. Jennings, "Using Electronic Interviews to Explore Student Understanding," submitted to 2004 PERC Proceedings.

Retention and Transfer from Trigonometry to Physics

Darryl J. Ozimek, Paula V. Engelhardt, Andrew G. Bennett[#], N. Sanjay Rebello

Department of Physics, 116 Cardwell Hall, Kansas State University, Manhattan KS, 66506
[#]Department of Mathematics, 138 Cardwell Hall, Kansas State University, Manhattan, KS 66506

Abstract. We examined the extent to which students retained and transferred various concepts from trigonometry to physics at the introductory college level. Online trigonometry homework assignments and pre- and post-instruction surveys in physics were our sources of data. Transfer was measured from a traditional as well as two contemporary perspectives. Students seemed to have more difficulty retaining and transferring the unit circle concept compared to others. Transfer was more evident when measured from some contemporary perspectives rather than a traditional perspective.

Keywords: physics education research, transfer of learning.
PACS: 01.40Fk

INTRODUCTION

Retention and transfer of learning are topics of interest for many educational researchers. There has been no significant prior research on retention and transfer from trigonometry to physics. The following research questions examine these issues.

1. To what extent do students retain their knowledge from trigonometry to physics?
2. To what extent do students transfer their knowledge from trigonometry to physics?

RELEVANT LITERATURE

Retention is the extent to which one can retrieve information from long term memory. The success of retrieval depends upon effective encoding that involves making associations with existing knowledge that can facilitate future retrieval. [1]

Transfer is the extent to which one can apply what was has learned in one situation to a new situation. Traditionally transfer has been measured by examining whether students can successfully apply what they have learned to new isomorphic problems. [2] Contemporary perspectives have expanded this view. Bransford and Schwartz [3] view transfer in terms of "preparation for future learning" (PFL). They focus on whether students can learn to problem-solve in a new context. Lobato's [4] "actor-oriented transfer" (AOT) model views transfer as the "personal construction of similarities" between the two contexts. She focuses on how the "actors" (or learners) see the two contexts as similar.

Trigonometry concepts can be understood and applied at different levels depending upon the problem context. Van Hiele [5] used a hierarchical system of levels to describe student understanding in geometry. Trigonometric concepts (sine, cosine, tangent and their relationships) have been described in terms of three Van Hiele Levels (VHLs) of understanding:

- VHL-I (Geometric): Relationships between the sides and angles of a right triangle.
- VHL-II (Unit Circle): Relationships between the angle of an arc of a circle and its projections.
- VHL-III (Function): As functions plotted on a 'f(x) vs. x' graph.

METHODOLOGY

Students in this study were enrolled in an algebra-based physics course and had previously taken a trigonometry course at Kansas State University. We used two sources of data:

(i.) Online homework assignment scores averaged over number of attempts in the trigonometry course.
(ii.) Scores on a multiple-choice survey containing mathematics and physics questions that required knowledge of trigonometry concepts.

CP790, *2004 Physics Education Research Conference*, edited by J. Marx, P. Heron, and S. Franklin

The survey was given on the first day of the physics course (pre-instruction) and again after they had completed the relevant topics (post-instruction) in the course. The survey contained 18 multiple-choice questions, organized into three groups – one for each VHL. Each group contained 'abstract' questions that tested trigonometry concepts devoid of a physics context paired with 'contextual' questions that required students to apply trigonometry concepts in a physics context. Figure 1 shows an 'abstract' question and an isomorphic 'contextual' question.

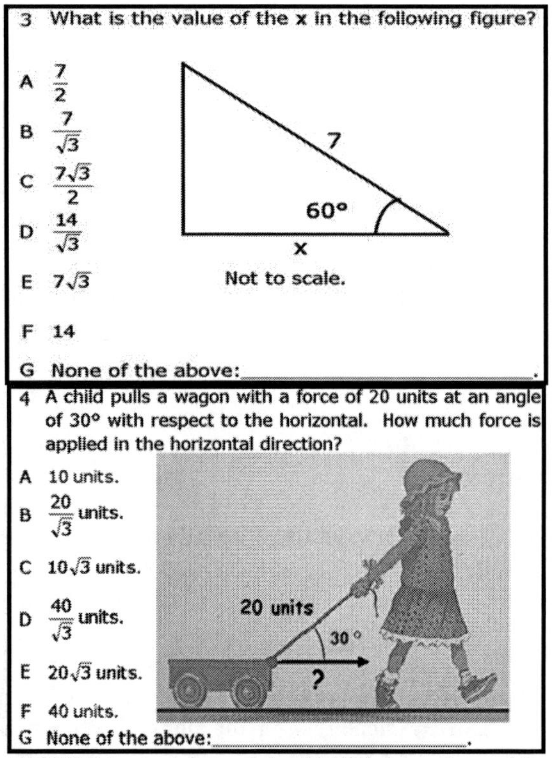

FIGURE 1. An 'abstract' (math) VHL-I question and its 'contextual' (physics) isomorphic question.

Research Q # 1: Retention

Student scores on the pre-instruction survey for the 'abstract' questions for each VHL were used to assess retention. We did not consider the 'contextual' physics questions because these questions require students to apply their knowledge in a new context and therefore measure transfer rather than retention. To account for the fact that a concept (VHL) that was not initially learned will not be retained, we divided the pre-instruction math score with the online homework score for that VHL. Thus, our retention metric was:

$$\text{Retention} = \frac{\text{'Abstract' (Math) Score on Pre-Survey}}{\text{Online Trig Score per Attempt}} \quad (1)$$

Research Q # 2: Transfer

Transfer was examined from a 'traditional' and two 'contemporary' perspectives (PFL and AOT). Each perspective carries a set of criteria and uses different measurements as summarized in Table 1.

TABLE 1: Perspectives of Transfer

Perspective	Criteria for transfer	Measure of transfer in this study
'Traditional'	Apply knowledge learned in a prior context to solve a problem in a new context.	Correlation between online trig score per attempt and 'contextual' physics score in pre/post survey.
'Contemporary' PFL	Apply knowledge learned in a prior context to *learn* to solve problems in a new context.	Correlation between online trig score per attempt and *gains* in 'contextual' physics questions.
'Contemporary' AOT	Recognize relations of similarity between the two contexts.	Correlations between 'abstract' and 'contextual' scores on the *same* survey.

'Traditional' Perspective

From the 'traditional' perspective transfer is measured as the ability to apply knowledge learned in a prior situation to a new situation. Transfer can be measured by comparing performance in the new situation (physics) with performance in the situation in which the knowledge was learned (trigonometry). We calculated correlations between students' online trigonometry scores with their scores on the 'contextual' physics problems on the pre- and post-instruction surveys at the same VHL. A statistically significant correlation between the average online trigonometry score and the average score on the physics survey questions at the same VHL was considered evidence of transfer.

'Contemporary' Perspective # 1: PFL

The 'contemporary' PFL perspective focuses on whether students' initial learning prepares them to *learn* how to problem-solve in a new context. In this study we examined whether students learning in a trigonometry course prepared them to learn physics. As per the PFL perspective we calculated correlations between students' online trigonometry scores and gains (pre- vs. post-instruction) on the 'contextual' physics scores at the same VHL. The rationale for using *gains* is that they are a measure of *learning* in the physics context. Thus, using gains to measure transfer is consistent with the PFL perspective which views transfer as the ability to learn in the new context.

'Contemporary' Perspective # 2: AOT

In the 'contemporary' AOT perspective, evidence of transfer is found when students create 'relations of similarity' between two situations, i.e. when they notice that two situations are similar in some way. As per this perspective we detect transfer through correlations between student performances on questions on the same survey. The fact that students' performance on two questions is significantly correlated implies that these questions have something similar about them from the students' perspective. This student-centered notion of perceived similarity is considered sufficient evidence of transfer.

As per the AOT perspective we calculated correlations between scores on the 'abstract' (math) and isomorphic 'contextual' (physics) questions at the same VHL on the same (pre- and post-instruction) survey. The rationale for using correlations between 'abstract' and 'contextual' question is that the degree of correlation between scores on an 'abstract' (math) problem and an isomorphic 'contextual' (physics) problem is a measure of the similarities perceived by students between these two problems.

RESULTS & DISCUSSION

Research Q# 1: Retention

To examine retention we used the metric described in equation (1) to calculate a retention score for each student at each VHL. The average (n=38) retention scores for each VHL are shown in Figure 2.

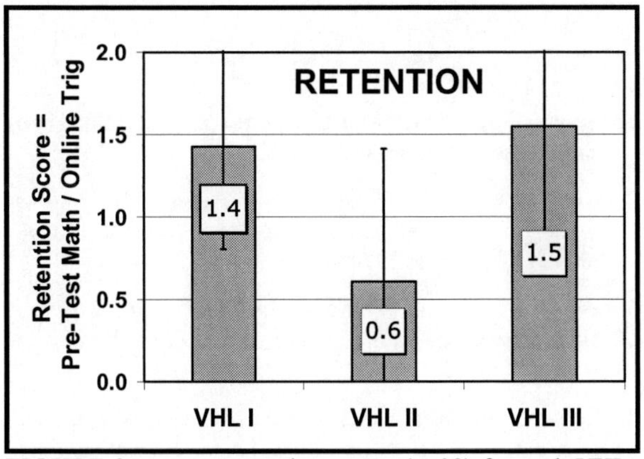

FIGURE 2: average retention scores (n=38) for each VHL-calculated as per equation (1).

Students appear to have greater difficulty retaining VHL-II concepts than VHL-I and VHL-III concepts. Differences between VHL-II and VHL-I or VHL-III at the $p<0.02$ significance level were detected. These results appear to indicate that the unit circle concept in trigonometry is more difficult for students to retain than the geometric or function concepts.

Research Q # 2: Transfer

As summarized in Table 1, transfer was analyzed using multiple perspectives. We describe below our results from each perspective.

'Traditional' Perspective

As per the 'traditional' perspective, we calculated correlations between average scores on online trigonometry assignments and scores on the 'contextual' physics questions (pre- and post-instruction survey) at each VHL. No statistically significant correlations were found at any of the VHLs for either the pre- or the post-instruction survey. Thus, no evidence of transfer was found from the traditional perspective.

'Contemporary' Perspective # 1: PFL

As per the PFL perspective, we calculated correlations (Table 2) between the *gains* (post-instruction – pre-instruction) on the physics survey at each VHL with the scores on online trigonometry assignments at the same VHL.

TABLE 2: Measuring transfer as from the 'contemporary' PFL perspective

Correlation between Physics Gain & Online Trigonometry Scores (n = 31)		
	r	p<
VHL-I	**0.36**	0.05
VHL-II	0.13	Not Significant
VHL-III	-0.15	Not Significant

A statistically significant ($p<0.05$) correlation (0.36) was found only for VHL-I. Thus, as per the PFL perspective, it appeared that students successfully transferred their learning at the geometric level (VHL-I), but not at the unit circle level (VHL-II) or the functional level (VHL-III).

'Contemporary' Perspective # 2: AOT

As per the AOT perspective, we calculated correlations (Table 3) between 'abstract' (math) and 'contextual' (physics) questions on the same survey for both the pre-instruction and post-instruction surveys.

TABLE 3: Measuring transfer from the 'contemporary' AOT perspective

Correlations between 'abstract' (math) & 'contextual' (physics) scores on the same survey				
	Pre–Survey (n = 43)		Post–Survey (n = 40)	
	r	p<	r	p<
VHL-I	**0.79**	0.001	**0.53**	0.001
VHL-II	-0.01	Not Sig.	0.20	Not Sig.
VHL-III	**0.76**	0.001	**0.39**	0.05

Statistically significant correlations were found for VHL-I and VHL-III on both the pre- and post-instruction surveys. Thus, as per the AOT perspectives, it appeared that students were able to dynamically transfer their knowledge from the 'abstract' (math) to the 'contextual' physics questions for VHL-I (geometric) and VHL-III (function), but not for VHL-II (unit circle).

CONCLUSIONS

Research Q # 1: Retention

Our retention results indicate that students had a significantly greater retention of the geometric and function concepts in trigonometry compared to the unit circle concept. The results are also anomalous with the assumption of hierarchical VHLs.

Research Q # 2: Transfer

As expected, the perspective of transfer that we adopted directly influenced whether we found evidence of transfer. We did not find any evidence of transfer from trigonometry to physics as per the 'traditional' perspective. However, when we broadened our perspective we found evidence of transfer as per the 'contemporary' PFL and AOT perspectives. We believe that this observation is not a weakness of our study; rather it underscores the importance of examining transfer from a variety of different perspectives.

Transfer was also found to be non-uniform across VHLs. Stronger evidence of transfer was detected for VHL-I (geometric) concepts than for VHL-III (function) concepts, and none was detected for VHL-II (unit circle) concepts.

Table 4 summarizes the retention and transfer results. Each check mark indicates evidence of transfer or retention. These results appear to converge on the conclusion that students have most difficulty in both retaining and transferring the unit circle concept (VHL-II).

TABLE 4: Summary of Results

VHL	I	II	III
Retention	✓		✓
'Traditional' Transfer			
'Contemporary' PFL Transfer	✓		
'Contemporary' AOT Transfer	✓		✓

FURTHER RESEARCH

Our results appear to be anomalous with the hierarchical nature of VHLs. Therefore, further research is necessary to analyze the nature of trigonometry knowledge, especially when applied in the context of physics.

We speculate that student difficulties with the unit circle concept are probably because they did not use this concept extensively in their trigonometry course. They completed only two online assignments on the unit circle compared with four assignments on the function concept. Further research will be necessary to investigate these speculations.

ACKNOWLEDGMENTS

Work funded in part by NSF grant DUE-0206943.

REFERENCES

1. Sousa, D.A., *How the Brain Learns*. 2nd ed. 2000, Thousand Oaks, CA: Corwin Press, Inc.
2. Reed, S.K., G.W. Ernst, and R. Banerji, *The role of analogy in transfer between similar problem states*. Cognitive Psychology, 1974. **6**: p. 436-450.
3. Bransford, J.D. and D. Schwartz, *Rethinking transfer: A simple proposal with multiple implications*. Review of Research in Education, 1999. **24**: p. 61-100.
4. Lobato, J.E., *How Design Experiments Can Inform a Rethinking of Transfer and Vice Versa*. Educational Researcher, 2003. **32** (1): p. 17-20.
5. Van Hiele, P.M., *Structure and insight: A theory of Mathematics Education*. 1986, Orlando, FL: Academic Press.

FREE-BODY DIAGRAMS: NECESSARY OR SUFFICIENT?

David Rosengrant, Alan Van Heuvelen, Eugenia Etkina

Rutgers University, New Brunswick, NJ

Abstract. The Rutgers PAER group is working to help students develop various scientific abilities. One of the abilities is to create, understand and learn to use for qualitative reasoning and problem solving different representations of physical processes such as pictorial representations, motion diagrams, free-body diagrams, and energy bar charts. Physics education literature indicates that using multiple representations is beneficial for student understanding of physics ideas and for problem solving. We developed a special approach to construct and utilize free-body diagrams for representing physical phenomena and for problem solving. We will examine whether students draw free-body diagrams in solving problems when they know they will not receive credit for it; the consistency of their use in different conceptual areas; and if students who use free-body diagrams while solving problems in different areas of physics are more successful then those who do not.

* *Supported in part by NSF grants DUE 0241078, DUE 0336713*

INTRODUCTION

An external representation is something that stands for, symbolizes or represents objects and/or processes. Examples in physics include words, diagrams, equations, graphs, and sketches. The positive role of multiple representations in learning has been suggested by many [1-5]. H. Simon said: "Finding facilitating representations for almost any class of problem(s) should be seen as a major intellectual achievement, one that is often greatly underestimated as a significant part of both problem solving efforts in science and efforts in instructional design." [6]. A free-body diagram (FBD) is one representation used by physicists while representing processes involving forces. Physics instructors teach students how to draw the diagrams and encourage student to use them in problem solving [7-9]. However, there is little evidence that those who draw free-body diagrams are more successful in problem solving than those who do not. We also do not know if students actually use FBDs beyond mechanics.

This manuscript investigates three questions: a) do students draw free-body diagrams while solving problems if they know there is no credit associated with this work; b) are students consistent in using free-body diagrams in different conceptual areas, and c) are students who use FBDs to solve traditional problems more successful then those who do not?

COURSE DESCRIPTION

This study was conducted in a two-semester large-enrollment algebra-based physics course for science majors. There were two 55-min lectures, one 55-min recitation and one 3-hr laboratory per week. The course followed the Investigative Science Learning Environment (ISLE) format [10]. Helping students learn to represent physical phenomena in multiple ways is one of ISLE's essential features.

In the course under study students learned to use motion diagrams, free-body diagrams, energy and momentum bar charts and so forth to reason about physical processes. As these representations are specific to physics, we call them physical representations [3]. Students also used pictorial and mathematical representations. The course had a specially designed packet that included, among other things, multiple representation tasks as separate problems [11]. In each lecture the instructor (AVH) discussed with the students how to approach several representation tasks and how to move from graphical representations to the mathematical. Then students worked individually and in small groups on similar activities. Feedback to the students was provided orally or via an electronic Personal Response System (PRS). Special multiple representations tasks occupied 40% of the recitation time. Drawing a physical representation was a part of the problem-

CP790, *2004 Physics Education Research Conference*, edited by J. Marx, P. Heron, and S. Franklin
 0-7354-0281-7/05/$22.50

solving strategy that AVH followed closely in lectures and students were encouraged to use in recitation and homework problems. Model solutions for the homework included a special step – drawing a physical representation of the problem situation.

FREE-BODY DIAGRAMS

While helping students learn to construct free-body diagrams we used the following strategy. Circle the object of interest in a sketch of the situation. Look for objects that touch the object or that interact with it at a distance. Represent each interaction with a force arrow. The number of the force arrows should equal the number of objects with which the object of interest interacts. Each force is labeled with two subscripts: for example: $\vec{F}_{\text{E on b}}$ would identify the force exerted by the Earth on the ball [6]. The relative length of the force arrow represents the relative magnitude of the force. The diagram must agree with other representations, for example the acceleration arrow in a motion diagram.

In lectures AVH consistently used FBDs to analyze phenomena and to construct the mathematical expression of Newton's second law in component form. In recitations students practiced drawing diagrams as separate exercises and then used them to write Newton's second law to solve problems. Model solutions for the homework problems emphasized free-body diagrams and their relationship to the component form of Newton's second law. In the second semester, free-body diagrams were used by AVH in electrostatics and magnetism to solve problems involving forces.

INSTRUMENTS/PARTICIPANTS

There were two midterms and one final each semester. The exams were predominantly multiple-choice with some hand-graded sections. We tried to devise specific multiple-choice problems that included several objects and would be difficult to solve without drawing free-body diagrams. The problem statements did not specifically ask for a diagram. The multiple-choice questions were graded with Scantron sheets. Students did not receive any partial credit for work or diagrams. We chose 5 exam problems from 4 exams for this study [see appendix].

We looked at FBDs that students drew on the exam sheets near the problem statement. If there was no diagram, we scored the response as 0. If a student drew the diagram but made mistakes in it (wrong forces, arrow length or mislabeled) we scored it as 1. If a student drew a diagram correctly with enough information to solve the problem, we scored it as 2. We related the scores to the students' answers.

We have complete data for 125 students. These 125 students came from an original sample of 200 students chosen randomly from 560 students in the first semester (10 students from each of the 20 recitation sections). Some did not take the second semester and some missed one of the exams reducing our final number to 125. The grade breakdown for the students in the study was almost identical to the breakdown for the class. Thus we believe that the sample was representative of the student population.

FINDINGS

Question (a) Do students draw free-body diagrams while solving problems if they know there is no credit associated with this work? To answer this question we counted all students who drew diagrams independently of whether the diagrams were correct or not [Table 1]. We found that 15% of students in the sample drew a diagram for each of the five problems chosen for the study (3 mechanics and 2 electrostatics); 72% of the students in the sample drew FBDs for at least one problem in mechanics and at least one in electrostatics; 12% drew free-body diagrams only for mechanics problems; 10% drew free-body diagrams only for electrostatics, and 5% never drew a free-body diagram [Table 2].

TABLE 1. Comparison of Students who Drew a FBD and those that did not (N=125).

Question	# who Drew a FBD	# who did not Draw a FBD
Mech. Exam 1 Qstn 1	94	31
Mech. Exam 1 Qstn. 2	63	62
Mech. Final	49	76
Electrostatics Exam 1	73	52
Electrostatics Final	91	34

TABLE 2. Consistency of Students Usage of FBD in Mechanics and Electrostatics (N=125).

Case	Percentage
Used in all 5 Problems	15
Used in at least one Mechanics & Electrostatics	72
Used only in Mechanics	12
Used only in Electrostatics	10
Never Used	5

Research question (b) Are students consistent in using free-body diagrams correctly? We found that 24% of the students drew a FBD correctly each time they attempted one, 26% drew correct diagrams for the majority of their attempts, 18% drew an equal number of correct and incorrect FBDs, 10% drew the majority of their FBDs incorrectly, and 2% got every diagram wrong. We put 14% of the students separately as they drew only one FBD for the five questions, thus we could not check for consistency. Another indication of consistency was that 5% of the students never drew a free-body diagram [Table 3].

TABLE 3. Consistency of Students Usage of FBD Associated with Correctness of Use (N=125).

Case	Percentage
All FBD Correctly	24
Majority of FBD Correct	26
Number Correct = Incorrect	18
Majority of FBD Incorrect	10
All Incorrect	2
One FBD Drawn	14
None Drawn	5

Research question (c) Are students who draw free-body diagrams to solve traditional problems more successful in solving the problems then those who do not? To answer this, for each exam problem we broke the students into 3 groups [Table 4]: students who drew a perfect free-body diagram; students who drew a diagram but made mistakes in it, and students who did not draw an FBD. Then we calculated the percent of the students in each group who chose the correct answer (first three columns). The fourth column represents the "success rate" for each problem. We measured the "success rate" of a problem as the percentage of students in the sample who chose the right answer for that problem. Table 4 provides additional data about the number of students who chose a correct answer divided by the number of students in each group.

DISCUSSION

There is little information in the literature [12] about student use of diagrams on tests in traditional courses (especially when diagrams are not required). Thus our study provides a contribution to the research of the role of free-body diagrams in successful problem solving. We realize the limitations of study: all data came from exam work only and there was no control group. We summarize our findings below.

Many students draw free-body diagrams: Students received no credit for their work on multiple-choice exam problems. Still, many of them drew FBDs. Table 1 shows that in most cases, there were more students drawing free-body diagrams than students who did not draw free-body diagrams (the only exception was the Mechanics final).

Students use free-body diagrams in more then mechanics situations. We found that students did use free-body diagrams while solving both mechanics and electrostatics problems. Roughly 3/4ths of the students we tracked used these diagrams while solving mechanic and electrostatic problems.

Students are consistent. Students are somewhat consistent in using free-body diagrams. Half of the students coded were consistently correct in using free-body diagrams.

Students who drew a correct free-body diagram were more likely to correctly solve the problem: We found that students who draw a free-body diagram correctly while solving a problem are much more likely to solve the problem correctly [Table 4]. We also found that drawing an incorrect FBD led to more incorrect solutions than having no diagram at all. In other words a wrong FBD is worse than no FBD at all. The distracters (possible answers due to incorrect reasoning process) for the electrostatics exam 1 problem should probably be revised so that students with incorrect diagrams are less successful.

TABLE 4. Comparison of Free-Body Diagrams Drawn to Answers Given (N = 125).

Exam Question [Rubric Score]	FBD Correct 2	FBD Incorrect 1	No FBD 0	Success Rate
Mechanics Exam 1 Question 1	61/79 = 77%	5/15 = 33%	18/31 = 58%	84/125 = 67%
Mechanics Exam 1 Question 2	45/51 = 88%	2/12 = 17%	19/62 = 31%	66/125 = 53%
Mechanics Final	12/22 = 55%	4/27 = 15%	24/76 = 32%	40/125 = 32%
Electrostatics Exam 1	49/55 = 89%	13/18 = 72%	33/52 = 63%	95/125 = 76%
Electrostatics Final	37/46 = 80%	14/43 = 33%	12/36 = 33%	63/125 = 50%

What are possible explanations of our findings? One can speculate why students did or did not draw free-body diagrams while solving a particular problem. We are currently conducting interviews to find whether it was just a habit developed in the course or students found diagrams helpful for solving difficult problems, and did not use them when they perceived problems as easy. A possible explanation for why drawing an incorrect diagram is related to a worse performance is that the students who had no FBDs could possibly draw them in their minds and did this correctly, or used some other approach to a problem, while those who did draw an incorrect diagram obviously misunderstood the physics. Obviously more studies are needed to rule out these explanations.

APPENDIX

Mechanics Exam 1, Question 1

A 100 kg fireman starts at rest and slides down a vertical pole with a constant downward acceleration of 4.0 m/s². The magnitude of the friction force that the pole exerts on the fireman is closest to:

Mechanics Exam 1, Question 2

Assume that the 100 kg fireman was traveling at a speed 6.0 m/s when he reached the floor. He managed to stop the center of his body in 0.40 m by bending his ankles and knees. What was the magnitude of the average force of the floor on his feet while stopping him?

Mechanics Final

Block A of mass 6.0 kg rests on a smooth table and is connected by a string that passes over an ideal pulley to block B of 4.0 kg. Block B is released from rest. Which answer below is closest to the time interval that block A, initially at rest, needs to travel 0.80 m?

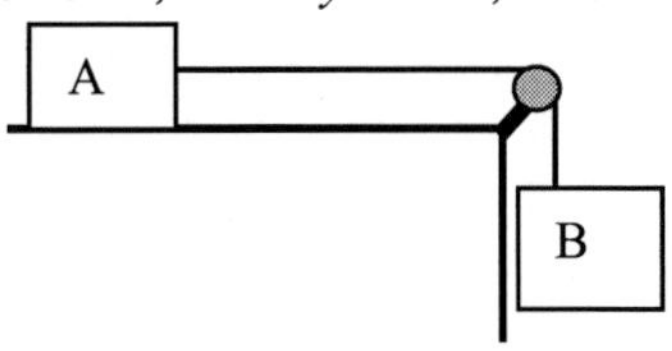

FIGURE 1. Picture with Mechanics Final exam question.

Electrostatics Exam 1

The three charged metal balls each have a charge of magnitude Q but of different signs, as shown. The positive direction is towards the right. Which expression below is the net electric force exerted on the right charged ball?

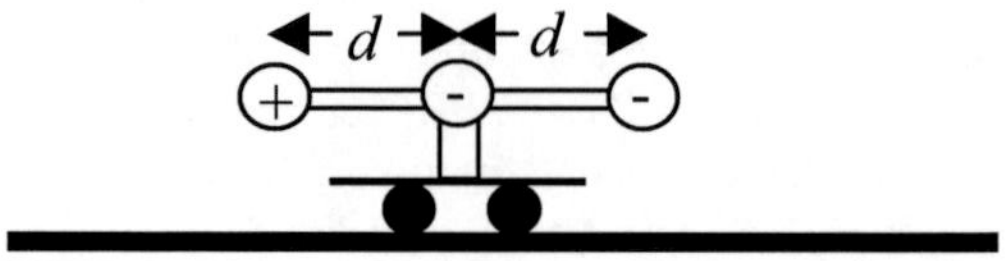

FIGURE 2. Picture with Electrostatics Exam 1 question.

Electrostatics Final

A small metal ball with +2.0 μC of charge hangs at the end of a vertical string. A second identical ball with -2.0 μC of charge hangs at the end of a second vertical string. The tops of the strings are brought near each other and the strings reach an equilibrium orientation (no longer vertical) when the balls are 3.0 cm apart. If the gravitational force of the Earth on each ball is 30 N, what is the tension in one of the strings?

REFERENCES

1. J.I. Heller and F. Reif, "Prescribing effective human problem solving processes: Problem description in physics," Cognit. Instruct. 1, 177-216 (1984).
2. J. Larkin, "Understanding, problem representations, and skill in physics," Thinking and Learning Skills, 2, 141-159 (1985).
3. A. Van Huevelen, "Learning to think like a physicist: A review of research-based instructional strategies," Am. J. Phys., 59, 891-897 (1991).
4. R. Dufresne, W. Gerace, and W. Leonard, "Solving Physics Problems with Multiple Representations," Phys. Teach. 35, 270-275 (1997)
5. D. Hestenes (1997), Modeling Methodology for Physics Teachers. In E. Redish & J. Rigden (Eds.) The changing role of the physics department in modern universities, AIP Part II. p. 935-957.
6. H.A. Simon, Models of discovery and other topics in the methods of science, Dordrecht:Reidel (1977).
7. D.P. Maloney, "Forces as Interactions," Phys. Teach. 28, 386-390 (1990).
8. A. Puri, "The art of free-body diagrams," Phys. Educ. 31, 155-157 (1996).
9. B. Lane, "Why can't Physicists draw FBD's?" Phys. Teach. 31, 216-217 (1993).
10. E. Etkina and A. Van Heuvelen, "Investigative Science Learning Environment: Using the processes of science and cognitive strategies to learn physics" Proceedings of the 2001 Physics Education Research Conference. Rochester, NY, 17-21 (2001).
11. A. Van Heuvelen and E. Etkina, Physics 203/204 Activity Guide, University Press, E. Brunswick, NJ (2003).
12. Harper, K. A., Investigating the Development of Problem Solving Skills during a Freshman Physics Sequence, Unpublished PhD dissertation, The Ohio State University, (2001).

Teachers' Investigation of Students' Self-Perceptions Regarding Physics Learning and Problem-Solving

Edit Yerushalmi*, Bat Sheva Eylon*, Rachel Seggev*

Weizmann Institute of Science, 76100 Rehovot, Israel

Abstract. Transfer is required in nearly every activity of problem solving. It spans from transferring procedures within a finite set of similar "end of the chapter problems" to applying problem solving strategies in completely unfamiliar problems. Students' self-perceptions, in the context of problem solving and learning, influence the success of instruction promoting transfer. Hence, teachers have to attend to such self-perceptions. We conducted a cooperative inquiry workshop to support teachers who modify their instruction in problem solving to better achieve transfer goals. As part of the workshop, the teachers raised the need to develop a questionnaire examining students' self-perceptions in the context of problem solving and learning in physics. The development of the questionnaire was supported by educational research, in a manner reflecting the teachers' motivation and time limits. In this paper, we describe the process of developing the questionnaire, present findings from a validation analysis of the questionnaire, and discuss its role in the teachers' professional development.

INTRODUCTION

Transfer is required in nearly every form of problem solving. It spans from transferring procedures within a finite set of similar "end of the chapter problems" to applying problem solving strategies in completely unfamiliar problems.

Students' self-perceptions regarding learning [1,2] and problem solving in physics, as well as their self-esteem [3], affect their responsiveness to instruction intended to develop transfer skills. For example, teaching systematic problem solving search strategies might be rejected by students who expect to engage in a fast, linear process for solving a problem, or by those who expect to search for the solution in a random manner [4]. Therefore, teachers who modify their classroom practice towards transfer goals need to address students' self-perceptions that were shaped in former and significantly different practice.

This issue troubled the participants of a teachers' workshop that focused on innovative ways to develop students' transfer skills in physics problem solving [5]. During the workshop, the participants initiated the development of a questionnaire to examine students' self-perceptions in areas related to the workshop theme. In this paper we describe the workshop context, how the workshop participants raised the need to study students' self-perceptions and the ways in which teachers and researchers cooperated in the development process. Then we examine the validation of the questionnaire and discuss how the development and analysis of such a questionnaire may help teachers to address students' self-perceptions.

DEVELOPING A QUESTIONNAIRE

Eight physics teachers participated in a year-long cooperative inquiry workshop. The teachers customized to their classrooms PER-based instructional materials designed to develop transfer skills in physics problem solving [6]. The instructional approach was characterized by explication and reflection on problem-solving processes used by experts. For example, students were required to explicitly present steps of a problem-solving strategy in their solutions, or, with guidance, to self-assess their test solutions.

A cooperative inquiry was carried out to ensure teachers' active involvement. It consisted of rounds in which each of the teachers, in turn, presented in a computerized forum an activity that had been tried out in the classroom. Next, a discussion was held with all participants to explore answers to questions raised by the presenters [5].

In the course of the workshop, the teachers repeatedly initiated a discussion about differences in students' learning styles, in their self-images and in

CP790, *2004 Physics Education Research Conference*, edited by J. Marx, P. Heron, and S. Franklin

their perceptions of learning. In the discussions a large spectrum of classifications was suggested. Teachers classified students according to their achievements as "strong" and "weak" as well as by personality traits.

For example, Teacher A raised the following question for discussion: *"There is a group of students who do not speak up in class. How can self-diagnostic activities be exploited to cause them to express themselves?"*

One of his colleagues, B, answered, *"A silent student is not necessarily a passive or weak student. There are those who think but do not express their thoughts, and those whose silence expresses lack of thinking and withdrawal from the lesson. The self-diagnosis activity can activate the passive student. It is worthwhile to cause the students to 'defend' their diagnosis. Maybe this can be promoted through a discussion between pairs of such students?"*

Teacher C raised the question, *"Why, in your opinion, were the weak students more critical with themselves in the diagnosis than the others?"*

Her colleague, D, answered: *"Self-diagnosis of one's own mistake influences self-confidence: on the one hand, learning from a mistake empowers; on the other hand, there are students who cannot accept failure and fall into a situation of denial or desperation."*

The teachers suggested that different traits of the students may interact (e.g. diagnostic ability with self-confidence; being withdrawn with passiveness or with low academic achievement, etc.). Following these discussions, the teachers suggested the preparation and administration of a questionnaire for students in order to characterize those traits that are relevant to the new teaching methods offered by the workshop.

There are a variety of research tools in the educational literature that can serve teachers in developing such a questionnaire. These tools examine traits such as learning styles (e.g. self-regulation, persistence, responsibility for studying); self-images (e.g. capability, anxiety) [7,8,9]. Since the workshop participants were neither acquainted with the literature, nor had time to explore it, both mediation and compromise were required. The workshop leader (a researcher in the field of science teaching) and the teachers collaboratively developed the questionnaire in several stages, taking different roles:

1. **The teachers** determined the structure and content of the questionnaire. They agreed on a 24-statement, self-reporting Likert-type questionnaire and proposed areas in which they wished to characterize their students: a. persistence; b. organization; c. cooperative work; d. reflectivity; e. concentration; f. self-confidence; g. anxiety; h. external vs. internal motivation; i. responsibility for learning; j. planning.
2. **The leader** imported resources. A pool of about 200 statements was gathered from relevant questionnaires found in the educational research literature.
3. **The teachers** constructed the questionnaire, selecting relevant statements from the aforementioned pool.
4. **The leader** prepared the preliminary version of the questionnaire.
5. **The teachers** pilot-tested the questionnaire with their students (N=139).
6. **The leader** revised and tuned the questionnaire through factor analysis.

This statistical analysis identified six highly-correlated groups of statements (factors). Two statements that did not pass the customary correlation threshold (0.35) with any grouping were sifted out. When the statements of each factor are considered together they characterize a certain personality trait. A group of experts was asked to describe and name each of the factors. The group included the workshop leader, a teacher-researcher, a workshop participant, a high-school principal and two physics curriculum developers. Differences were discussed until agreement was reached. The various experts chose almost identical names for each factor and clarified the identity of each factor. Statements from the questionnaire appear in Table 1.

TABLE 1. Factors in the general population.

Factor 'A': The persistent, self-regulated searcher:
1. I try different approaches until I reach the solution.
7. When I have difficulty in solving a problem, I search for the point where I went wrong.
11. I try to find a connection between what we learned in class and what is written in the book.
12. (marginal) I ask myself questions on the material we just learned in order to ascertain that I understood it.
20. I don't leave the problem until I find a solution to it.
22. When I approach a problem, I plan what I am going to do.
Factor 'B': The determined student:
13. I am confident about my ability to meet the requirements of the subject of physics.
17. What I am learning in physics is important to me.
21. A high grade in physics is very important to me.
Factor 'C': The self-confident student:
10. I am confident of my ability to understand whatever they teach me in physics class.
13. I am confident of my ability to meet the requirements of the subject of physics.
14. I am very worried about exams (-).
15. Eventually I will always be able to find a solution, even to difficult problems.
18. I feel stressed when I have to work in an unfamiliar way (-).
Factor 'D': The student striving to improve:
2. I check my work to ascertain that I have done it correctly.
5. I prefer to learn systematic approaches to problem solving,

rather than just practicing by solving many problems (-). 9. Sometimes in class, or while doing homework, I start thinking about other things (-). 19. If I don't do well on an exam I try to learn from my errors.
Factor 'E': The self-securing student: 3. When I am studying for an exam I prepare a list for myself of the important points in the material. 16. When I encounter a difficulty, the first thing I do is to ask the teacher for help.
Factor 'F': The individualist: 4. When I get a hard problem I try to solve it together with a friend (-). 6. Studying can only be done alone because teamwork interferes with the pace of personal progress. 8. I don't ask many questions in class so as not to disturb the course of the lesson (-).

The (-) sign represents opposite polarization

VALIDATING THE QUESTIONNAIRE

How valid is this instrument? The content validation was achieved through the process of developing the questionnaire. We studied the construct and criterion validity of the questionnaire via a) analysis of sub-samples; b) comparison of students' answers in the questionnaire to students' achievements in an activity developed by the workshop teachers.

Content Validation

The questionnaire was constructed by teachers in response to a need they had raised. Thus it reflected the teachers' conception of relevant student traits. The consensus in the group of experts on the meaning of the various factors constitutes additional content validation of the questionnaire.

Construct Validation

The following analysis was carried out on the improved version of the questionnaire using a sample of 15 classes (N=352). First, a factor analysis was carried out on the whole sample (see Table 1 showing the titles (in italics) and statements of each factor). Second, factor analyses were carried out on several sub-samples defined through various criteria (e.g. sex, class level, etc.). Then, the factors obtained for the whole sample served as a reference set and were compared with those obtained for the particular sub-sample in the following manner: for every factor in the whole sample, factors were sought in the partial sub-sample that almost coincided with it. The differences between the composition of factors in the whole sample and in the sub-samples were interpreted and judged on the basis of the relevant literature. As an example, we present two analyses carried out for the sub-samples of boys (N=245) and girls (N=101).

- Factor A includes statements 1,7,11,20 for the boys and statements 1,2,(-)5,16,20,22 for the girls. The girls' factor differs significantly from the original A factor in the whole sample. Instead of reflective statements (7, 11) one gets the hard work and self securing statements (2,(-)5,16).
- Factor C includes the original statements for the boys and statements 10,12,13,(-)14,15,20 for the girls. Statements 12 (assuring understanding) and 20 (persistence) are added to the girls' factor.

That is, for the girls, the "persistent, self regulated searcher" factor transforms to the "persistent, hard working and self-assuring searcher" factor, and the self-confidence factor is associated with a determined approach to studying and to solving problems. These results are consistent with those reported by Sroufe et. al. [10] who claim that parents, teachers and even boys and girls themselves attribute high achievements by girls in mathematics to hard work, whereas those of boys are attributed to talent. If a girl has self-confidence in her ability to meet the requirements of the subject, and to solve even difficult problems without fear, it is not taken for granted and "natural" as it is with boys. The girl demands from herself a commitment to reach the solution. To summarize, the findings in these sub-samples are in accord with the developmental theories regarding differences between boys and girls. These results contribute to the construct validity.

Criterion Validation

The teachers expected that the students' actual achievements in an activity requiring transfer would correlate with the students' profile of traits. We focused on one activity, developed by a teacher, to study the relationship between students' traits and their achievements. The activity dealt with self-diagnosis of a problem solution on an exam. The students had to identify the execution of central stages in the problem solution (description, definition and presentation of sub-problems, checking the answer), and to pinpoint mistakes in the application of the relevant principles of physics. The teacher carried out the activity in his class. We developed a rubric to assess students' achievements in self-diagnosis. The rubric judged students' achievements in locating, categorizing, explaining and correcting mistakes. The teacher administered the questionnaire and characterized the students' profiles. A "profile" stands for the distribution of intensities of the various factors. Since the sample was small (n=10) the profile averages were calculated for high and low achievers in the self-diagnostic activity (Table 2). It was found that the low achievers had higher intensity of the factors "C-self-confident" and "E-self-securing". These findings correspond to the teacher's expectation that students who are very confident and students who believe that they work "by the book" would have a lesser tendency to diagnose themselves. These findings contribute to criterion validation of the questionnaire.

TABLE 2. Students' Profiles vs. Achievements

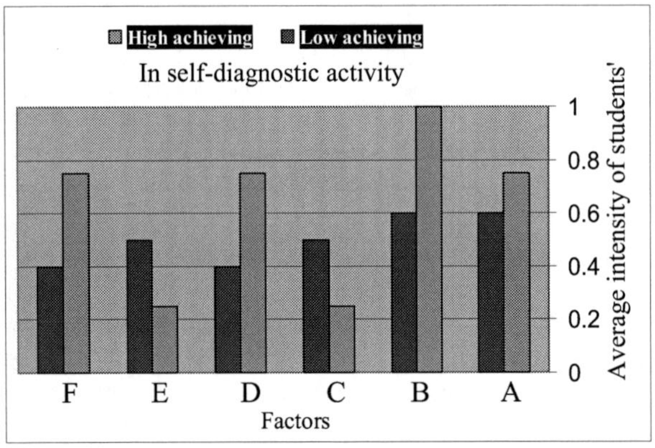

DISCUSSION AND SUMMARY

This paper describes a workshop in which teachers attempted to promote transfer in problem solving by their students. They realized the importance of students' traits, in realms such as self-efficacy and self-monitoring, to their goal. The teachers suggested designing a questionnaire that would allow them to identify relevant student traits. The teachers cooperated with researchers in developing the questionnaire: The researchers imported resources from the professional literature, while the teachers judged the relevance of these resources to their practice. In constructing the questionnaire, the teachers set aside academic achievements and focused on learning practices, perceptions and attitudes.

The questionnaire can serve the professional development of teachers by enabling them to re-formulate and sharpen their intuitive distinctions regarding traits of students in the following ways: a) The questionnaire reflects an agreement between the teachers regarding which student traits are relevant to fostering self-monitoring in solving problems; b) the factor analysis develops a common and well-defined professional language; and c) this professional language can be used to discuss and address traits of students that affect transfer in problem solving.

To summarize, the construction of the questionnaire can raise teachers' awareness to relevant traits of their students and lead towards student-centered teaching. The findings of this study suggest that: a) Students' profiles, as defined by the factor analysis, are correlated with students' achievements in problem-solving activities requiring transfer. b) Findings from studying the composition of factors in sub-samples (e.g. boys vs. girls) are in accord with theory. These findings contribute partially to the validation of the questionnaire and justify using it as a professional development tool.

ACKNOWLEDGMENT

We wish to express our gratitude to the teachers who participated in the workshop: Miryam Hershko, Arkady Mishna, Zeev Nemtsov, Korina Polinger, Philip Rojnikovski, Rachel Seggev, Irina Weissmann and Berl Zaltsman.

REFERENCES

1. E.F. Redish, J.M. Saul and R.N. Steinberg, "Student expectations in introductory physics", *American Journal of Physics*, **66** (3), 212-224, (1998).
2. D. Langley and B. Eylon, "Probing High School Physics Students' Views and Concerns about Learning Activities", *International Journal of Science and Mathematics Education*, (Accepted 2005).
3. A. Bandura, "Self-efficacy: Toward a Unifying Theory of Behavioral Change", *Psychological Review*, **84**, 191-215, (1977).
4. A. H. Schoenfeld, "Learning to Think Mathematically: Problem Solving, Meta Cognition, and Sense Making in Mathematics" in *Handbook of Research on Mathematics Teaching and Learning* edited by D. A. Grouws, Macmillan, New York, 1992, pp. 334-370.
5. E. Yerushalmi and B. Eylon, "Teachers' Approaches to Promoting Self-Monitoring in Physics Problem Solving by Their Students", GIREP Meeting Proceedings, Spain, 2001.
6. D. P. Maloney, "Research on Problem Solving: Physics" in *Handbook of Research on Science Teaching and Learning* edited by D. L. Gabel, Macmillan, New York, 1994, pp. 327-356.
7. M. Jerusalem and R. Schwarzer, "Self-efficacy as a Resource Factor in Stress Appraisal" in *Self-efficacy: Thought Control of Action,* edited by R. Schwarzer, Hemisphere, Washington, DC, 1992, pp. 195-213.
8. P. R. Pintrich and E. V. De Groot, "Motivational and Self-Regulated Learning Components of Classroom Academic Performance", *Journal of Educational Psychology*, **82**(1), 33-40 (1990).
9. Zimmerman and M. Martinez-Pons, "Construct Validation of a Strategy Model of Student Self-Regulated Learning", Journal of Educational Psychology, 80(3), 284-290 (1988).
10. L. A. Sroufe, G. R. Cooper and G. B. DeHart, *Child Development Its Nature and Course*, McGraw-Hill, New York, 1996.

Conceptual Underpinnings of Students' Ability to Understand Reflections from a Plane Mirror

Karen Cummings and Edward Grillo

Southern Connecticut State University, 501 Crescent St. New Haven, CT 06515

Abstract. In this paper we explore students' pre-instruction knowledge of several conceptual and procedural pieces of knowledge that we believe are prerequisite to one's ability to generate correct light ray diagrams and understand image formation by a plane mirror. The research population is an algebra-based, introductory physics class of about 50 students at a medium-sized, urban, public university. Both individual interviews and written free response questions were used to gather data.

PRIOR FINDINGS

In 1986, Fred M. Goldberg and Lillian C. McDermott published a careful study of the difficulties students have applying their knowledge of geometric optics to novel situations involving image formation by a plane mirror.[1] Goldberg and McDermott probed students' functional knowledge within this domain using four tasks. They asked students about these tasks either in interviews or with written questions. Three of the tasks involved a dowel placed in front of a plane mirror. The students were asked to predict if (and where) an image would be visible for various observer locations. They were also asked to explain their reasoning. The forth tasked involved a small mirror held in front of the student. The student is asked if there is anything that she or he can do in order to see more of their image in the mirror.

Goldberg and McDermott found that many students fail at these tasks, both pre- and post-instruction. That is, they found that most students could not determine the location of an image in a plane mirror for shifting observation points nor could they determine whether observers at different locations would be able to see the reflection at all. Goldberg and McDermott also identified a common, largely incorrect student conception, the "line of sight" conception, which is depicted in Figure 1.

Seminal to this study, Goldberg and McDermott also found that many students could not draw correct ray diagrams. Even more notable, they found that students often could self-correct initially incorrect predictions if they were encouraged to draw ray diagrams and could do so correctly. These last two findings, stated immediately above, were the motivation for this study.

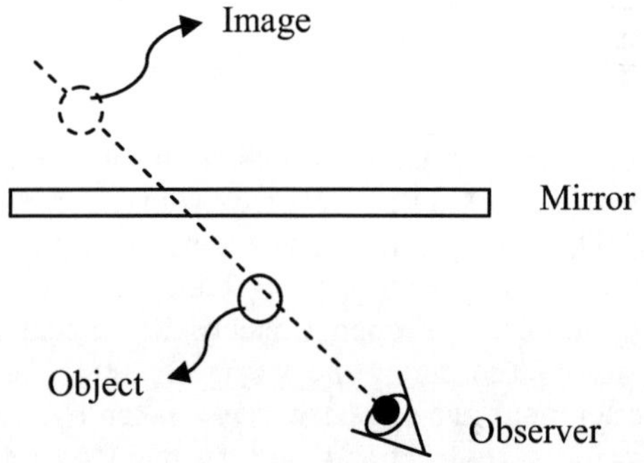

FIGURE 1. The "line of sight approach," depicted here, is a largely incorrect but common student approach to determining the location of the image formed by a plane mirror. The image is believed to lie somewhere along the observer-object line of sight. (This is correct if the observer-object line of sight happens to be perpendicular to the mirror.)

INTRODUCITON

The Goldberg-McDermott findings are helpful in that they clearly establish the difficulty students have applying physical principles from geometric optics in explanation of image formation by a plane mirror. In addition, their work supports our own theoretical belief that learning to draw correct ray diagrams is of fundamental importance if students are to develop

CP790, *2004 Physics Education Research Conference*, edited by J. Marx, P. Heron, and S. Franklin

functional knowledge of geometric optics. However, when we attempted to apply the Goldberg-McDermott findings to our own instructional practices we were left with several significant questions. Hence, this study was a preliminary attempt to answer a cluster of research questions related to students' inability to draw correct ray diagrams within the domain of reflections from a plane mirror.

Primarily, we wanted to know what piece or pieces of conceptual and/or procedural knowledge are missing or misunderstood when students fail to generate correct ray diagrams for the Goldberg-McDermott tasks. This is largely unaddressed by the Goldberg-McDermott paper. For example, do college-level, general-education students know that light is required for image formation? Do they realize that in general objects reflect light that falls on them from other sources? What are their intuitions about the reflection of light in a plane mirror?

STUDENT POPULATION

This study took place at Southern Connecticut State University, a public university with approximately 12,000 students in an urban setting. The student population studied was an algebra-based introductory physics class of approximately 50 students. The study took place during the spring of 2004. The course is a terminal one-semester course taken by computer science students and biology majors who are not pre-medical students. High school physics or a conceptual physics course is a prerequisite.

RESEARCH TOOLS AND METHODOLOGY

The data reported here were gathered as part of a study that employed four different sets of written, open-ended questions. The question sets were given to all students present during four fairly evenly spaced class periods. Individual student interviews were preformed periodically to clarify, verify and extend the information gained via the written responses. This approach is similar to the "hybrid" interview-written response approach used by Goldberg and McDermott in their earlier study. In each case the question sets were *not* returned to the students and the correct answers were *not* directly discussed. The first question set was given prior to any instruction in geometric optics. This "pre-instruction" assessment is the focus of this paper.

FINDINGS

With Good Eyes, You Can See Anywhere.

One of the most fundamental pieces of knowledge required for an understanding of image formation by a plane mirror is that at least some light must be present. We wondered if our population of students understood this. We probed it by asking Question #1, which is shown below along with student answers.

> ***Question #1:*** *Suppose that you are in a totally closed room with only a light bulb and a mirror. There are no windows and no doors in the room. The light bulb is placed in front of the mirror but is NOT turned on. So, there is no light in the room. Will you see an image of the light bulb reflected by the mirror?*
> **Student Responses:**
> **82 % No**
> **18 % Yes**

Among the small group who incorrectly answered "yes" there were often statements made that "one's eyes will adjust in time and then you will be able to see (even in a room with no light at all)".

Although the number of students using this line of reasoning was small in this study, some researchers believe that they see a higher prevalence of this type of faulty reasoning in other college level populations.[2]

Another preliminary conceptual piece required for understanding image formation in plane mirrors is knowledge that there are light rays reflected from nearly all objects and that this is required if an object is to be imaged. This idea interested us. We hypothesize that some students may have stumbled from the start of the Goldberg-McDermott tasks because they fail to see a dowel as a source of light rays. After all, it is wooden and so neither directly produces light nor is highly reflective. In order to probe student pre-instruction conceptions we asked Question #2 shown below with answers.

Wood Doesn't Reflect Light

There are two noteworthy results from our study of Question #2. First, a fairly large percentage of students (62%) did not recognize a wooden table as a source of reflected light. This is, of course, a fundamental problem in understanding image formation in most real situations.

> **Question #2**: *An object can be a "source" of light in several ways. For example, an object might produce, reflect, scatter, or transmit light from an outside source. In each case, we will call the object a "source" of light. Suppose that you are in a closed room with only a light bulb, mirror and a wooden table present. The bulb is turned on, so there is light in the room.*
>
> **Student Responses are shown in bold.**
>
> *Is the bulb a source of light? If so how?*
> **100% Yes, it produces light**
>
> *Is the mirror a source of light? If so how?*
> **81 % Yes, it reflects light**
> **11 % Yes, it reflects light and thereby produces more light or amplifies the existing light**
> **8% Other**
>
> *Is the table a source of light? If so how?*
> **38% Yes, it reflects light**
> **62% No**

Reflection is a Light Production Process

The second noteworthy result was a student conception of the mirror as an "amplifier of light" or reflection as a light production process. For example, one student wrote in free response and confirmed verbally that she believed "*the mirror's reflection produces a light*". Another said "*The mirror reflects the light from the bulb and multiplies its intensity.*" These students also drew (incorrect) ray diagrams in response to Question #3 (discussed on the following page) that were consistent with this belief. An example is shown in Figure #2.

Although the concept of reflection as a light production process was not commonly expressed in our pre-instruction study (11%), it did seem to exist in some form in the minds of many of the students at various points of time. For example, later in the instructional/assessment process, when students were asked to draw ray diagrams for a two-eyed observer located next to an object (as opposed to behind it), several students "developed" the idea of multiple reflective rays from a single incident ray in order to justify their predicted image location. An example diagram is shown in Figure #3.

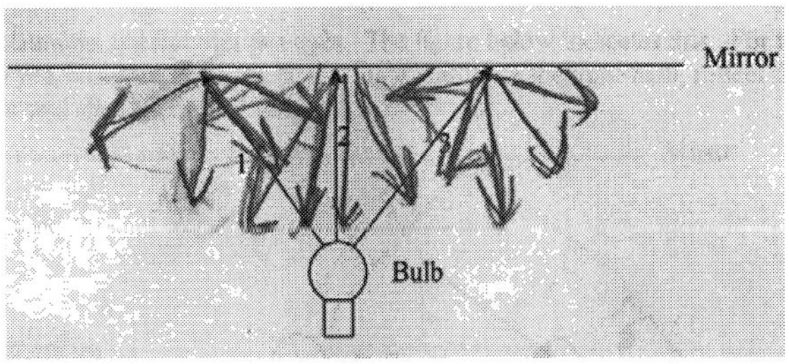

FIGURE 2. A student response to Question #3, discussed on the next page. This student's diagram is consistent with her assertion that reflection "produces light".

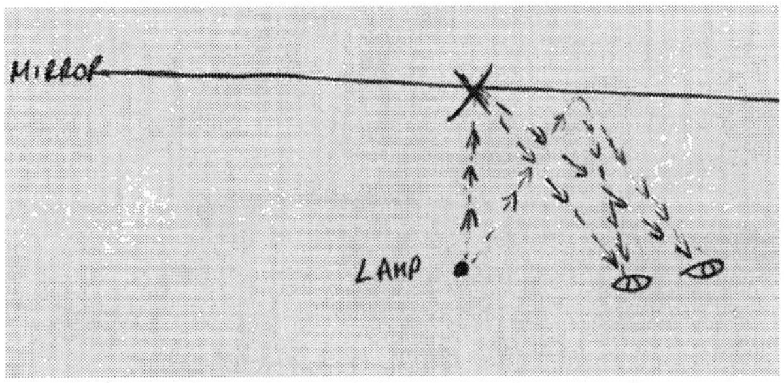

FIGURE 3. Another student diagram indicating that multiple reflective rays can be produced by a single incident ray. This student did not express that belief during pre-instruction questioning. Rather, he seemed to "develop" this conception in response to novel, more difficult tasks. The "X" in the diagram marks the location the student predicts for the image.

It Goes Out the Way It Came In.... Only Opposite

We had encouraging finding as well. Our students often appeared to have helpful, correct intuitions, preconceptions or remnants of knowledge from previous courses regarding the reflection of light by a plane mirror. The same is true for the path light travels in the process of image formation by a plane mirror. For example, 90% of the students questioned before instruction knew that light had to travel from the object to the mirror and then to the observer's eye in order for a reflective image to be seen.

Many students could draw correct diagrams indicating the reflection of light from a plane mirror, as probed with Question #3 shown below. In addition, 50% of students came up with a correct "law of reflection" although it was often stated in non-technical terms. For example, one student said "*Whatever angle the light hits the mirror is the angle it will reflect at.*"

Question #3: *A light bulb emits rays of light in all directions. Consider the two dimensional situation shown in the figure below. A bulb is place in front of a mirror. Three rays of light coming from the bulb hit the mirror as shown. Draw appropriate rays of reflection for each of these three rays. Please label them R1, R2 and R3.*

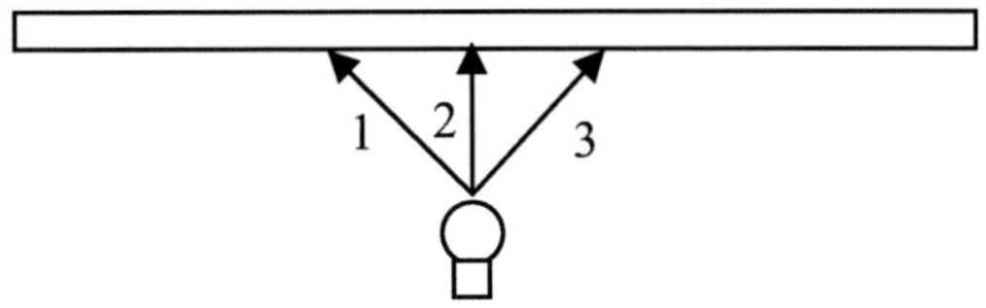

Student Responses:
55% Correct Diagram
8% More than one reflected per incident ray
8% Reflected rays enter in the mirror (as in refraction)
5% Right angle between incident and reflected
24% Various Other Incorrect Diagrams

The Image Forms in Your Eye ...and in Front of the Mirror

The simplest Goldberg-McDermott tasks involved an object placed in front of a mirror. The students were asked to identify where the image would be located. We asked the same question of our students. Table #1 shows the breakdown of responses for the two groups.

Stated Image Location	Goldberg Population	Cummings Population
Behind the Mirror	70%	16%
On the Mirror Surface	25%	8%
In front of the Mirror	---	21%
In the observer's eye	---	18%
"In" the Mirror	---	11%
In the observer's eye and in/on the mirror	---	18%
Other	5%	8%

TABLE 1. Percentage of students indicating various image locations for an object placed in front of a plane mirror

There are two obvious differences between the responses in this study and those in the Goldberg-McDermott study. First, far fewer of our students knew that the image was located behind the mirror. Second, 36% of our students wanted to discuss, or even focus on, the formation of the image in the observer's eye. This was not even mentioned in the Goldberg-McDermott study.

But That's Different

Evaluating students' answers to Question #3 in light of their ability to state a correct law of reflection, we found that many students simultaneously hold conflicting conceptual beliefs. For example, 25% of the students who stated the angle of incidence is equal to the angle of reflection drew diagrams that were inconsistent with this. Only 42% of the students gave clearly consistent answers to where an image is located (on the mirror, behind the mirror and so on) when asked to answer using words and then using diagrammatic representations. The idea that students might have "mixed-models" or context dependent beliefs is not surprising. It is consistent with what we know from the domain of mechanics.[3]

CONCLUSIONS

We have found that to varying degrees this population of students has conceptions regarding light and image formation by a plane mirror that include the idea that reflection is a light production or amplification process, light is not necessary for vision to occur (if you have good eyes), and "dull" objects like wood don't reflect light. In response to this last finding, we suggest that it might be prudent to start instruction with a lit light bulb as an object. This may help limit the "cognitive load" on students as they begin learning about image formation. In general, this study indicates that instruction for general education students likely needs to include clear demonstrations of the fact that nearly all objects reflect at least some light. Based on the results shown in Table #1, we also believe that many students need a chance to address their existing knowledge regarding image formation in the eye so that they can shift their attention to external images. In addition, Table #1 indicates that instructors should be careful not to reinforce a likely misunderstanding that the reflection is "in the mirror" through causal use of language.

REFERENCES

1. Goldberg, F.M. and McDermott, L.C., The Physics Teacher, 11, 472-480 (1986).
2. See for example, M. Schneps and P. Sadler, A Private Universe, Santa Monica, CA, Pyrimid Film and Video, 1989
3. L. Bao and E.F. Redish, AJP (PERS) 69, (7), S45-S53, 2001.

A Web-based Tool For The Analysis Of Concept Inventory Data

Joseph P Beuckman[1], Scott V Franklin[2] and Rebecca S Lindell[1]

[1]*Southern Illinois University Edwardsville*
[2]*Rochester Institute of Technology*

Abstract. "FOCIA" stands for Free Online Concept Inventory Analyzer. FOCIA, our new web-based tool will allow teachers and researchers in any location to upload their test data and instantly receive a complete analysis report. Analyses included with this tool are basic test statistics, Traditional Item Analysis, Concentration Analysis, Model Analysis Theory results, pre and post test comparison, including the calculations of gain, normalized change and effect size. The tool currently analyzes data from the Lunar Phases Concept Inventory (LPCI), the Force Concept Inventory (FCI), the Astronomy Diagnostic Test (ADT), the Force and Motion Concept Inventory (FMCE) and generically, any multiple choice test. It will be expanded to analyze data from other commonly utilized concept inventories in the PER community and, from user-designed and uploaded tools. In this paper, we will discuss the development of this analysis tool including some technical details of implementation and a description of what is available for use. Instructors and researchers are encouraged to use the latest version of the analysis tool via our website, http://www.sciedures.org.

Keywords: FOCIA, Web Analysis, concept inventory, item analysis, concentration analysis, model analysis
PACS: 01.40.Fk

INTRODUCTION

The last few years in education research have brought about somewhat sophisticated experimental analysis methods for interpreting student assessment data. The new methods typically require tedious, repetitive calculations that lend themselves readily to automation. At present, the methods are typically implemented by a combination of hand-manipulation of the data in spreadsheet software, mathematics software and graphics software. In FOCIA, we automate as completely as possible this analysis including several experimental data analysis techniques, as well as more traditional test analysis techniques and to provide wide access to our new tool with the overarching goals:

- To give educators and researchers instant access to the data analysis methods
- To accelerate the development of these and other new methods of assessment.

This paper presents the tool with some technical description of its implementation, examples of output and future directions.

FOCIA TECHNICAL DESCRIPTION

FOCIA makes sophisticated analyses of common and experimental test data and reports the results all via the World Wide Web. We use PHP as our programming language. PHP (recursive acronym for "**PHP**: **H**ypertext **P**reprocessor")[1] is a general purpose scripting language that works on a web server to provide HTTP (**H**yper **T**ext **T**ransfer **P**rotocol) content to a user's computer and is readily integrable with relational databases.

Several types of stored data are required as inputs to perform item, concentration and model analysis. There are the concept space structure, the concept inventory structure and content and the inventory result (classroom) data on which to perform the analysis.

The data comprising the concept space and concept inventory are relatively static. Changes to a concept space or concept inventory are only made by a researcher after analysis of potentially many datasets of class response data. To provide a portable storage

CP790, *2004 Physics Education Research Conference*, edited by J. Marx, P. Heron, and S. Franklin

format for concept spaces and inventories, an XML (**EX**tensible **M**arkup **L**anguage)[2] document type was devised to store and communicate them in a human-readable "ASCII text" format (Fig. 1).

```
<dimension name="Period of the Moon's orbit around the Earth" ID="0">
   <model name="null" ID="0"></model>
   <model name="approximately one month" ID="1" correct="yes"></model>
   <model name="less than one month" ID="2"></model>
   <model name="greater than one month" ID="3"></model>
</dimension>
```

FIGURE 1. The XML document format for marking up concept spaces includes, for example, this tag to describe a dimension of the LPCI[3] concept space and the dimension's four different conceptual models.

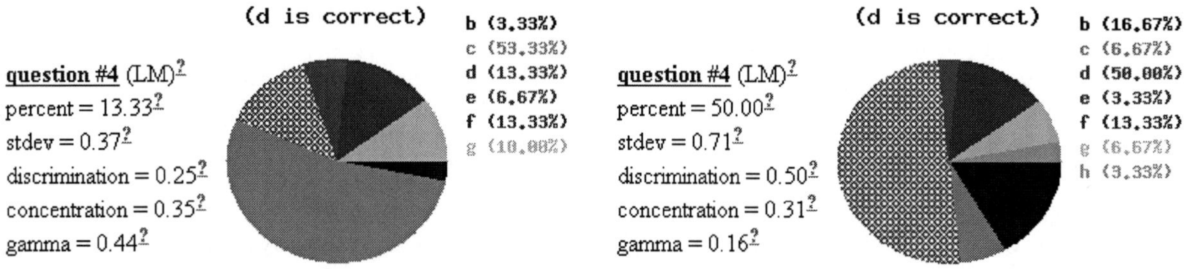

FIGURE 2. An example of the side-by-side response distribution data and graphics for pre-instruction versus post-instruction analysis for a particular item (in this case question #4). The superscript question marks at each figure hyperlink to an explanation of how that figure was calculated.

To facilitate efficient access to the concept space and inventory during the automated analysis, we store these relational data items in a minimized, SQL (**S**tructured **Q**uery **L**anguage)[4] database.

Inventory response data is parsed from the user's delimited text files into arrays in program memory and stored in the user's account on the website as a FOCIA "dataset." The analysis is then performed on the response data using the information from SQL queries of the concept space and inventory database tables.

An object-oriented programming methodology is appropriate here to organize the structures of inventory result and analysis data. PHP classes for a "Record" (a single student's response vector to the inventory) and an "Analyzer" (a group of Records with its analysis data) encapsulate the arrays and functions having to do with each. Several other classes implement portions of the interface and processing within the program.

USING FOCIA

We currently require users of FOCIA to register only so that they have a password-controlled account to maintain a separate, private space for their classroom data.

To use the automatic analysis, the user creates a "dataset" by uploading classroom result data for one of the inventories (or the generic multiple choice type) available within FOCIA. They then select which analysis results they want to be reported via an HTML form. Analysis is initiated with a button on the form.

The analysis is typically completed within a few seconds and result begin transfer to the web browser. The requested analysis results are organized and returned in a printable hypertext report. Numeric data are available as well as popular graphical representations (Fig. 3 and 4) and graphical representations unique to this program (Fig. 2).

ANALYSES CURRENTLY AVAILABLE

The tool currently analyzes data from the Lunar Phases Concept Inventory (LPCI), the Force Concept Inventory (FCI), the Astronomy Diagnostic Test (ADT), the Force and Motion Concept Inventory (FMCE) and, in a basic way, data from any multiple choice test.

The most difficult part of automating this analysis is properly importing the user class data. We have a model for this parser in place that works for a wide variety of text data file formats.

Automated item analysis[7] for each item in a given inventory is currently available. Functions have been written to calculate difficulty, standard deviation of the mean and discrimination. A pie-chart showing response distribution for each item is available (Fig 2). It should be noted that item analysis can be performed on any multiple-choice test regardless of whether the test is a concept inventory with concept space and item to dimension mappings defined. Item analysis can be performed on ANY test data by selecting a "generic" option when creating its dataset.

Concentration analysis[8] for each item has been implemented. Functions exist to calculate item concentration and gamma (concentration deviation) and report item score and concentration in the "L-M-H" coding system explained by Bao and Redish[8]. An automatically generated chart shows score vs. concentration for every item in an analyzed inventory (Fig. 3).

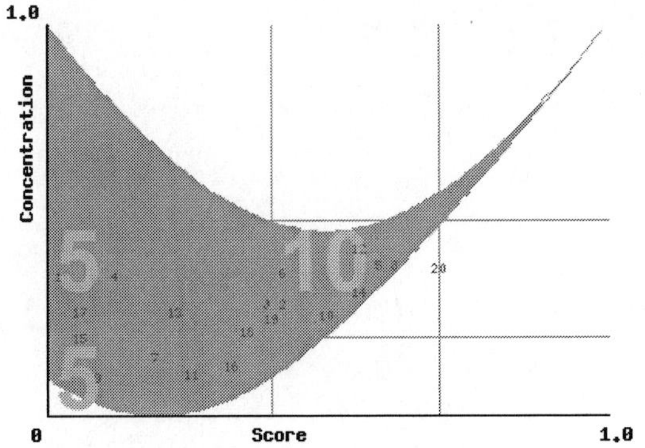

Figure 3. This plot of score versus concentration shows data from the 20 questions of the LPCI[3] for anonymous sample data.[6] The plot is divided into sections according to the L-M-H encoding[8] and the number of data points in each section is displayed.

Performing model analysis[9] requires several ordered manipulations of vectors containing the probabilities that a student within the dataset will choose a given concept model for each dimension in the concept space. Functions are in place to calculate the required probability vectors, then normalize to the "probability amplitude" vectors and produce the "density matrices." The eigen-analysis of the density matrix is performed and analyzed for model use consistencies in the dataset. For binary comparisons between viable pairs of preferred models, "model plots" are generated programmatically (Fig. 4).

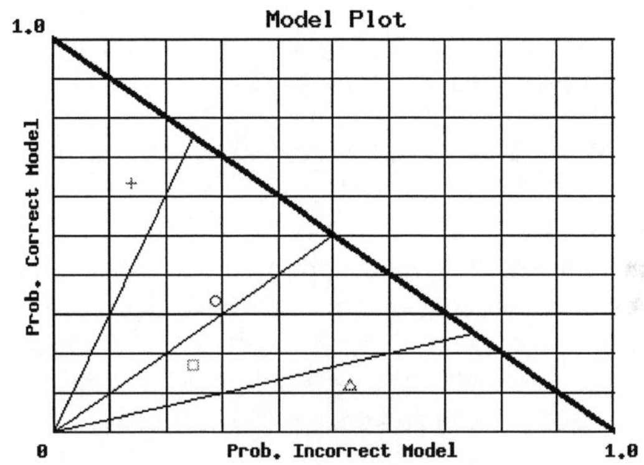

Figure 4. This example model plot[5] shows results in each probability region from four separate instances of an inventory.

A mechanism is provided for performing two analyses side by side. This could be used to compare arbitrary datasets or to analyze a class' performance pre-instruction versus post-instruction. When this type of analysis is selected, the program provides assistance with matching student records from the two datasets being analyzed. Plots of pre versus post instruction scores for each student can be generated as well as Hake plots[10] and combined pre and post score versus concentration plots. A list of raw pre and post scores for each student is generated and each student in the list is hyperlinked to full pre and post instruction response listings for immediate examination.

FUTURE WORK

Eventually, we will allow users to upload independently developed concept spaces and inventories in the designated XML format. Users would be able to smoothly develop their own inventories in this manner. We envision a system that would allow users public and private data areas for analysis and sharing results. This will require careful attention to privacy issues regarding the classroom datasets.

As mentioned, the parser code for class data files is a complicated area of the program as we want to be able to accept a wide variety of input formats. A means will be required for the user to efficiently ascertain whether their data has been imported correctly. It may be necessary to require of the user very specifically formatted data files. This is what we aim to avoid. We would like to include the ability to interpret Microsoft Excel format files in addition to the text data files we currently accept. We are studying the types of file formats most commonly used by testing services.

When users have functional access to the working analysis tool, we will focus on furthering the analyses available. We will expand and improve the current analyses and add other analyses as they become available or are requested, especially with regard to concept hierarchy analyses, another subject of our research. We will continue to improve the interface and add popularly requested features.

ACKNOWLEDGMENTS

We would to like to acknowledge the Physics, Astronomy, Chemistry and Biology Education Research (PAC_BER) group at Southern Illinois University Edwardsville for the valuable insight they have contributed to this project. In addition, we would like to thank the SIUE College of Arts and Sciences and Department of Physics for providing the financial support for this paper.

REFERENCES

1. What is PHP?. <http://www.php.net/> Accessed May 2004

2. Introduction to XML. <http://www.w3schools.com/xml/xml_whatis.asp> Accessed June 2004

3. Lindel, Rebecca S. and Olson, James P. (2002) Developing the Lunar Phases Concept Inventory

4. Melton, J. SQL: The Standard And The Language. <http://www.opengroup.org/public/tech/datam/sql.htm> Accessed June 2004

5. Bao, L. Mathematical Features of Model Analysis

6. Note that one data point falls outside of the gray region. This is because the gray region represents the possible results for a Multiple Choice Single Response (MCSR) tool with exactly 5 possible responses for each item. The LPCI items vary in number of possible item responses.

7. Kehoe, Jerard (1995) Basic Item Analysis for Multiple-Choice Tests, *ERIC/AE Digest*

8. Bao, L. and Redish, E. F. (2001) Concentration Analysis: A Quantitative Assessment of Student States, Accepted for publication in PERS of *Am. J. Phys.*

9. Bao, L. and Redish, E. F. , Model Analysis: Modeling and Assessing the Dynamics of Student Learning, submitted to *JPER*

10. Hake, R. R. (1998). Interactive-engagement versus traditional methods: A six-thousand-student survey of mechanics test data for introductory physics courses. American Journal of Physics, 66(1), 64-74.

Feedback With Web-based Homework And PADs

Scott Bonham

Department of Physics and Astronomy
Western Kentucky University

Abstract. The Physics Applets for Drawing (PADs) allow students to interactively make graphs and other physics diagrams on the Web and have them evaluated. PADs are able to evaluate qualitative as well as quantitative drawings and to give customized feedback. These features greatly expand the range of exercises possible in a web-based homework system and make the latter more able to support research-based curricula. While feedback is an important component in learning, it is a challenge to provide enough feedback so that students do not become stuck and frustrated while at the same time not so much that it enables students to avoid thinking, particularly in an on-line environment. In this paper six different approaches to computer-based feedback are discussed along with how PADs could be used for different approaches. Participants are invited to discuss and make suggestions as to how PADs could be best used to support research-based curricula.

INTRODUCTION

Two significant trends in physics instruction are Physics Education Research (PER) and the use of web-based homework. The latter is often driven by economic and logistical motivations, but could support a key principle of PER-based curricula: providing immediate, formative feedback to the student. The challenge is to provide the right level of feedback.

If insufficient structure and feedback are provided, weaker students will flounder and become frustrated. If too much structure and feedback are provided, students can use it as a crutch and avoid thinking. Balancing the structure/feedback needs of different students in the classroom is a significant challenge for a good human instructor. Web-based implementation exacerbates the challenge, reasons including limited communication (no facial expressions) and the difficulty of anticipating all possible responses. Over the years different approaches have been used for computerized feedback. Some general approaches will be discussed below with an eye towards lessons that can be learned and applied to the PADs project.

The Physics Applets for Drawing (PADs) project[1] is sponsored by the National Science Foundation for a suite of Java Applets for making qualitative and quantitative graphs, vector diagrams, and other diagrams on the web. PADs can be used in static web pages or within web-based homework. They are currently supported by WebAssign. PADs have an internal grading engine with the ability to give conditional feedback customized to a particular exercise. I would like to engage in discussion with other members of the PER community on how to best use these features to support PER-based curriculum. The first curricular application was to code homework questions from RealTime Physics Module 1[2] (which uses many graphs) into WebAssign,[3]. WebAssign is reviewing the questions and obtaining permissions to add RealTime Physics 1 as a supported textbook.

TYPES OF FEEDBACK

Different approaches have been used over the years to try to provide correct levels of computer feedback. They can be divided into six major categories with different advantages and disadvantages.

Poll

The poll type offers no direct feedback about what is right or wrong. Examples include Just-in-Time-

CP790, *2004 Physics Education Research Conference*, edited by J. Marx, P. Heron, and S. Franklin

Teaching[4] (JiTT) and Peer Instruction.[5] In this approach the instructor eventually gives feedback in the discussion of student answers, resolving the structure/feedback dilemma by passing the buck to the instructor. Since the instructor must provide feedback, it may be delayed and it may not meet the needs of a particular student.

Right Or Wrong

This is a common mode for web-based homework because of its ease of implementation. The system simply tells the student whether the response is correct or not, with no other information. Students can learn from their own mistakes as they figure out the error that was made, but a student working alone without enough resources to identify the error may spend hours in frustration. The straight right/wrong feedback may also lead some students to focus more on getting things "right" compared to understanding.

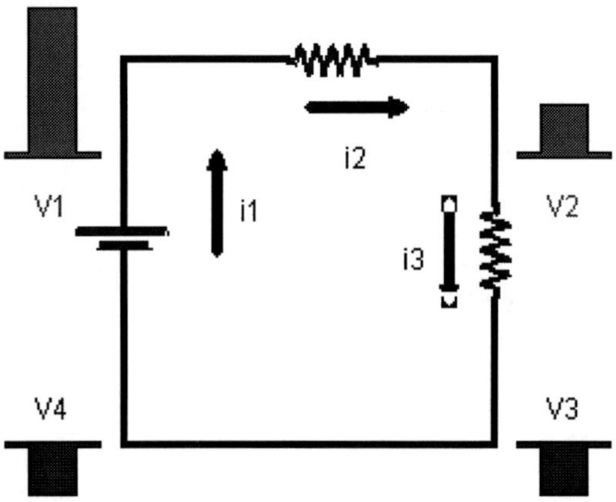

FIGURE 1. An example of a qualitative circuit PADs exercise. The student adjusts the bars to represent the voltage at each point, and the arrows to indicate direction and magnitude of current through each element. The applet will grade correct any diagram in which V2 is halfway between V1 and V4, V3 is the same as V4, and the currents are clockwise and the same magnitude. This example uses VectorPAD, illustrating both vector and bar forms of objects.

Help On Demand

This approach makes additional help available to students when they ask for it. However, there needs to be some sort of disincentive for using help when the student doesn't really need it. The first web-homework system I used, CyberProf,[6] often included hint and help links in the text of the exercise that students could expand for additional suggestions and pointers. In practice, many students would expand all of the hints and helps in the page before even reading the problem. CyberProf's replacement, the Tychos[7] system, provides additional help not in statements but through a series of extra questions and steps the student goes through if they chose; help is available at the cost of having to answer additional questions. Another approach is to assess a point penalty for asking for extra help. However, I heard of students at my school partially defeating this penalty in Mastering Physics[8] by one person in a group sacrificing the point and then passing the hint on to the rest of the group.

Reactive Help

This approach provides help when students submit an incorrect answer. A simple approach (employed by CAPA[9]) is to simply provide a hint from a list. Students have been known to randomly submit wrong answers until they had received the whole list of hints provided.[10] A more sophisticated and challenging approach is for the system to try to figure out the cause of the mistake and provide appropriate feedback. A student worker of mine told me about a light refraction problem in which Mastering Physics kept telling him there was a sign error when in reality an additional index of refraction had been left out.[11]

Simulated Results

Students receive feedback from simulated results by observing a simulation of the result. An example might be an object moving according to the equation of motion that was entered. Examples of computer programs that use this are Graphs and Tracks[12] and Electric Field Hockey.[13] This type of feedback tends to emphasize the physics in the problem and requires the student to judge for himself/herself whether the goal was achieved, and what errors would have lead to an undesired result. However, not all problems easily lend themselves to this type of feedback.

Structured guidance

The alternative to trying to help students figure out their mistakes is to help them do it correctly in the first place. This may be less discouraging as it focuses on what is right instead of what is wrong. It is also easier to code the exercise, being easier to anticipate all the responses the student might make at one particular point than in trying to reconstruct the error from a final answer. The simple approach is to break a large, complex exercise into a series of smaller exercises, an example being the program Freebody[14]. A more sophisticated approach is to give initial guidance and then allow the student some flexibility in taking larger or smaller steps through the material. As the students' skills progress, the assisting structure needs to be reduced. The challenge is to provide enough guidance and feedback for the weaker student without putting a straightjacket on a stronger one, especially on exercises with multiple valid approaches to solving. The more sophisticated approach reduces the straightjacket problem at the cost of a more complex exercise/program that requires more effort to write.

FEEDBACK IN PADS

The internal grading engine of PADs provides the exercise author great flexibility in establishing what counts as correct and providing various levels of feedback. This is necessary for the wide range of possible applications of PADs, from qualitative to quantitative exercises, from kinematics graphs to force diagrams to voltage bar charts, all of which have different criteria to be evaluated. The goal of the discussion here is to simply give a sense of what is possible to do in PADs, and the interested reader is referred to additional information on the website[1] for additional information.

The key aspect of the PADs approach to grading is the ability to nest criteria and values inside each other, providing a way to construct complex evaluations and messages with a set of simple statements. For example, the value of V2 in figure 1 should be halfway between that of V1 and V4. PADs has a statement that instructs it, "Return the set of the objects that are labeled 'V1,'" and another, "Return the height of all objects in this set." Combined, it instructs the applet to return the height of all objects labeled 'V1.' A similar statement can be made to obtain the height of V4. These can be nested in yet another statement that says, "Add these two values together and divide by 2," to get the average of V1 and V4. This can then be nested in another statement that compares two values (e.g. the average and the height of V2) and returns one message if they differ by less than a tolerance, and a second if they do not. This is illustrated in Figure 2. The messages themselves can be compound statements of nested criteria and values that depend on almost any aspect of the current status of the diagram.

Selection criteria can depend on almost any property of an object—labels, position, height, width, magnitude, direction, style, size, color, others—or any combination thereof. Both numerical and string comparisons are available. A wide range of mathematical operations can be specified, along with store/recall functions and limited conditional operations. Not only can the grading criteria and feedback messages depend on any combination of nested statements, they can also be used to specify properties of the objects themselves, such as displayed label, style (e.g. vector, bar, line, etc.) or color. Note that in Figure 1, the color of the bars depends on whether the height is positive or negative.

Possible Modes For Using PADs

PADs can be used in most of the feedback modes described above. A PAD can return whether or not it is correct, the feedback message, or both. For a Poll exercise, the feedback would simply describe the student work, e.g. on a simple graph it might return "Increasing line," "Horizontal line," or "Decreasing line". In a Right or Wrong exercise, the messages specified could be either empty strings or omitted altogether. Figure 1 illustrated a Reactive response. Feedback can range from very general ("Double check all the voltages.") to very specific ("Current i2 should have a value of 1.5 A."). Simulated exercises are enabled through inter-PAD communication. GraphPAD can plot how a student moved an object in MotionPAD or vice versa. Structured Guidance can be achieved by specifying a task (e.g. construct a freebody diagram) that requires a number of steps and providing the student a button to periodically click. Each time the button is clicked, the PAD determines where the student is in the process based on the configuration of objects, and the feedback message provides instructions for the next step.

LOOKING TO THE FUTURE

What are the most fruitful ways to use PADs to support research based curricula? The overriding question of this paper is how to best provide feedback for all students, from the weakest to the strongest. I invite discussion on that as well as more specific questions, including:

Polling systems:, PADs could be used in class poll situations. It would be possible to conduct Interactive Lecture Demonstrations[14] (ILDs) in which the class results of student graphs were collected and viewed before the experiment. However, this would require students to have some sort of computer access in the classroom, as well as the software for such a system to be developed, both of which could require substantial resources. How valuable would this be to undertake?

Feedback and coding: Complex exercises with good feedback may be more beneficial for student learning. However, they require significantly more effort to code (can take several days compared to less than an hour). What is the value of some complex exercises with good feedback versus many simple, limited feedback ones?

Tying in with curricula: Many research-base curricula have similar activities, such as making position and velocity graphs. One could develop generic problems that could be used for a variety of curricula to make more available. However, generic may not fit perfectly with a curriculum and could lead to uncertainty on what to use when. How important is it to have problems written for a particular curriculum?

Appropriate scaffolding: A valuable method for teaching skills is scaffolding, initially providing extra support to the learner, which is gradually removed. One approach would to place high-feedback exercises at the beginning of an assignment, followed by similar, low-feedback exercises. Another would to be to spread these across several assignments. Another would be to assign only low-feedback exercises, but make high-feedback ones available for students who are struggling. There are advantages and disadvantages for different approaches.

Problem-solving: Problem-solving is a major goal of physics instruction. Teaching students to take the time to carry out important intermediate steps is common to most efforts to teach problem solving. PADs allow students to carry out and receive feedback on such steps, such as drawing free-body diagrams. Implementation in a web-based system raises tricky issues. Does one make the student choose what kind of diagram before they get the applet for drawing? If so, how does one grade the energy diagram for an exercise that requires a force diagram? However, simply handing the student an applet for making a force diagram could provide a major clue as to how to solve it. Does the student get to choose variable names? How does one get them to enter equations? Do we need a PAD for that?

CONCLUSION

The PADs project is an on-going effort to bring important capabilities to web-based homework. Graphical responses and customized feedback enable it to support curricula based on Physics Education Research. However, there are important questions to be discussed as to the best way to achieve this goal

ACKNOWLEDGMENTS

This work is sponsored in part by the National Science Foundation under grant DUE-0231353.

REFERENCES

1. Bonham, S. W., *Physics Applets for Drawing*, Western Kentucky University, *http://www.wku.edu/pads*
2. Sokolof, D. R., Thornton, R. K., Laws, P. W., *RealTime Physics Module 1: Mechanics*, New York: John Wiley & Sons, 1999.
3. Martin, L., Titus, A, *WebAssign,* North Carolina State Unviersity: 2004, *http://www.webassign.net.*
4. Novak, G. M., Patterson, E. T., Gavrin, A. D. et al., *Just-in-Time Teaching: Blending Active Learning with Web Technology*, Upper Saddle River: Prentice-Hall, 1999.
5. Mazur, E., *Peer Instruction*, Upper Saddle River: Prentice-Hall, 1997.
6. Hubber, A. P., *CyberProf,* How-Why Inc., 2003, *http://www.howhy.com/home/.*
7. Kane, D. Tychos, University of Illinois, 2004, *http://www.physics.uiuc.edu/research/per/Tycho.html*
8. Pritchard, D., *Mastering Physics*, Pearson Education, 2004, *http://www.masteringphysics.com/*
9. Kashy, E., *CAPA: A Computer-Assisted Personalized Approach*, LON-CAPA, *http://www.lon-capa.org*
10. Heller, P., personal communication, 1999.
11. Rice, I., personal communication, 2004.
12. Trowbridge, D., *Graphs and Tracks*, Raleigh: Physics Academic Software, 1996.
13. Chabay, R. W., *Electric Field Hockey*, Raleigh: Physics Academic Software, 1996.
14. Oberem, G., *Freebody*, Raleigh: Physics Academic Software, 1996.
15. Sokoloff, D. R., Thornton, R. K., *The Physics Teacher*, **35**, 340 (1997)

Computer Problem-Solving Coaches

Leon Hsu and Kenneth Heller

University of Minnesota, Minneapolis, MN 55455

Abstract. Computers might be able to play an important role in physics instruction by coaching students to develop good problem-solving skills. Building on previous research on student problem solving and on designing computer programs to teach cognitive skills, we are developing a prototype computer coach to provide students with guided practice in solving problems. In addition to helping students become better problem solvers, such programs can be useful in studying how students learn to solve problems and how and if problem-solving skills can be transferred from a computer to a pencil-and-paper environment.

INTRODUCTION

The ability to solve problems in a variety of contexts is becoming increasingly important in our rapidly changing technological society. In particular, good problem-solving skills are critically important for scientists and engineers, who use these skills to create new knowledge and to apply existing knowledge to the real world.

Because an introductory course in physics is a pre-requisite for study in nearly all science and engineering fields, it is an ideal venue for teaching problem solving. However, studies have shown that the majority of students emerge from such courses having made little progress toward developing good problem-solving skills [1]. One obstacle to students' learning effective problem-solving strategies is the difficulty and expense of providing good coaching, i.e., supplying students with an environment where they receive guidance and feedback while they solve problems. Even when the instructor and textbook model good problem-solving techniques, students often continue to use previously developed weak strategies and not those that are modeled [2].

In this paper, we present a framework for combining previous research in student problem solving with research in student-computer interactions to develop a practical means of providing every student with effective coaching in problem solving. We then describe our ongoing efforts to implement these ideas by designing a prototype of a computer problem-solving coach for students. Because this type of coaching focuses on the underlying student problem-solving process, it has little in common with existing computer homework systems that provide hints for specific problems. The work described in this paper is concerned with the practicality of building a software coach that interacts with students. After several prototypes have been constructed, we intend to test their efficacy first in small pilot studies of a few students and then in large-scale classroom trials.

RESEARCH BASE

Data exists on curricular interventions designed to help students become better problem solvers [3]. Each of the successful interventions had three common features: (1) explicit teaching of a problem-solving framework, (2) modeling of the use of the framework by the instructor, and (3) students using the framework when solving problems.

All of the problem-solving frameworks are based on the strategy developed by Polya [4]. The Minnesota problem-solving framework, shown in Fig. 1, is a typical example. Since real problem solving is rarely linear, Fig. 1 is meant only to outline the basic stages through which a solver might loop multiple times, and not to imply that problem solving can be reduced to a linear algorithmic process.

CP790, *2004 Physics Education Research Conference*, edited by J. Marx, P. Heron, and S. Franklin

Minnesota problem-solving framework

1. Focus the problem
   - Draw a picture illustrating the situation
   - Determine the question to be answered
   - Choose which physics principle(s) to use
2. Describe the physics
   - Draw physics diagrams
   - Determine target quantity(ies)
   - Write down quantitative relationships
3. Plan the solution
   - Select equation containing the target quantity
   - Identify other unknowns in equation
   - Solve a sub-problem to find each unknown
   - Check units
4. Execute the plan
   - Calculate value of target quantity(ies)
5. Evaluate the answer
   - Check if answer is properly stated
   - Check if answer is unreasonable
   - Check if answer is complete

FIGURE 1. The Minnesota problem-solving framework.

In the curricular applications, the instructor models good problem-solving techniques, usually during lectures, making explicit reference to the framework, and students are required to use the framework in their own problem solutions [5]. This strategy for teaching problem solving stems from the cognitive apprenticeship approach [6], in which the instructor models and makes explicit the cognitive processes necessary for performing a task and students are then expected to perform similar tasks with guidance before performing them on their own. In a standard physics class, students may have opportunities to practice solving problems in recitations, where they can receive feedback from both their peers and instructors [7,8].

Built into the framework for successful problem solving are the basic cognitive functions of deciding, implementing, and assessing. At each step in the solution process, the solver must *decide* on an action, *implement* it, and *assess* whether the implementation is adequate. Expert problem solvers perform these functions automatically, as an adult might tie a shoe. Novices, however, must be deliberate in their performance in order to succeed.

Students usually focus on implementing and rarely make deliberate decisions or assess their performance. This failure to make deliberate decisions often results in students' invoking inappropriate or incorrect knowledge, in not recalling useful knowledge they do have, and in applying procedures incorrectly. This process leads not only to incorrect problem solutions, but also to a failure to learn from one's mistakes. A good problem-solving aid should make these basic cognitive functions explicit to the student and enable the student to practice each one with feedback.

COMPUTER COACHES

One fundamental limitation of the curricular methods described previously is the reliance on interactions with peers or an instructor to achieve targeted coaching. Social interaction is likely a necessary part of an efficient learning process but does not allow for enough guided practice for most students. In many large lecture classes, students have less than one hour each week of organized practice solving problems in the supportive environment of a small group discussion section with a knowledgeable instructor.

The availability of powerful personal computers has led researchers to try to exploit their capabilities to provide students with individualized guidance and feedback [9]. Reif and Scott [10] developed a set of computer coaches called Personal Assistants for Learning (PALs) designed to teach students how to apply Newton's motion law, $\mathbf{F}_{\mathbf{net}} = m\mathbf{a}$. A pilot study of the PALs found that such coaches could help students improve their ability to apply Newton's motion law to solving problems.

Our goal is to build on previous work to develop computer coaches (called PS-PALs or Problem-Solving PALs) that can help students learn to solve the full range of problems they encounter in a typical introductory physics course. PS-PALs differ from PALs in that the previous coaches by Reif and Scott were designed to help students apply Newton's motion law to a problem for which it was already known that such an approach would lead to a successful solution. PS-PALs are designed to coach students in the more difficult and useful problem-solving practice of decision making: it is the student who must decide which physics principle(s) (e.g., kinematics, Newton's laws, conservation of energy or momentum, etc.) are necessary for solving the problem, in addition to applying the principle(s) correctly.

Following the literature on designing effective cognitive tutors [9], PS-PALs are based on a task analysis of the thought processes required to use a systematic problem-solving framework and give students guided practice performing these thought processes while solving problems. PS-PALs also give students immediate feedback on errors and gradually reduce the amount of help given to students, requiring them to learn to work more independently.

Pedagogy

PS-PALs incorporate all of the previously described characteristics of good problem-solving aids by interacting with the student in three different modes. The first two modes are based on the reciprocal-teaching strategy [11].

In the first mode, the PS-PAL acts as a coach, performing the functions of deciding and assessing, while the student implements. The PS-PAL displays the framework shown in Fig. 1 and graphically highlights each step as it guides the student through solving the problem by breaking the complete solution down into smaller questions. For example, the answers to the first series of questions lead to the construction of a useful picture illustrating the situation, including the relevant objects shown at important time instants and various kinematical and dynamical quantities associated with the objects. The picture aids the student in visualizing the problem situation and also may help to decide which physics principle(s) might be useful in solving the problem. Thus, the PS-PAL *decides* on the step to perform, the student *implements* the step by answering PS-PAL's questions, and the PS-PAL *assesses* the student's answer. If it is wrong, the PS-PAL gives feedback to help the student correct his or her work. A similar method is used to guide students in performing other problem-solving tasks, such as choosing and elaborating the physics principle(s) to use to solve the problem, writing down the quantitative relationships corresponding to the principles, and solving for the desired quantities.

In the second mode of reciprocal teaching, the roles are reversed. The student acts as the coach, *deciding* on the action to be implemented by the PS-PAL. The computer *implements* the action (but may deliberately make a mistake) and the student then *assesses* PS-PAL's implementation and makes any necessary corrections. This mode gives students practice in making deliberate decisions during the problem-solving process and in assessing work. The mistakes the PS-PAL is programmed to make are ones that are commonly found in student solutions.

The advantages of the reciprocal teaching strategy are (1) the instruction is highly interactive so that the student is constantly engaged mentally, (2) the cognitive processes of deciding, implementing, and assessing are made explicit and practiced in the context of solving a problem, (3) the PS-PAL models both the use of the problem-solving framework and good problem-solving performance for the student, and (4) the student receives individualized guidance and feedback, ensuring effective practice.

The third mode of interaction used by PS-PALs emphasizes the cognitive apprenticeship stage known as fading, slowly withdrawing the scaffolding of the computer program enforced problem-solving framework. The instructional strategy is based on learning from well-studied examples [12]. After students have received guided practice solving some problems as described above, the computer asks the student to solve a similar problem on paper, without any help. It then asks the student a series of questions designed to detect and diagnose any errors. If the student's solution is incorrect, the PS-PAL provides the student with a series of increasingly detailed hints, beginning with a hint to look back at a similar problem that was previously solved with the PS-PAL's guidance (the well-studied example). Students have a chance to correct their solution after each hint and thus receive the minimum amount of help necessary to solve the problem. This fading of assistance is designed to help students develop the ability to work independently.

Guidance and feedback

The guidance and feedback provided by the PS-PALs are primarily process-oriented hints but include some content-oriented hints specific to the particular problem. This is designed to focus students' attention on the use of their own mental framework to solve problems and differs from other web-based homework assistance systems such as WebAssign and Mastering Physics, in which a large fraction of the hints are content-oriented and specific to a particular problem. The intent of this type of coaching is that process-oriented hints emphasize the generality of physics and promote the development of an underlying problem-solving process that helps students transfer their skills to the solving of other problems [9].

Problems

The problems presented by the PS-PALs are "context rich" problems [13]. Such problems (1) are challenging enough that students must use an effective problem-solving framework to reach a solution, (2) require students to make decisions on how to proceed with the solution, (3) have a context and motivation that could help trigger cognitive connections for a student, (4) require students to visualize the situation, and (5) are mathematically straightforward to solve from basic principles. Traditional textbook problems can often be solved without the use of an expert-like problem-solving framework, so context-rich problems are employed to motivate the use of the framework shown in Fig. 1.

Programming

We use Authorware, a commercial programming language developed by Macromedia as the development language for PS-PALs because (1) Authorware is the product of a large established company that continues to develop and make improvements to the language, (2) Authorware programs can be delivered over the web and run on both Macintosh and Windows computers, (3) Authorware allows for easy incorporation of graphics, sound, and movies, and (4) Authorware is designed to be easily learned by even non-programmers, so that one can focus on instructional design without worrying about programming details. Using Authorware allows for eventual adoption and adaptation of the tutorials by individual instructors. PS-PALs are a test-bed for this type of computer coaching, which probes how much this technology can achieve through good design and the use of effective instructional strategies and without resorting to the complexities of artificial intelligence [14,15] that may be possible in the future.

Assessment

Thus far, we have constructed a PS-PAL prototype for each of the three modes of interaction and are testing them with students both to assess their usability and to gauge students' potential interest in using such computer coaches. All of the students who volunteered to test the programs were enthusiastic about them and said they thought such coaches would help them to become better problem solvers. These students were also able to point out instances where the computer interface was confusing or the feedback provided by the PS-PAL was unclear.

Once we are satisfied that the programs are acceptable to a large fraction of students, our plan is to develop a set of PS-PALs for eight to ten physics problems that can be solved using kinematics, Newton's laws, and/or conservation of energy. We have chosen this set of principles because it represents the point in most introductory physics courses at which problem-solving decisions become challenging for students. Because PS-PALs log all student interactions with the computer, we are able to study the difficulties students have during the problem-solving process and thus refine the computer coaching structure. We will also test the PS-PALs for educational impact with students in an introductory physics course. The results of that assessment will guide our future work with the PS-PAL computer coaches.

ACKNOWLEDGMENTS

This work was supported by NSF DUE-0230830 and by a University of Minnesota Grant-in-Aid award.

REFERENCES

1. Reif, F., *Am. J. Phys.* **63**, 17-32 (1995).
2. Larkin, J., McDermott, J., Simon, D., and Simon, H. A., *Science* **208**, 1335-1342 (1980).
3. Maloney, D. P., "Research on problem solving: Physics," in *Handbook of Research on Science Teaching and Learning*, edited by D. L. Gabel, New York: Macmillan, 1994, pp. 327-356.
4. Polya, G., *How to Solve It,* Garden City, NY: Doubleday, 1945.
5. Leonard, W.J., Dufresne, R.J., and Mestre, J.P., *Am. J. Phys.* **64**, 1495-1503 (1996).
6. Collins, A., Brown, J.S., and Newman, S.E., "Cognitive Apprenticeship: Teaching the Crafts of Reading, Writing and Mathematics," in *Knowing, Learning and Instruction*, edited by L. Resnick, Hillsdale, NJ: Erlbaum, 1989, pp. 453-494.
7. Heller, P., Keith, R., and Anderson, S., *Am. J. Phys.* **60**, 627-636 (1992).
8. Johnson, M., *Am. J. Phys.* **69**, S2-S11 (2001).
9. Anderson, J. R., Corbett, A. T., Koedinger, K. R., and Pelletier, R., *J. Learn Sci.* **4**, 167-207 (1995).
10. Reif, F. and Scott, L. A., *Am. J. Phys.* **67**, 819-831 (1999).
11. Palincsar, A. L., and Brown, A. L., *Cogn. Instruct.* **1**, 117-175 (1984).
12. Zhu, X. and Simon, H. A., *Cogn. Instruct.* **4**, 137-166 (1987).
13. Heller, P. and Hollabaugh, M., *Am. J. Phys.* **60**, 637-644 (1992).
14. Anderson, J. R., Boyle, C. B., and Reiser, B. J., *Science* **228**, 456-462 (1985).
15. VanLehn, K., Lynch, C., Taylor, L., Weinstein, A., Shelby, R., Schulze, K., Treacy, D., and Wintersgill, M., in *Intelligent Tutoring Systems: 6th International Conference*, edited by S. A. Cerri, G. Gouarderes, and F. Paraguacu, Berlin: Springer, 2002, pp. 367-376.

Using Electronic Interviews to Explore Student Understanding

DJ Wagner[*,†], JJ Rivera[*], Fran Mateycik[*], and Sybillyn Jennings[§]

[*]*Department of Physics, Rensselaer Polytechnic Institute, Troy, NY 12180*
[†]*Department of Physics, Grove City College, Grove City, PA 16127*
[§]*Department of Psychology, Russell Sage College, Troy, NY 12180*

Abstract. This paper reports on methods used to probe student understandings of optical fibers and total internal reflection (TIR). The study was conducted as part of the expansion and improvement of web-based materials for an innovative introductory physics course. Initially, we conducted face-to-face Piaget-style interviews with a convenience sample. Our next step was to interview students taking the course at Rensselaer. Physical limitations necessitated that this be done from a distance, so we conducted "e-interviews" using a Chat Room. In this paper we focus on the e-interview experience, discussing similarities to and differences from the traditional face-to-face approach. In the process, we address how each method informs us about students' activation of prior experiences in making sense of unfamiliar phenomena (*e.g.*, "transfer of learning").

INTRODUCTION

Since 1999, Rensselaer has offered The Science of Information Technology (ScIT) to a total of 158 academically diverse students [1]. This innovative course teaches the scientific principles behind the operation of computing systems. For example, in ScIT we apply total internal reflection (TIR), a phenomenon commonly covered in traditional physics courses, to optical fibers, a technology eminently relevant to the students [2].

Part of our ongoing curriculum development process is to assess student preconceptions and construct a diagnostic test that distinguishes between preconceptions associated with different levels of comprehension. The Physics Education Research (PER) literature abounds with studies of preconceptions [3], but few (an exception is ref. [4]) have examined how students approach topics addressed in ScIT: *e.g.*, TIR, propagation of signals, and semiconductors. We chose to study how students describe refraction and optical fibers.

Piaget [5] developed the method of critical exploration to see what thoughts lay behind a child's initial answer to a question. Such clinical interviews have been widely accepted as an effective means of eliciting learners' reasoning and conceptual frameworks [5-7]. The PER community has its own strong history [8-10] of using such interviews to effectively probe student comprehension about a physics-related concept.

We initially conducted face-to-face interviews with a convenience sample of 12 volunteers with diverse physics experience. When such face-to-face interviews with ScIT students were not possible, we conducted "e-interviews." (The use of e-interviews to guide the development of web-based curricular materials seemed particularly apt.) In this paper we discuss our experience with both interview formats.

METHODOLOGY

Interviews have traditionally been conducted using face-to-face (f2f) dialogue. The interaction is recorded and later transcribed. In the summer of 2002, we used this format to interview a convenience sample of 12 volunteers (Group A, Subjects A1 to A12). The interviewees included five in-service teachers who had formerly taken introductory physics, five physics REU participants, and two physics faculty. These interviews

CP790, *2004 Physics Education Research Conference*, edited by J. Marx, P. Heron, and S. Franklin
 0-7354-0281-7/05/$22.50

provided a "broad picture" of how individuals from a variety of backgrounds think about optical fibers.

Our next step was to interview members of our target audience: students enrolled in ScIT (Group B, Subjects B1 to B22). All interviewees had taken high school physics; many had also taken college physics courses, as summarized in Table 1.

TABLE 1. Physics Background of Interviewees.

Highest Completed Physics Course	Grp A	Grp B
High School Physics	1	9
College Physics I	0	8
College Physics II	4	3
Upper-Level Coll. Phys.	5	1
Physics Faculty	2	0

Face-to-face interviews with students enrolled in ScIT were not possible, since the interviewer and students were in different regions of the country. Additionally, Rensselaer's tradition of innovative use of technology in and out of the classroom, along with the amount of time spent by college students in Chat Rooms and using Instant Messaging (IM) [11], suggested that ScIT students would be fairly comfortable using electronic communication. We therefore conducted "e-interviews", using the Chat Room tool of WebCT. All 22 ScIT students were interviewed in this manner before instruction; post-instruction e-interviews were conducted with 8 of those 22.

ScIT students signed up for interview times in one of WebCT's four chat rooms. At the appointed time, the interviewer and the interviewee "met" by logging into the chosen Room. Like their f2f counterparts, our e-interviews started with participant identification and verification of informed consent. Initial questions about students' backgrounds helped build rapport and establish comfort with the environment before turning to the content-laden questions.

Both sets of interviews were conducted by the same interviewer (a physicist); a cognitive psychologist "observed" many interviews to gauge their efficacy. The content portion of all interviews began with the question "What does an optical fiber do?" Subsequent questions included, "How might the fiber keep the light from escaping out the sides?" "What can you tell me about refraction?" and "How does TIR occur?" In the tradition of Piaget, our interviews were dynamic and driven by the participants' responses, so the exact set of questions depended on those responses.

Consecutive interviews were scheduled in different Chat Rooms. Each Chat Room opens up in its own browser window, creating individual "examining rooms." For example, the interviewer could wrap up an interview with one student in Room 1 while starting the preliminaries with the next student in Room 3. The separate browser windows, along with the scroll bars and automatic transcription of the conversation, assisted the interviewer in keeping track of the "tone," as well as the specific content, of each interview.

FINDINGS

Our purpose in this paper is to report on our experiences using the two interview formats and offer some observations for those who might consider the e-interview approach. We are not attempting to show either the superiority of one method or the equivalence of the two; indeed, such direct comparison of these two disparate formats is akin to the proverbial comparison of apples and oranges. We have grouped our observations around three aspects of the interview experience: descriptions of targeted physics topics, how the different modes of communication affected the social interaction, and practical considerations.

Targeted Physics Topics

We found that the physics descriptions gleaned from the e-interviews were consistent with those observed in the f2f interviews. Analysis of those descriptions is reported elsewhere [12]. Here we present specific examples of descriptions found in the two formats.

Consider the following excerpts from e-interviewee B11 and f2f-interviewee A8.

> **B11**: "Light [in an optical fiber] is prevented from escaping too early, by using mirrors to make sure the light bounces through the cable and doesn't escape. ... [the mirrors] are made of glass and located throughout the cable."

> **A8**: "if uh, you want the light to go in different directions inside of the tube you would ... Somehow screw, uh, put mirrors on the sides of the tube at a certain angle so when the light goes, uh, enters from this side ... it would be reflected here (points at diagram)."

In both formats subjects successfully conveyed an image of mirrors embedded at points within a fiber

F2f-interviewee A5 and e-interviewee B10 each drew upon past experiences to *reject* a mirrored-surface model of fibers. A5 recalled "those things you get at like Disney World ... there's like a light that you see at the end of the strings, but the strings themselves don't look lit up." B10 also recounted his experience with character wands which "looked similar to a flashlight with a battery inside ... when you turned them on, there were all of these wires coming out of the top of the wand, and the ends would glow different colors." Although A5 expressed her experiences vocally while B10 typed his, both were able to effectively communicate the memories they activated.

Social Interaction

Social interaction in face-to-face interviews utilizes multiple communication channels. The f2f subject uses not only words, but also tone of voice and gestures to convey meaning. Additional communication may occur through drawing or interaction with demonstration equipment [8]. For example, we provided a scratch pad in the f2f interviews; it was used in all but one of those interviews. Furthermore, the f2f-interviewer may discern the subject's emotions and motivational state by reading facial expressions and postures. These visible channels of communication are simply not an option in e-interviews; their absence changes the nature of the social interaction, as described below. In our specific experience, this change did not lead to a substantive difference in descriptions of the targeted concepts provided by interviewees.

Consider the descriptions of Disney World wands provided by subjects A5 and B10. A5, in her f2f interview, used gestures to illustrate how the fibers sprouted out of the character's head and wiggled. B10's description required a few more words but conveyed a similar description of his past experience with optical fibers. Similarly, B11 and A8 were both able to convey a model of fibers with mirrors embedded in them, even though e-interviewee B11 did not have the scratch pad provided in A8's f2f interview.

Participants in the e-interviews created work-arounds to constraints they experienced. Subject B18 "drew" with keyboard symbols: "I think it would bend toward that wall -- so it enters like '\', then inside is more towards '---' (but not, necessarily), and then when it leaves, it's back to '\'." Subject B21 drew off-line: "[light bends] towards the normal -- Changed my mind, away from the normal -- I drew a picture." The interviewer commented on the lack of a scratch pad in 6 of the 30 e-interviews when visualizing what students described was particularly difficult. Some success was achieved through questions such as "Can you describe what you're visualizing?" (One student replied, "I'm visualizing the beam that goes through the boundary to bend closer to parrallel [sic] with the boundary.") Still, the absence of a scratch pad in e-interviews did sometimes limit the extent of probing.

The Chat Room interviewer can glean some insight from the language and format used. Lack of upper-case letters, the inclusion of IM jargon [13] such as "lol" or "iirc", and the use of emoticons such as :) and :-& can indicate how comfortable the participant is or qualify a statement. Twelve of the 22 ScIT students e-interviewed used IM jargon, and the interviewer used jargon or emoticons in almost every e-interview, to put the subjects at ease. Subjects frequently chose to type out their qualifications and hesitations – phrases such as "i don't know the specifics, but ...," "umm -- i could make a stab in the complete dark," and "I think I am confused -- Let me re-phrase my answer" were common throughout the e-interviews. Also common was the use of question marks at the end of statements to indicate hesitancy. The interviewer suggested the use of a symbol (#) to signify the completion of a turn, since pauses were easily misinterpreted.

Gestures, facial expressions, and tone of voice may be misinterpreted or get lost in a transcription of an f2f-interview. E-interview participants cannot use anything except their words and are therefore likely to describe their thoughts more explicitly. And the e-interviewer does not have to worry about her unconscious reactions influencing an interviewee's responses. We harbor no illusions, however, that the text of a Chat Room transcript can provide the same detailed information as can a video of a f2f interview. Emotions/reactions in an e-interview arise from the interviewee's conscious effort; unconscious (or unacknowledged) reactions are not easily identifiable in an electronic conversation. Additionally, the e-interview is not a "controlled" environment; *e.g.*, the subject's doorbell could give rise to a pause that the interviewer interprets as the subject thinking hard about a question. The importance of these constraints will, obviously, depend upon the information sought in the study.

Practical Considerations

Researchers conduct interviews to collect information, but the information collected is not the only factor one must consider. Time and resource constraints, along with the willingness of subjects to participate, play a crucial role in the success of a study.

E-interviews are convenient. They can involve participants from across the country – even from around the world. Times can be arranged without regard to building access or office hours. (Ours were often held between 9 PM and midnight, a convenient time for the participants to be on-line.) When necessary, the interviewer can overlap interviews by using multiple Chat Rooms, decreasing the necessity of quickly wrapping up an interesting interview because the next interviewee showed up.

While the "contact" time of e-interviews is indeed longer than that required by f2f interviews, it may be offset by the increased convenience, negligible transcription time for the interviewer, elimination of travel time for the interviewee, and the potential to make constructive use of the time waiting for the other participant to type a response. Our e-interviews had a median time of around 38 minutes, roughly six times that of our f2f interviews, although this is a tentative comparison because of the difference in samples. Only one e-interviewee truncated the interview because of time constraints; others seemed not to mind. One subject replied, "I'm not going anywhere," when the interviewer apologized for the interview length; another commented, "this is very convenient for me."

SUMMARY

We conducted interviews primarily to gain insight into how students conceptualize a very specific set of topics, to drive curriculum and assessment tool development. We believe both types of interview experience contributed significantly to this goal. Even with the limitations of e-interviews, students were able to dialogue comfortably with the interviewer about their understanding of these concepts. Interviewees in both formats provided comparable explanations of the physics topics. And subjects in both types of interview conveyed how they were using past experiences to make sense of optical fibers.

For the purposes of our interviews, the lack of a scratch pad posed the greatest hurdle. Some options for future e-interviews are to post a set of drawings on the web that can be accessed and used for reference by both parties, or to use WebCT's Whiteboard Tool. (The Whiteboard allows electronic images to be created and viewed in real time by all participants.) We anticipate that the advent of Tablet PCs will allow interviewers and subjects to share images drawn with an electronic pen, but it will be a while before that promising technology becomes ubiquitous.

The e-interview experience provided us with access to students not otherwise available. Was the information we gathered the same as the information that could have been gathered in a face-to-face interview? Of course not. But the content-specific information in which we were most interested appears consistent with that gathered in face-to-face interviews driven by the same set of questions. Other interviewers should use their time and resource constraints, awareness of their participants' familiarity with technology, along with consideration of the information they are seeking, when deciding whether to conduct e-interviews.

ACKNOWLEDGMENTS

The Rensselaer portion of this work was supported in part by NSF CCLI Program under grant # DUE-0089399. Special thanks to Leo Schowalter, the ScIT Advisory Committee, and all the interviewees.

REFERENCES

1. 35% of Rensselaer's ScIT students have been Information Technology majors, 20% were Computer Science majors, 17% were graphic arts majors, 10% were Engineering or Physics majors, 8 % were in other sciences, and 8% were studying Architecture or Management.
2. All materials are web-based, and many can be found at the ScIT website: http://www.rpi.edu/dept/phys/ScIT.
3. See, for example, E. F. Redish and L. C. McDermott, "Resource Letter PER-1: Physics Education Research," *Am. J. Phys.* **67** (9), 755-767 (1999).
4. M. C. Wittmann, R. N. Steinberg, and E. F. Redish, "Investigating student understanding of quantum mechanics: Spontaneous models of conductivity," *Am. J. Phys* **70** (3), 218–226 (2002).
5. J. Piaget, *The child's conception of the world*, trans. by J. & A. Tomlinson (Totowa, NJ: Littlefield, Adams,& Co. 1926/1972).
6. E. Hunt and J. Minstrell, "A Cognitive Approach to the Teaching of Physics," in *Classroom Lessons: Integrating Cognitive Theory and Classroom Practice*, edited by K. McGilly (Cambridge, MA: MIT Press, 1994).
7. H. Ginsburg, *Entering the child's mind* (New York: Cambridge University Press, 1997).
8. L.C. McDermott, "Research on conceptual understanding in mechanics," *Physics Today*, **37** (7), 24-32 (1984).
9. I. A. Halloun and D. Hestenes, "Common sense concepts about motion," *Am. J. Phys.* **53** (11), 1056-1065 (1985).
10. E. F. Redish and R. N. Steinberg, "Teaching Physics: Figuring Out What Works," *Phys. Today* **52** (1), 24-30 (1999).
11. See, for example, "The Internet Goes to College," by the Pew Internet and American Life Project, found at www.pewinternet.org/pdfs/PIP_College_Report.pdf
12. F. Mateycik, D. Wagner, J. Rivera, and S. Jennings, "Student Descriptions of Refraction and Optical Fibers," submitted to 2004 PERC Proceedings.
13. "lol"= "laugh out loud," and "iirc" = "if I remember correctly." Emoticons are faces on their sides: ":)" is a smile and ":-&" means the typist is "tongue-tied." See http://www.netlingo.com/ for more.

Time to completion reveals problem-solving transfer

Rasil Warnakulasooriya & David E. Pritchard

Dept. of Physics & Research Laboratory of Electronics
Massachusetts Institute of Technology
Cambridge, Massachusetts 02139.

Abstract. Two equally skilled groups of students taking introductory mechanics solve related physics problem pairs in reverse order with respect to each other, using the web-based Socratic tutor, MasteringPhysics. In tutorial problems containing help in the form of requestable hints, descriptive text, and feedback, twice as many students were able to complete problems correctly in real-time compared to problems that did not provide any help (end-of-chapter problems). The prepared group in a given related pair was able to solve it in ~15% less time on average compared to the unprepared group. Furthermore, the prepared group requests ~7% fewer hints on average than the unprepared group. We conclude that shorter completion times and problem-solving transfer are facilitated through tutorial problems.

INTRODUCTION

In this paper we provide evidence of learning and problem-solving transfer in physics as seen by time to completion curves using the web-based homework tutor, MasteringPhysics. The time to completion curves depict the fraction of students who have completed a given physics problem correctly versus the logarithmic time for the completion of that problem.

A study of time to completion of physics problems is enabled by the general mastery learning [1] pedagogy of MasteringPhysics and the availability of the log files it generates. Such studies are crucial for maximizing learning for a given amount of student time on task. In MasteringPhysics mastery learning is implemented within a Socratic dialogue where students are provided with hints and simpler sub-problems upon request, and are given feedback when wrong answers are proposed.

THE STUDY

The data in this study involve students from the fall 2003 semester taking the introductory Newtonian mechanics course at the Massachusetts Institute of Technology (MIT). The students (~ 430 in total) were divided into two equally skilled groups based on their MasteringPhysics homework assignment grades for the first six weeks of the semester. Our study involves these two groups solving related problem pairs in opposite orders relative to each other with no intervening problems. That is, the "first" problem in a given pair to one group is the "second" to the other and vice versa. We will call the group that solves a given problem in a pair first, the unprepared group and the group that solves the same problem second (having solved its related problem first), the prepared group.

The problems in a pair are related in the sense that they both involve the same concepts and methods. Seven problem pairs in the concept domains of torque, linear momentum conservation, energy conservation, friction, angular kinematics and dynamics, rigid body rotation, and Newtonian gravitation were studied. Since these problems involve requestable hints and wrong answer responses, we will call them problems with help or tutorial problems. Five end-of-chapter (EOC) problems [2] that do not contain any hints or wrong answer responses were also administered through MasteringPhysics in the domains of energy conservation and rigid body rotation.

The main variable of interest in this study is time to completion, which is defined as the time interval between a student's first opening of a problem and his/her submission of the completed problem. No time intervals for interactions in between (even for logins and log-offs) are accounted for. Completion is defined

CP790, *2004 Physics Education Research Conference*, edited by J. Marx, P. Heron, and S. Franklin

as finishing all the main parts of a given problem correctly (finishing sub-parts when hints are requested is not required). We chose to bin the data in time intervals of 0.5 using a logarithmic time scale (a factor of 1.6 per bin in real time).

RESULTS AND CONCLUSIONS

1. **Three groups:** We identify three major groups of students in completing a given problem by plotting the rate of completion against logarithmic time (see Fig. 1). The three groups occur, consistently separated by a local minimum, in all fourteen tutorial problems.

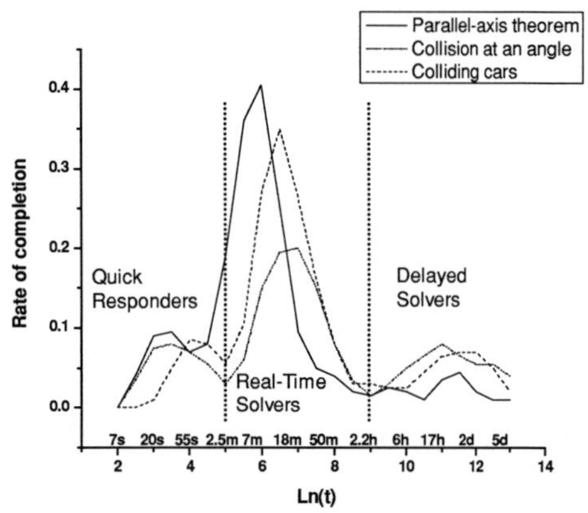

FIGURE 1. The rate of completion graphs for three representative tutorial problems reveal three distinct peaks. In the order of increasing time, these peaks correspond to the quick responders, the real-time solvers, and the delayed solvers.

The "quick responders" (~10% of the students in the course) are able to finish the problem within 2.5 minutes. These students typically do not request any help (hints) or submit wrong answers. The peak that occurs after about 2.2 hours, we identify as "delayed solvers." These students request help within MasteringPhysics and submit incorrect answers (see Fig. 2). We believe that the majority of both the quick responders and the delayed solvers seek assistance outside MasteringPhysics. This may indicate intellectual dishonesty on the part of quick responders and/or help from human tutors for the delayed solvers. Indeed, further analysis show that the more quick responses a student submits, the more his/her score is below average in the paper-based final examination.

The most interesting group from the view point of problem-solving, we believe, is the "real-time solvers" – they complete the problem in the interval from 2.5 minutes to 2.2 hours, generally using hints and feedback (For two gravitation problems this peak is shifted towards longer times and occur between 7 minutes to 6 hours). This group corresponds to the prominent central peak in rate of completion graphs. Since these students seem to struggle with the problem by giving wrong answers and requesting hints within reasonable time intervals we believe that they solve problems in real-time (see Fig. 2).

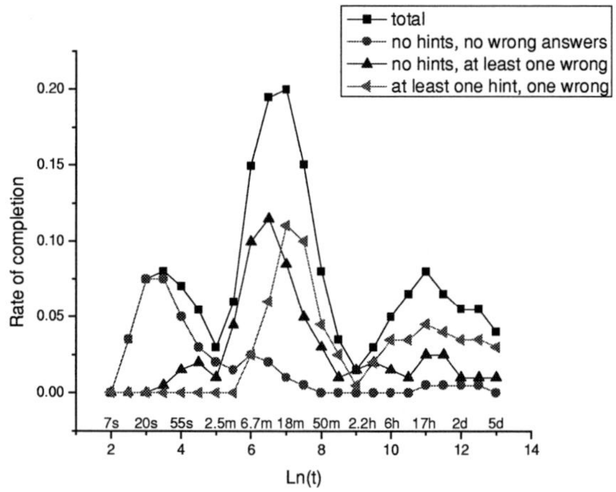

FIGURE 2. The break-down of the rate of completion graph for a typical tutorial problem. The real-time and the delayed solvers make mistakes and ask for help.

The PER research community (and others) defines "problem" (as opposed to "exercise") fairly consistently – the route to solution is not immediately apparent to the student. We therefore feel these questions are "problems" for the real-time solvers since the central peak mainly consists of students who give wrong answers and/or request hints (at the expense of a penalty of their homework grade), and only a small minority answer without either of these indications of struggle.

2. **Sigmoidal shape for "real-time solvers":** The time to completion graphs for real-time solvers (which are the integral curves of the rate of completion curves) yield S-shaped (sigmoid) curves for all problems with help (see Fig. 3). This observation is interesting since the sigmoid curve plays a prominent role in growth models [3] and time to completion of other psychological tasks [4,5] from sentence completion [6] to pigeon learning [7]. These curves are important for short term and long term objectives. Short term, they indicate whether certain instructional strategies will help more students obtain the correct solutions with less expenditure of time, and enable measurements of knowledge transfer per unit time. Longer term, they will seriously constrain models of human problem solving which must be able to predict the detailed shape observed (as is already the case for pigeon learning [7]).

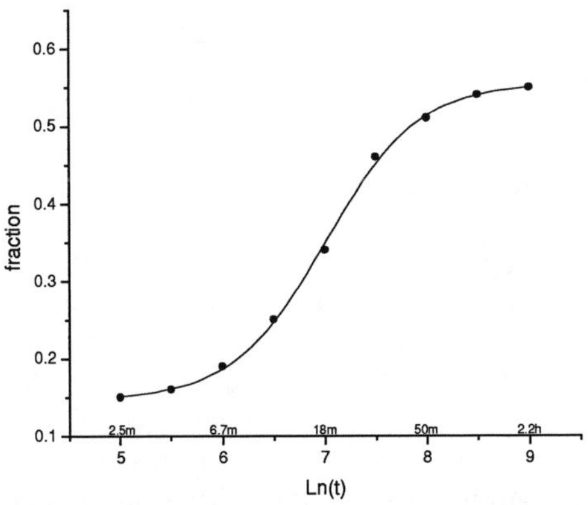

FIGURE 3. The integral of the rate of completion curve of a typical tutorial is sigmoidal or S-shaped. This shape is a characteristic of psychological tasks.

3. **Logistic and Boltzmann fit best:** The high signal to noise ratio of our S-shaped time to completion curves for real-time solvers invites a more detailed fit. For this, we picked three sigmoidal functions even though we have no a priori reason (e.g. they do not result from plausible "growth models" of problem solving):

Logistic $$f_L(x) = \frac{a-b}{1+(\frac{x}{x_C})^p} + b$$

Boltzmann $$f_B(x) = \frac{a-b}{1+e^{(x-x_C)/d}} + b$$

Gompertz $$f_G(x) = ae^{-e^{-k(x-b)}}$$

These were considered as both functions of time (x = t) and functions of logarithmic time [x=Ln(t)]. We fitted the differentials of the above functions to the rate of completion curves in order to eliminate the interdependency among the data taken from different times, and to remove the effects of the quick responders and delayed solvers. This also allowed us to use the reduced chi-square (χ^2/ν) goodness-of-fit test as well as the more sophisticated second order Akaike information criterion (AIC_c) [8,9]. We find that the Logistic and the Boltzmann functions using x=Ln(t) are both good fits (χ^2/ν=1.15) and are preferred for our data (i.e. ΔAIC_c >4) relative to the Gompertz or to any function of linear time.

In order to see if our finding of the preference for x=Ln(t) has precedents, we reanalyzed the human word recognition data in sentence completion reported by Tulving et al.[6]. These data, when plotted against a logarithmic time, were generally fitted better by a linear model (evidence ratio > 100) than with any of the above sigmoidal functions. We also find, when considering the data in linear time (as originally reported by Tulving et al.), that the Gompertz model seems more likely (ΔAIC_c >4) than the linear, the Logistic, or the Boltzmann. The only precedent we have found for the superiority of Ln(t) is in computer adapted testing [10] where a reanalysis of the data seems to favor this independent variable. In this context we note that it is for complex tasks like the problems given in our study but unlike the simple tasks in [4-7].

4. **Lack of hints reduces "real-time solvers."** We find clear differences in time to completion curves between tutorial problems (which contain hints and feedback to specific wrong answers) and the EOC problems (which do not). The S-shape typically becomes linear in EOC problems for the real-time solver interval. In Fig. 4 we provide an example of this effect for an EOC problem and compare with a tutorial problem both involving the concept of energy conservation. We find that the fraction of real-time solvers is 67±3% for 12 problems with help but is only 33±5% for five EOC problems (see Fig. 4). We feel that these problems are sufficiently equal in difficulty that this factor of two difference almost certainly results from the presence of tutoring.

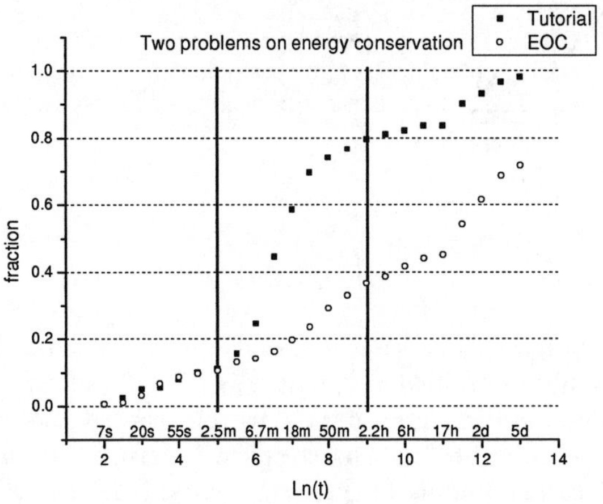

FIGURE 4. Twice as many students are able to solve a given tutorial problem with hints and wrong answer responses in real-time (2.5 min. – 2.2 hrs) compared to an end-of-chapter (EOC) problem.

5. **Problem-solving transfer to related problems:** We find that learning and problem-solving transfer [11] takes place from one related problem to another in the sense that the prepared group have a time advantage of 14.6±2.2% on average over the

unprepared group on the same problem. This effect is seen across all the fourteen problems thus providing robust statistical evidence. We arrive at this result by fitting the differentials of the above sigmoidal functions to the rate of completion data and identifying the time corresponding to the peak rate for the real-time solvers (see Fig. 5 for an example). This method also reveals that the median time on a typical problem is about 15 minutes, which is quite realistic (there are only a few students at the margins of the time interval 2.5 minutes to 2.2 hours). We also find that averaging over six pairs of problems, the fraction of prepared students requesting hints is 7±2% lower than the unprepared students. In other studies [12] we have identified quicker time to solution and use of fewer hints with better performance on the final exam.

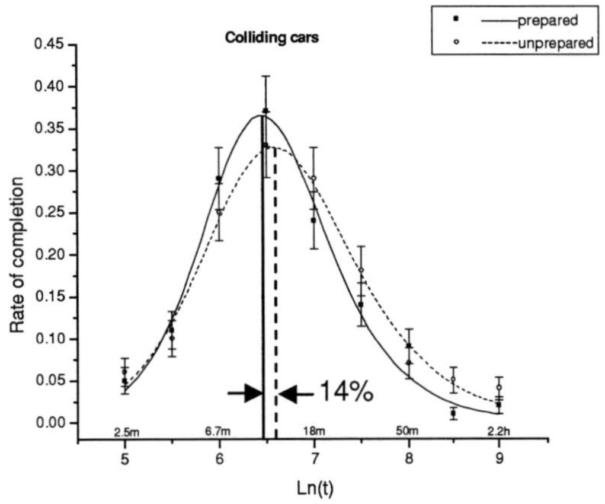

FIGURE 5. The prepared group is able to solve a given tutorial problem in less-time compared to the unprepared group. The average advantage in time seen across fourteen problems is 14.6%.

DISCUSSION

We have demonstrated a methodology of extracting useful information about time on task in an uncontrolled homework environment – namely, measuring the fraction complete against the variable logarithmic time for problem completion. Identifying the real-time solvers we have demonstrated that their time to completion curve follows the sigmoidal shape against logarithmic time rather than linear time as observed in other psychological tasks. Furthermore, we have shown evidence that the problems with tutorial help can be solved in a reasonable time by twice as many students as end-of-chapter problems. We also find robust evidence of problem-solving transfer as evidenced by the reduction in time to solution having solved a related prior problem across fourteen problems covering seven concept domains. The effectiveness of the tutorial problems can be attributed to the hints and wrong answer responses leading to learning in much the same way Simon and Reed [13] have observed in the context of the "cannibals and missionaries" problem.

We note that our results are not biased since the effects we see are reciprocal; that is the prepared group for a given problem in a pair is the unprepared group for its related problem and vice versa. The various comparisons we have made in this study are general in nature in the sense that the problems are related only by content and the relevant conceptual domain. Also, the tutorial problems and the EOC problems are not isomorphs of each other. Since our larger goal is to investigate and establish best pedagogical practices in web-based tutoring, a study designed to look at the isomorphic problems with and without hints and wrong answer responses is being carried out. The findings from these studies will be reported in the future.

[We thank the National Science Foundation for providing assistance through its grant 0231268.]

REFERENCES

1. B. S. Bloom, *All our children learning*, McGraw-Hill, New York, 1981.
2. H. D. Young and R. A. Freedman, *University physics*, Pearson, San Francisco, 2004.
3. K. P. Burnham and D. R. Anderson, *Model selection and multimodel inference: a practical information-theoretic approach*, Springer, New York, 2002.
4. C. L. Hull, *Principles of behavior: An introduction to behavior theory*, Appleton-Century-Crofts, New York, 1943.
5. D. L. Schnipke and D. J. Scrams, *Journal of Educational Measurement* **34**, 2132 (1997).
6. E. Tulving, G. Mandler, and R. Baumal, *Canadian Journal of Psychology* **18**, 62 (1964).
7. A. Machado and P. Guilhardi, *Journal of the Experimental Analysis of Behavior* **74**, 25 (2000).
8. H. Akaike, "Information theory and an extension of the maximum likelihood principle," in *Second international symposium on information theory*, edited by B.N. Petrox, F. Caski, Budapest, 1973, pp. 267-281.
9. N. Sugiura, *Communications in Statistics, Theory and Methods* **A7,** 13 (1978).
10. L. F. Hornke, *Psicológica* **21,** 175 (2000).
11. R. E. Mayer, *Learning and instruction*, Pearson, New Jersey, 2003.
12. D. E. Pritchard and R. Warnakulasooriya, *EdMedia* (2005) – To appear.
13. H. A. Simon and S. K. Reed, *Cognitive Psychology* **8**, 86 (1976).

How We Think About And Prepare To Teach Physics

Dewey I. Dykstra, Jr.

Physics Department, Boise State University, Boise, ID 83723-1570

Abstract. We have been preparing physics teachers in the same manner for many decades. Yet, physics education research reveals for some observers disturbing evidence little or no change in understanding the phenomena occurs as a direct result of physics instruction from elementary school through the college years. The apparent compatibility between these learning results and prevailing paradigm enables the construction of a description of the paradigm. If it can be demonstrated that there is even just one alternative paradigm from which much more effective alternative pedagogical practice is derived, are we not obligated to change how we prepare to teach physics?

INTRODUCTION

"...a physics major has to be trained to use today's physics whereas a physics teacher has to be trained to see a development of physical theories in ... students' minds."[1]

In recent decades those interested in science teaching and learning have investigated very carefully science learning. The rise in attention to an explanation of the origins of knowledge and evidence supporting the explanation has inspired a focus in physics education research (PER) on students' conceptions of the phenomena studied in physics.

EVIDENCE ABOUNDS...

The "modern" era of this research began in the 1970's, but articles on students' conceptions about phenomena have been uncovered as far back as 1903. In the 1980's several bibliographic efforts on students' conceptions were coalesced into one bibliography.[2] Maintained by Reinders Duit at the University of Kiel, it now contains 6,314 entries.

This body of research supports one of the more robust findings in educational research:

The general outcome of science instruction is ***no significant change*** *in students' conceptions of the phenomena studied in the classroom.*

It is not possible in the space available to do more than show a few examples and invite the reader to consult the bibliography:[2]

Kinematics-acceleration: "The conceptual difficulties with acceleration that were encountered by the students in our study appeared to be very persistent. Often, ...the procedures used by a particular student were the same before and after instruction. ... A significant number of students from a wide variety of courses confused the concepts of velocity and acceleration. ... At the completion of instruction, fewer than half of the students demonstrated sufficient qualitative understanding of acceleration as a ratio to be able to apply this concept in a real situation."[3]

Electric circuits: "We have examined students' explanations of an extremely simple electric circuit, one that involved only three major components. ...many students were unable to interpret the circuit correctly. ... One suspects, therefore, that a significant proportion of students in physics courses will have this type of difficulty. ...the misconception persisted in some students who had been through a calculus-based course in electricity which included five experiments on electric circuits."[4]

RECENT EXAMPLE IN DETAIL

To sample more of the population in order to find out how widespread these conceptions might be, multiple-choice format diagnostic instruments were developed. One diagnostic is the Force and Motion

CP790, *2004 Physics Education Research Conference*, edited by J. Marx, P. Heron, and S. Franklin

Conceptual Evaluation (FMCE).[5] This diagnostic is designed to discern the nature of a student's conception or "physical theory" concerning the explanation of motion in terms of force, not merely to determine whether a student knows the "correct" answer.[6]

An analysis of the FMCE responses of students in standard algebra-trig-level and calculus-level introductory students from state and private institutions of higher education across the U.S.A. over a dozen years was conducted by Dykstra[7,8]. The analysis of this data was guided by that of Thornton.[6]

Using a more coarse scale of analysis than Thornton, two keys to the responses in the FMCE were generated. One key is the set of choices consistent with a *person-on-the-street* (*pots*) explanation of motion in terms of force, *i.e.*, velocity goes as the force. The other key is a set of choices consistent with a *Newtonian-like* (*N*) explanation of force, *i.e.*, acceleration goes as the net force. Comparing a student's responses to each key yields a score in the range of 0 – 15 indicating the degree to which the student's responses match the corresponding explanation of motion in terms of force.

TABLE 1: **Results of Traditional Instruction**

Traditional, *Content-Driven* Instruction

Algebra-Trig Level Intro Physics

Year	Term	n	Effect Size (st dev) *N*	*pots*	Normalized Gain <g>	Loss <L>	Averages Pre *pots*	*N*	Post *pots*	N
West Coast Public Univ. A										
1990		99	0.59	-0.47	0.14	-0.09	10.1	1.5	8.5	3.3
"Prairie State" Public Univ.										
2002	SP	112	0.66	-0.40	0.13	-0.06	10.3	0.9	9.0	2.7
Calculus Level Intro Physics										
North East State Public Univ.										
1998		72	0.47	-0.30	0.15	-0.06	9.6	1.7	8.5	3.5
West Coast Public Univ. B										
1999	W	87	0.60	-0.62	0.22	-0.30	9.3	2.6	6.5	5.4
1999	SP	73	0.38	-0.36	0.12	-0.01	9.1	2.3	7.6	4.0
2000	SP	115	0.59	-0.50	0.15	-0.10	9.2	2.4	7.2	4.8
West Coast Private Univ.										
2000	SP	38	0.54	-0.08	0.09	0.05	9.8	0.6	9.6	1.9

TABLE 1 gives some of the results of this analysis. Effect size (Cohen's d) is the difference between the pre and post scores in units of pooled standard deviations. Normalized gain is the fraction of the possible increase in the *N* score. Normalized loss is the fraction of the possible decrease in *pots* score.

From the average pre-scores in TABLE 1 these typical students, who have all had instruction on motion and force at least once before, are still answering in a manner consistent with the *pots* explanation. In the whole sample of students in this on-going study, now numbering nearly 1150 students, there have been only 5 whose pre-responses earned an *N* score high enough to be convinced they were using a *Newtonian-like* view of motion and force. With the second or third treatment of traditional, *content-driven* instruction in this study, while there is some change, the change is still very small after the semester's instruction by Ph. D. physicists.

EXPLAINING THE EVIDENCE

The prevailing paradigm apparently accounts for this long-standing state of affairs. If not, physics teaching and physics teacher preparation would have changed long ago. This enables us to develop descriptions of the standard paradigm on a phenomenological basis.

In the prevailing paradigm, the nature of the enterprise of physics teaching can be characterized in the following way[7]:

Physics teaching is the presentation of the official canon of physics by approved methods for the benefit of those who deserve it.

As with other core elements of a paradigm, it is taken as given and so, goes unchallenged.

The word "deserving" means that a student has met two criteria in order to "learn" physics. (1) The student must be capable. (2) The student must work diligently. Not all people are capable of this feat because "physics is hard." Failure to meet one of these two criteria is one of the general explanations why most students do not understand physics at the end of the semester. The other general explanation relates to the teaching of physics.

In this enterprise, in order to prepare to teach physics first one must know the official canon. Girded with the official canon, one needs be able to apply approved methods of presentation. Thus, degree programs for the preparation of physics are designed. Take the standard degree program in physics. Trim a little away to make room for the methods class and a chance to practice the methods in student teaching. There is no attention to the necessary fundamental issue Niedderer indicates for the preparation of a physics teacher: the ability "*...to see a development of physical theories in ...students' minds.*"

If the teacher presents the canon properly, the "deserving" student will "get" it. The canon is a commodity transferable from teacher to student under suitable circumstances. This is founded on a realist view of the nature of knowledge which postulates that: "...the objective existence of physical reality ... can be known to our minds...with an ever growing precision by the subtle play of theory and experiment."[9] The nature of knowledge, taken as a given foundational aspect of this paradigm, is whatever it is, independent of human beings. By clear implication this knowledge can exist outside the mind in order to be transmitted by physical means from teacher to student and in order to have an independent existence.

Having presented the established canon by accepted methods, the responsibility of the teacher has been met. If the necessary characteristics for student success satisfactorily account for the "learning" outcomes of physics teaching, there is no need to look any deeper. Thus, we can understand the lack of research in students' understandings over literally centuries until recent decades and resistance to the findings of such research. No wonder that the findings from PER have had so little effect on physics teaching. In the standard paradigm, we are apparently doing about as good a job as possible at physics teaching.

AN ALTERNATIVE PARADIGM

In the standard paradigm TABLE 1 is an example of what we can expect of the most deserving students, the science and engineering majors, in courses given by instructors who know the established canon and use approved methods. It should then be the case that students less likely to be "deserving" would not be able to show any more change in understanding than the science and engineering majors.

Let us take a look at students more likely to be the "less deserving." The first six rows of results in TABLE 2 are from non-science, non-engineering majors in a course that satisfies general university requirements in science. The data was collected using the FMCE and analyzed in the same way. These students experienced a pedagogical practice rooted in a different paradigm. Clearly, the standard paradigm belief that there is a limited set of "deserving" students is contradicted by these results. The instructional materials used were developed for the Motion and Force Modules of the Powerful Ideas in Physical Science (PIPS) Project of the American Association of Physics Teachers.[10]

Comparing TABLE 1 and TABLE 2, it is quite clear that the shift in the response patterns is considerably greater. The revised PIPS instruction average effect size of 2.5 standard deviations and 61% normalized gain in the *N* scores for the college students far exceeds that of the science and engineering majors in TABLE 1. Clearly the notion that the *status quo* in physics teaching is about as good as we can expect is totally discredited by this repeated example of a different pedagogy based on an alternative paradigm.

TABLE 2: Results of Alternative Paradigm Instruction

Student Understanding-Driven **Instruction**

Conceptual Physics--Standard PIPS Instruction

			Effect Size (st dev)		Normalized		Averages			
					Gain	Loss	Pre		Post	
Year	Term	n	*N*	*pots*	<g>	<L>	*pots*	*N*	*pots*	N
Boise State University										
1999	SP	97	2.00	-1.90	0.49	-0.68	9.4	0.9	3.0	7.8
1999	FL	93	1.78	-1.80	0.48	-0.63	9.9	0.8	3.6	7.6
Conceptual Physics--revised PIPS Instruction										
Boise State University										
2000	FL	90	2.50	-2.20	0.59	-0.66	9.3	0.8	2.5	9.2
2001	SP	87	2.40	-2.40	0.62	-0.74	9.8	0.8	2.2	9.6
2002	FL	66	2.62	-2.19	0.57	-0.72	9.4	0.8	2.2	8.8
2004	SP	69	2.31	-1.85	0.67	-0.78	7.9	2.2	1.8	10.8
High School Level--Standard PIPS Instruction										
2001	FLa	23	6.30	-5.40	0.89	-0.95	11.3	0.6	0.6	13.3
2001	FLb	24	6.10	-3.70	0.86	-0.93	10.6	0.9	0.8	13.1

PREPARING TEACHERS IN AN ALTERNATIVE PARADIGM

We arrive back at Niedderer's statement: "*a physics teacher has to be trained to see a development of physical theories in ...students' minds.*"[1] The phenomena studied in physics courses are what the physical theories are about. Sadly, most physics courses are not studies of the phenomena, but drill-and-practice at the canon. They do not train a person to see the development of explanatory knowledge in students' minds.

A physics teacher candidate needs to be meta-cognitive about the development of his own explanatory knowledge and to interact with others about this meta-cognition. The future teacher needs to observe students to discern evidence of their explanatory knowledge. The teacher candidate should become familiar with the PER research literature. The physics teacher candidate should also practice strategies intended to maximize the chances of disequilibration in students and to facilitate their development of new explanatory schemes. Also, the teacher candidates should study the historical development of physical theories and the philosophical analysis of the nature of knowledge and its development.

In this alternative paradigm, the emphasis is on the nature of understanding. The students' understandings, physical theories in student minds, are the primary objects of study and manipulation; not the phenomena, the apparatus, lab skills, mathematics or the canon. The teacher must be able to construct in her own mind effective models of the students' understandings and to pick examples of experience the students will decide are challenges to their existing understandings.

PREPARING PHYSICS TEACHERS

To meet the challenge, we need to develop courses and degree programs to bring out the ability to see the development of physical theories in students' minds. Two examples of courses to this end do exist.

A course sequence with an intense focus on the development of more powerful understanding and the ability to clearly, rigorously describe experiences within this understanding is offered at the University of Washington by the Physics Education Group (PEG).[11] The course sequence is Phys 407–409: Physics by Inquiry.[12 - 13]

At Boise State University a course is being developed to introduce teacher candidates to the "the development of physical theories" in the minds of students. The course, PHYSCI 497/597: Alternative Conceptions in Science, engages teacher candidates in two major activities. Teacher candidates study articles in the bibliography[2] to develop a catalog of descriptions of conceptions their future students are likely to have when they come to class. They also study student work: responses on diagnostics, posters developed by students describing their ideas, writing assignments, electronic course discussions, etc. to discern evidence of student conceptions and of change in their theories.

CONCLUSION

Conventional, *content-driven* physics instruction fails at leaving students with deeper understandings of the phenomena they study. *Student-understanding driven* pedagogy does result in significant change in student understanding. Its existence places on us the burden of responding in our own practices and in how we prepare the physics teachers who come after us.

We need to assist physics teachers at developing the skills, habits, and points of view that enable the ability "to see the development of physical theories" in the minds of students. Courses to this end are being developed. More such courses need to be established and built into whole degree programs, not only for physics teachers, but also for all teachers.

ACKNOWLEDGMENTS

Without students willing to share their ideas none of this work nor the data would have been possible. Without the opportunity to interact with the PER community what quality there is in this work would be seriously diminished. On several occasions funding from NSF provided support as some of the data was collected and ideas herein were formulated.

REFERENCES

1. Niedderer, H., "What Research can Contribute to the Improvement of Classroom Teaching," in *Proceedings of the International Conference on Physics Teachers' Education*, Dortmund, University of Dortmund, 1992, p. 151. Available at url: <http://didaktik.physik.uni-bremen.de/niedderer/download/8dortmph.pdf>
2. Duit, R., Students and Teachers Conceptions in Science. url: <http://www.ipn.uni-kiel.de/aktuell/stcse/stcse.html> (2004)
3. Trowbridge, D. & McDermott, L. *American Journal of Physics* **48**(3), 242 – 253 (1981).
4. Fredette, N. & Clement, J., *Journal College Science Teaching* **10**(5), 280 – 285 (1981).
5. Thornton, R. & Sokoloff, D. *American Journal of Physics* **66**(4), 338 – 352, (1998).
6. Thornton, R., "Conceptual Dynamics: Following changing student views of force and motion," in *The Changing Role of Physics Departments in Modern Universities: Proceedings of ICUPE*, edited by E. F. Redish & J. S. Rigden, College Park, MD: American Institute of Physics, 1997. Abstract online at http://ase.tufts.edu/csmt/html/abstracts/icupe_cd.html.
7. Dykstra, D., "Why teach kinematics?: An examination of the teaching of kinematics and force, Part I" submitted to the Physics Education Research Supplement of the American Journal of Physics, (2002). Available at url: <http://www.boisestate.edu/physics/dykstra/WTK1.pdf>
8. Dykstra, D., "Why teach kinematics?: An examination of the teaching of kinematics and force, Part II" submitted to the Physics Education Research Supplement of the American Journal of Physics, (2002). Available at url: <http://www.boisestate.edu/physics/dykstra/WTK2.pdf>
9. de la Torre, A. & Zamorano, R., *American Journal of Physics* **69** (2), 103 (2001)
10. American Association of Physics Teachers, *Powerful Ideas in Physical Science—A Model Course*, College Park, AAPT (2002)
11. PEG, <http://www.phys.washington.edu/groups/peg/> (2004)
12. McDermott, L, Am. J. Phys. **69**(11), 1127 - 1137 (2001). Downloadable from <http://www.phys.washington.edu/groups/peg/oersted.html>
13. PEG, <http://www.phys.washington.edu/groups/peg/pubsa.html> (2004)

Author Index